Companion Animal Behaviour Problems

Prevention and Management of Behaviour Problems in Veterinary Practice

FSC
www.fsc.org
MIX
Paper from responsible sources
FSC® C022174

Companion Animal Behaviour Problems

Prevention and Management of Behaviour Problems in Veterinary Practice

---

Edited by

Rachel Casey,
Sarah Heath

and

Helen Zulch

CABI

CABI is a trading name of CAB International

CABI
Nosworthy Way
Wallingford
Oxfordshire OX10 8DE
UK

Tel: +44 (0)1491 832111
Fax: +44 (0)1491 833508
E-mail: info@cabi.org
Website: www.cabi.org

CABI
WeWork
One Lincoln St
24th Floor
Boston, MA 02111
USA

Tel: +1 (617)682-9015
E-mail: cabi-nao@cabi.org

Please note that throughout the book the term 'referral' has been used when discussing a veterinarian seeking to delegate the behavioural support of their patient to a behaviour practitioner. It is accepted that this process is likely to differ depending on whether the behaviour practitioner is a veterinarian or a non-veterinarian, as well as the jurisdiction in which the veterinarian operates. It may take place as a formal referral or it may be a less formal delegation of care for the patient's behavioural needs. Whatever process is followed, the point that the text aims to make is that a close working relationship between a patient's primary care and/or specialist care veterinarian and the behaviour practitioner is key to the appropriate behavioural care of the patient.

Please also note that the advice given throughout this book is considered appropriate for the majority of animals. However, it remains the responsibility of the veterinarian and behaviour practitioner to tailor handling, management and treatment approaches for the individual pet based on the evidence available to them at the time.

A catalogue record for this book is available from the British Library, London, UK.

References to Internet websites (URLs) were accurate at the time of writing.

ISBN-13: 9781780643458 (hardback)
9781780643465 (paperback)
9781780643472 (ePDF)
9781800621312 (ePub)

DOI: 10.1079/9781800621312.0000

Commissioning Editor: Caroline Makepeace
Editorial Assistant: Kate Hill
Production Editor: Marta Patiño

Typeset by SPi, Pondicherry, India
Printed and bound in the UK by Severn, Gloucester

Contents

List of Contributors

David Appleby MSc CCAB

David is an ASAB (Association for the Study of Animal Behaviour), Certificated Clinical Animal Behaviourist (CCAB). Until retirement from clinical practice in June 2021 he was the visiting clinician and lecturer to Helsinki University Veterinary School. He continues as Co-coordinator of the postgraduate course in Clinical Animal Behaviour and Welfare, Odisee University of Applied Sciences, Belgium. He is a member of the Fellowship of Animal Behaviour Clinicians (FABC).

Trudi Atkinson Dip.AS.(CABC)

Trudi qualified as a veterinary nurse in 1986 and gained accreditation as an ASAB (Association for the Study of Animal Behaviour) Certificated Clinical Animal Behaviourist (CCAB) in 2003. She is a founder member of the Fellowship of Animal Behaviour Clinicians (FABC) and, until retiring in 2021, she ran a clinical feline behaviour practice covering the south west of England. She is a member of International Cat Care's feline wellbeing panel and continues to work with iCatCare on their feline behaviour courses. She is also the author of *House-Soiling Problems in Domestic Cats* and *Practical Feline Behaviour – Understanding Cat Behaviour and Improving Welfare.*

Julie Bedford PGCE PGDip(CABC) CCAB

Julie has run an animal behaviour/training consultancy for 35 years. She is a visiting university lecturer, expert witness, supervising clinician, staff trainer at leading animal charities and advises on welfare policy and strategy. Julie has extensive experience of handling, advising on and resolving behavioural and training problems in companion animals and regularly competes in various dog sports with her own dogs. In 2015, The *Daily Mirror* and RSPCA acknowledged her work within rescue and rehoming by awarding her the Animal Hero Superstar award.

Emily Blackwell BSc PhD FHEA CCAB

Emily graduated with a First Class honours degree in Zoology from University of Leicester, before completing her PhD in the aetiology, prevention and treatment of separation-related behaviour in dogs at University of Bristol. Emily is a Senior Lecturer at Bristol Veterinary School, with over 20 years' experience of coordinating and executing research into companion animal behaviour and welfare, playing a key role in advancing the knowledge base in this field.

While there is a plethora of advice and behavioural modification protocols available to dog owners, from sources such as the internet and magazines, few have received scientific scrutiny and Emily was one of the first pioneering researchers to scientifically validate treatment protocols for separation-related behaviour problems. Emily's research informs her work as a Clinical Animal Behaviourist, certificated by the Association for the Study of Animal Behaviour, and she is a Founder Member of the Fellowship of Animal Behaviour Clinicians.

Jonathan Bowen BVetMed MRCVS DipAS(CABC)

Jonathan graduated in veterinary medicine from the Royal Veterinary College in 1992. After several years in general practice he developed an interest in behaviour and completed the postgraduate diploma in companion animal behaviour counselling at Southampton University. He runs the behavioural medicine referral service at the RVC and is an honorary lecturer in behavioural medicine. He has contributed to a number of books, including the current *BSAVA Handbook of Behavioural Medicine* and *The Welfare of Animals in Animal-Assisted Interventions*. Jonathan is a regular speaker at national and international conferences, including the Canine Science Forum, International Veterinary Behaviour Meeting, World Veterinary Association Congress, Southern European Veterinary Congress, and North American Veterinary Congress (VMX). He is actively involved in research, with recent publications on the human–animal bond, the impact of the COVID pandemic on pets, and the connection between physical health and behaviour. Apart from his academic affiliation with the RVC, Jonathan is also a member of the scientific committee of the Chair for Animals and Health at the Autonomous University of Barcelona.

Caroline Bower BVM&S MRCVS Dip(CABC)

Caroline is a member of the British Veterinary Association (BVA), British Small Animal Veterinary Association (BSAVA) and the British Veterinary Behaviour Association (BVBA).

Emma Brown BVSc PGDip MRCVS CCAB

Emma works as a Veterinary Behaviourist taking canine and feline cases on referral. Emma is an RCVS Advanced Practitioner in Companion Animal Behaviour. She is a Certificated Clinical Animal Behaviourist (CCAB), a certificated member of the Fellowship of Animal Behaviour Clinicians (FAB Clinicians) and an Animal Behaviour and Training Council (ABTC) Registered Veterinary Behaviourist.

Rachel Casey BVMS DipECAWBM(BM) PhD MRCVS

Rachel is Director of Canine Behaviour and Research at Dogs Trust, the UK's largest dog charity. She is an RCVS and an EBVS® European Veterinary Specialist in Behavioural Medicine and an EBVS® European Veterinary Specialist in Animal Welfare, Science, Ethics and Law. She has research interests in canine welfare science, including identification of risk factors for undesired behaviours. Rachel also oversees community programmes aimed at preventing behaviour problems, and the behavioural rehabilitation of dogs in rehoming centres.

Claire Hargrave MSc BSc(Hons) PGCE CSci CChem MRSC CCAB

Claire is a self-employed ASAB Certificated Clinical Animal Behaviourist (CCAB), working from her clinic in Wales. Claire is currently the Honorary Secretary of the Fellowship of Animal Behaviour Clinicians and lectured on the Animal Behaviour and Welfare masters' programme at Newcastle University.

Sarah Heath BVSc PgCertVE DipECAWBM(BM) CCAB FRCVS

Sarah is both an RCVS and an EBVS® European Veterinary Specialist in Behavioural Medicine, as well as a Certified Clinical Animal Behaviourist. She was made a Fellow of the Royal College of Veterinary Surgeons in recognition of her work in establishing behavioural medicine as a veterinary discipline.

Stephanie Hedges BSc(Hons) RVN ASAB CCAB

Stephanie is a Registered Veterinary Nurse (RVN), an ASAB (Association for the Study of Animal Behaviour) Certificated Clinical Animal Behaviourist (CCAB), and a founding member of the Fellowship of Animal Behaviour Clinicians (FABC). By virtue of the latter, she is recognized as a Clinical Animal Behaviourist by the Animal Behaviour and Training Council (ABTC). She is also a Director of and treasurer for ASAB (Accreditation) Ltd. She studied Applied Animal Behaviour at the University of Southampton and Veterinary Nursing with Business Management with Warwickshire College.

Emma Magnus BSc (Hons), MSc

Emma has a single honours degree in Zoology, a joint honours in Psychology and Sociology, and a Masters in Applied Animal Welfare and Behaviour, from Edinburgh University. Emma worked in the field of Pet Behaviour Counselling for 15 years, and was a member of the APBC, with CCAB status. She has several papers and books published within the field of small animal behaviour, and recently undertook a qualitative piece of research into the meaning of grief for pet owners. Emma (now Lightfoot) currently works in the field of personnel and wellbeing.

Anne McBride BSc(Hons) PhD Cert Cons FRSA ABTC-CAB

Anne gained her degree and PhD from University College London and is a senior lecturer at the University of Southampton. She was founder and director of the Masters in Companion Animal Behaviour Counselling at Southampton. The first British academically recognized qualification, this course ran from 1994 to 2015, drawing students from several continents. Anne was a Trustee of the Animal Behaviour and Training Council (ABTC), the UK regulatory body for trainers and behaviourists, from 2016-2022. Currently, she is a member of the ABTC Programme Recognition Committee, and an ABTC listed Clinical Animal Behaviourist. She is also Chair of the Association of Pet Behaviour Counsellors. Anne is proud to be an Honorary Life Associate Member of both the British Small Animal Veterinary Association and the British Veterinary Nursing Association, and a Fellow of the International Society for Anthrozoology. A passionate advocate of the power of interesting and relevant education to improve One Welfare, she teaches and writes extensively about human–animal interactions and animal behaviour, including small mammals, for audiences of different ages and backgrounds (as both Anne McBride and E.A. McBride).

Francesca Riccomini BSc(Hons) BVetMed CCAB MRCVS PGDip(CABC)

An experienced general practitioner, Francesca specialized in companion animal behaviour counselling. No longer professionally active, Francesca was a member of International Cat Care's Behaviour Advisory Panel and is now a retired ASAB accredited Certified Clinical Animal Behaviourist, a retired member of The Fellowship of Animal Behaviour Clinicians and an honorary life member of the Association of Pet Behaviour Counsellors. Francesca wrote extensively on her subject for a range of publications aimed at veterinary professionals and pet owners. She was a regular lecturer for several organizations, including the BSAVA, BVNA, APBC, independent CPD providers and rescue centres and is also the author of several books about cats and feline behaviour.

Reducing the impact of clinic-related stress on patients and the important role veterinary professionals play in helping owners to prevent and resolve problems associated with their pets' behaviour were among Francesca's particular professional interests.

Amanda L. Roshier BSc(Hons) PhD PGCHE PGCert MA(higher education) MSc(behaviour counselling) SFHEA

Mandy is an Associate Professor in Anatomy and Behaviour in the School of Veterinary Medicine and Science at the University of Nottingham. She was awarded a BSc (Hons) in Equine Science from the University of Bristol before completing a PhD in biomedical engineering at the University of Nottingham, Institute of Biomechanics. Her PhD investigated the use of diagnostic ultrasound as a method to diagnose soft tissue injuries of the neck following a whiplash injury. During this period of study, Mandy also demonstrated human anatomy to medical students. An interest in the public engagement of science throughout this period led to presenting her research to Parliament, schools and national science exhibitions. She joined the School of Veterinary Medicine and Science in March 2006 as part of the foundation team developing the new vet school. She studied an MSc in Companion Animal Behaviour Counselling at the University of Southampton, where her dissertation investigated veterinary support and client expectations of behavioural medicine in first opinion practice. As an advocate of integrating behaviour into veterinary practice, Mandy joined the British Veterinary Behaviour Association as a committee member in 2014.

David Ryan PGDip(CABC) CCAB

David has a postgraduate diploma in Companion Animal Behaviour Counselling (2002) and is accredited as a Certificated Clinical Animal Behaviourist by the Association for the Study of Animal Behaviour (2008). For 26 years, until 2007, he was employed as a Home Office Accredited Police Dog Handler and Handling Instructor. He has worked as a pet behaviour counsellor on veterinary referral (2002–2013) and lecturer on Newcastle University's MSc in Applied Animal Behaviour and Welfare (2011–2017). He was Chair of the Association of Pet Behaviour Counsellors (2009–2012), is an Animal Behaviour and Training Council registered Clinical Animal Behaviourist and Expert

Witness (2010–), and a founder member of the Fellowship of Animal Behaviour Clinicians (2020–). He now specializes in legal consultations on allegedly dangerous dogs, and in 2018 was requested by The House of Commons Environment, Food and Rural Affairs Committee to provide evidence to their inquiry into Dangerous Dogs: Breed Specific Legislation.

Caroline Warnes BVSc MSc CCAB MRCVS

Caroline qualified as a Veterinary Surgeon from the University of Bristol in 1985 and worked for 10 years in general veterinary practice, during which time she became particularly interested in companion animal behaviour. Caroline gained an MSc in Companion Animal Behaviour Counselling from the University of Southampton, and went on to become an ASAB accredited Certificated Clinical Animal Behaviourist and an RCVS Advanced Practitioner in Companion Animal Behaviour. Until she retired at the end of 2020, Caroline ran a companion animal behaviour referral practice based near Swindon in Wiltshire, alongside various other activities over the years, including helping to run a veterinary practice and being an external lecturer at Southampton and Bristol Universities. Caroline is currently a retired member of the Fellowship of Animal Behaviour Clinicians, still active in their Vet practice liaison subcommittee.

Clare Wilson MA, VetMB CCAB MRCVS PGDip(CABC)

Clare is a veterinary surgeon and an Association for the Study of Animal Behaviour (ASAB) Certificated Clinical Animal Behaviourist (CCAB). She works as a locum in general veterinary practice and also provides a behavioural referral service in Warwickshire. In addition to clinical work, she provides Continuing Professional Development for the veterinary profession and has been, and continues to be, involved in various behavioural organization committees. She is a Founder Member of the Fellowship of Animal Behaviour Clinicians (FABC), which is a membership organization that supports independent accreditation of clinical behaviour practitioners.

Helen Zulch BVSc(Hons) DipECAWBM(BM) CCAB MRCVS

Helen is a veterinarian and Royal College and European Specialist in Veterinary Behavioural Medicine. She is Associate Professor in Clinical Animal Behaviour in the School of Life Sciences, University of Lincoln, UK, and an Honorary Assistant Professor at the University of Nottingham School of Veterinary Medicine and Science. Helen has worked in the field of companion animal behaviour for over 20 years, lecturing on undergraduate and postgraduate programmes, consulting for pet owners and engaging with research across a range of areas in animal behaviour, learning and cognition. She has a particular interest in the interface between physical health and behaviour, as well as in the prevention of the development of problem behaviours in pets. She has published in the scientific literature, presented lectures and seminars at international conferences and events, authored books and book chapters and was instrumental in developing the Life Skills for Puppies programme as well as the Life Skills for Puppies Train the Trainers course.

1 Introduction

Helen Zulch

Over the past few decades, the field of companion animal behaviour has grown exponentially, with research flourishing and the field of behaviour consulting rapidly expanding. Qualifications in this field are now available and encompass both postgraduate veterinary specialisms and non-veterinary undergraduate and postgraduate degrees and accreditations. This, together with a greater awareness within the pet-owning public of the availability of behavioural support for their pets, means that, in many parts of the world, it is now common for owners of pets to seek both preventive as well as remedial behavioural advice.

In human health, the interrelationship between physical, emotional and psychological health is well recognized; for example, the World Health Organization (WHO) states in the preamble to their constitution: 'Health is a state of complete physical, mental and social well-being and not merely the absence of disease or infirmity.' In animals the interrelationship between physical and psychological welfare and health and behaviour has also been recognized for a number of years, largely in production animal species (for example, see Weary *et al.*, 2009; Phythian *et al.*, 2016), but more recently links have been investigated and highlighted in companion animals too (for example, Bécuwe-Bonnet *et al.*, 2012; Mills *et al.*, 2020). This is one of the main reasons that those working in the field of clinical animal behaviour believe strongly in the need for the veterinary and behaviour professions to work closely together for the best possible outcomes for pets and their owners.

In addition to the importance of taking a holistic approach to health and behaviour, the veterinary profession, as represented by veterinarians and nurses in first-opinion practice, is well placed to offer advice on appropriate pet acquisition a well as early-life advice to protect against the development of later behaviour problems (Gazzano *et al.*, 2008). Ensuring good communication between owners and the practice and promoting the practice as behaviourally aware and concerned can also assist in the early identification of behaviour concerns such that appropriate interventions can be recommended. Finally, ensuring that the practice puts behavioural awareness at the heart of their work can go a long way to preventing the development of problems associated with visits to the vet. Volk *et al.* (2011) presented evidence that cat owners are more reluctant to take their pets to the vet for routine treatment if their cat exhibits distress during the process. Reducing the distress of pets attending veterinary practices not only benefits the pet, whose care is likely to be more regular and potentially more successful, but is also by association likely to reduce owner stress and can lead to stronger client/practice relationships.

All of these factors led to the conception of this book, the key aims of which are to provide

DOI: 10.1079/9781800621312.0001

those in first-opinion veterinary practice with information to:

- Help pet owners to better understand their pets' behaviour in order to ensure that needs are provided for and thus problems are less likely to develop.
- Enable them to offer owners preventive behavioural advice in order to support the long-term behavioural health of their patients.
- Assist them in the early detection of behaviour problems in pets that attend the practice and ensure that they are comfortable in the best approach to take to remediate problems that are identified.
- Assist practices in developing environments and operating procedures that help to create a positive experience for pets and owners visiting the practice.
- Assist practices in fostering positive and collaborative working relationships with their local behaviour practitioners.

The book is arranged in three parts. The first part aims to highlight important aspects of the natural behaviour of a range of common companion animal species and show how this may be at odds with the domestic environments in which they are often kept. This section aims to provide information to veterinary professionals that is both up to date and welfare compatible, in order that this can be conveyed to clients, thus helping to set appropriate expectations of normal behaviour and needs, leading to improved pet well-being.

The second part of the book focuses on prevention of the development of behaviour problems in pets. This encompasses advice on how to raise young animals in a manner that best equips them to live in the modern human-centric world and builds resilience to deal with its stressors. It also examines how the practice can create an environment where veterinary visits are as positive as possible for pet and owner. This is key in both reducing the risk of pets developing a fear of the practice and veterinary handling, which can compromise both immediate and long-term care and welfare, and in creating an environment where owners feel supported in the behavioural care of their pets.

Part three offers support for the practice when there is a need to address a client's concerns regarding their pet's behaviour. Included in this part are chapters covering the following.

- Creating an environment where owners are encouraged to raise their concerns, knowing that they will be empathetically and non-judgementally supported.
- Appropriately handling challenging patients in order to be able to offer a good clinical service for general care as well as in order to rule in or out the presence of a medical condition contributing to the behaviour problem.
- Assessing the risk of a medical condition causing or contributing to the behaviour presented.
- Offering first aid advice which will support an owner to reduce risk and improve welfare in the interval between the problem being identified and referral (internal or external) for a full behaviour consultation.

This chapter also deals with the difficult subject of decision making regarding outcomes, for example when a problem poses a significant welfare or safety risk and decisions around relinquishment or euthanasia need to be considered. Finally, this section covers aspects of referral to behaviour specialists and the fostering and maintaining of relationships between client, practice and behaviourist for the best long-term interest of the client and of course their pet.

This book is not a textbook describing detailed assessment and treatment protocols for specific problem behaviours, nor is it an exhaustive academic discourse on the research underpinning the field of companion animal behaviour. The authors' aim is to ensure that veterinary professionals feel comfortable offering preventive and first aid advice for the pets in their care as well as creating a safe and supportive environment for the discussion of behaviour problems and an empathetic and compassionate place for the handling of pets who are brought into the practice environment. All those involved in its writing hope that it will serve as a key reference for veterinary professionals as well as practitioners and students in the field of companion animal behaviour. We trust that it will play an important role in their ability to offer the best possible support to the pets and owners with whom they come into contact.

References

Bécuwe-Bonnet, V., Bélanger, M.C., Frank, D., Parent, J. and Hélie, P. (2012) Gastrointestinal disorders in dogs with excessive licking of surfaces. *Journal of Veterinary Behaviour* 7(4), 194–204.

Gazzano, A., Mariti, C., Alvares, S., Cozzi, A., Tognetti, R. and Sighieri, C. (2008) The prevention of undesirable behaviours in dogs: effectiveness of veterinary behaviorists' advice given to puppy owners. *Journal of Veterinary Behaviour* 3(3), 125–133.

Mills, D.S., Demontigny-Bédard, I., Gruen, M., Klinck, M.P., McPeake, K.J. *et al.* (2020) Pain and problem behavior in cats and dogs. *Animals* 10(2), 318.

Phythian, C.J., Michalopoulou, E., Cripps, P.J., Duncan, J.S. and Wemelsfelder, F. (2016) On-farm qualitative behaviour assessment in sheep: repeated measurements across time, and association with physical indicators of flock health and welfare. *Applied Animal Behaviour Science* 175, 23–31.

Volk, J.O., Felsted, K.E., Thomas, J.G. and Siren, C.W. (2011) Executive summary of the Bayer veterinary care usage study. *Journal of the American Veterinary Medical Association* 238(10), 1275–1282.

Weary, D.M., Huzzey, J.M. and Von Keyserlingk, M.A.G. (2009) Board-invited review: using behaviour to predict and identify ill health in animals. *Journal of Animal Science* 87(2), 770–777.

World Health Organization (WHO) Constitution. Available at: https://apps.who.int/gb/bd/pdf_files/BD_49th-en.pdf#page=6 (accessed 1 July 2022).

Part I

The Influence of the Domestic Environment on Behaviour

2 Dog Behaviour

David Ryan

The Ancient Dog

With advances in genetics and genomics, it has become evident that the domestic dog is descended from the same ancestor as today's grey wolf. However, whilst similarities in dog and wolf behaviour can still be found, today's domestic dog is by no means merely a tamed wolf.

Mitochondrial DNA evidence points to the domestication of dogs from the wolf-type ancestor as probably taking place less than 20,000 years ago, but studies suggest considerable variation in dates. The current failure of DNA analysis to provide an exact date is caused by difficulties in accurately interpreting what the DNA 'clock' is telling us. Dates for the divergence of all species and subspecies are measured by mutations in their DNA, but this can become confused since it appears that domestication (or other periods of environmental change or population bottlenecks) speeds up the rate of mitochondrial DNA mutation. This produces an artificially high number when compared with wild species and pushes back the estimation of the time of the event. It can again become difficult to be exact when, as is thought to be likely in the case of the dog, the domestication process takes place more than once, in more than one location. Interbreeding between dogs resulting from different centres of domestication further confuses the picture.

The archaeological record shows that the bones of wolves and humans are found together in sites that are half a million years old, but these may arise just from the association with or even the capture and containment of wild individuals and are not evidence of domestication. For domestication to occur, the behaviour of the wild animals has to adapt to fit the new domestic environment, otherwise subsequent generations simply revert to the wild type.

The first solid archaeological evidence of domestication appears when the teeth and skull of domestic dogs are seen to be smaller than those of the local wolves and the apparently deliberate positioning of skeletons suggests a relationship between human and dog. This kind of evidence starts to appear at settlement sites dating from about 12,000 years ago.

There are many pieces of the archaeological and DNA jigsaw that identify dogs in different parts of the world at different times, and evidence is emerging that there may have been earlier independent domestication events that subsequently died out, but currently it is fair to conclude that the domestication of dogs probably took place in more than one location, around 20,000 years ago, and was well underway by 12,000 years ago (Bradshaw, 2011).

Whilst this gives us a rough date, it does not explain how and why wild wolves changed their shape and behaviour to become dogs. These

 Companion Animal Behaviour Problems: Prevention and Management of Behaviour Problems in Veterinary Practice (eds. R. Casey, S. Heath and H. Zulch)
DOI: 10.1079/9781800621312.0002

questions can probably never be answered with absolute certainty, but we can speculate as to the most likely turn of events.

A major feature of our domestic dogs that is not seen in wild canids is the retention of puppy characteristics into adulthood, known as 'neoteny'. This restriction of adult development applies to both physical form and behaviour. Adult domestic dogs retain a strong tendency to play and the ability to approach unfamiliar dogs and people without fear after relatively little experience. They can also exhibit predatory behaviour towards non-prey items. In undomesticated canines this behaviour is restricted to puppies and is lost after the juvenile period. Likewise, the head shape of many varieties of dog is characteristic of the early developmental stages of wolf puppies, notably short muzzles on comparatively large heads.

The plasticity of both shape and behaviour demonstrated here has most likely enabled domestication and the subsequent development into the huge variety of modern breeds.

Long-term research with silver foxes in Russia (Trut, 1999) has shown that it is possible to obtain dog-like behaviour within a few generations by breeding for a reduced flight distance, tolerance of humans, or 'tameness'. Along with this come other puppy-like characteristics, such as a tendency to whine and bark. A tiny minority of the foxes also appeared with piebald coat colouring, dropped ears and lifted tails. Although we cannot extrapolate too much from this, it is interesting to note that these dog-like features emerge (albeit in around only 1% of these foxes) quickly and appear to represent what might be considered to be an early stage of domestication.

Humans and our waste products provide an ecological niche that animals can exploit and, around 20,000 years ago, a reduced flight distance would have been useful in allowing wolves to take food from humans, village waste or at hunting kill sites. The tendency not to run away but to be able to stay and feed for longer confers a survival benefit in this environment. Add to this the human tendency to take young animals as pets, or perhaps keep them as the original 'ready-meal' (or both), and the situation arises where both species benefit.

Many modern feral dogs continue to exist like this all over the world, living on the edge of human society, taking advantage of our waste and handouts. Where culture allows, dogs are still eaten as food and the most successful part of the evolution of the domestic dog is that pups are taken from their dams to be kept as pets by humans. We rear their pups for them.

The light-bulb moment in the 'are dogs wolves?' discussion comes when we examine what happens when domestic pet dogs return to the wild. They do not behave like wolves, but like feral dogs.

When most domestic animals escape or otherwise return to the wild the behaviour of their offspring tends to revert to that of the wild ancestor, albeit over a number of generations. When domestic dogs go feral, they do not revert to behaviour as seen in modern wolves, but to that of scavenging pariahs or feral dogs. This suggests that the domestication process started from a pariah-dog-type rather than from the wolf-type; that the feral dog was domesticated into the companion animal rather than the wolf being domesticated into the dog; and that the baseline behaviour of the domestic pet dog is the feral pariah dog, rather than that of the wolf.

In order to examine how much like a wolf our pet dogs behave, we need firstly to understand how wolves behave.

Wolves

The unit of grey wolf society is the 'pack' and usually consists of an unrelated, monogamous, reproducing sire and dam, and their offspring. The pack cooperates in hunting, territorial defence and care of the pups.

Female wolves come into season once each year and this coincides with a seasonal increase in testosterone in the reproducing male. Both males and females have a distinct biological breeding season – male testicular volume in June is less than half of that in the mating season in December. This maximizes breeding potential whilst minimizing the likelihood of social competition and conflict for the rest of the year.

It is usual for pups of both sexes to remain with their parents for two years or more, as further litters are born, and the relationship between the breeding pair and their offspring is one of mutual cooperation for genetic benefit.

Whilst with the pack, there is cooperation in hunting and care of the offspring and all members contribute by either 'babysitting' whilst others hunt, or by regurgitating food for stay-at-home pups and nannies. After two or more years, some wolves leave the family pack to find a mate, ensuring genetic diversity in the population.

Outright aggression is expensive in genetic terms. Fights injure and disable, so family disputes tend to be resolved through the use of social posturing and expressive body language. Rather than the populist view of wolves as fighting each other for supremacy in the pack, the wolf family unit cooperates to the best advantage for all. There is no genetic benefit in offspring challenging their parent for the right to rule the pack and mate with their own parent.

That is not to say that wolves will not attack itinerant wolves that stray into their territory, or members of other packs that might intrude. Territories are marked and guarded and any interlopers that are not driven away may be killed by the resident pack. Only when conditions force them to come together, such as in times when food sources are very concentrated, will wolves tolerate the proximity of non-pack members.

Dogs

Unlike the pack structure of the wolf, the unit of the feral dog is 'one dog' or, at the most, 'one female with dependent pups'. Feral dogs do not generally cooperate in packs as wolves do, although they do congregate in resource-rich areas, leading to a looser association with dogs with which they share a territory. Whilst they will defend their territory, it is difficult to tell if it is a joint effort or a number of individuals that happen to be doing the same thing in response to the same stimulus (Pal, 2011; see also Bonanni *et al.*, 2010).

Feral dogs are highly promiscuous, with females mating with multiple males during their biannual season. Male dogs are reproductively active all year round and, although dogs will fight for access to a bitch in heat, the fight is for precedence only.

Occasionally a male dog will help the dam with her pups, by pairing up with her and sometimes regurgitating food for them, although this is not always the case. Older siblings take no part in puppy rearing. It is of course impossible for the male dog to know if he is the sire of any of the pups the dam carries, and the pair-bonds seem to dissolve quite easily, with other males in the territory possibly pairing with the female at her next season.

Rather than utilizing the model of extended family care as a reproductive strategy as for the wolf, the dog has a strategy of increased numbers of puppies and investing less in each one. Although mortality is higher, overall numbers of puppies born are higher. Genetic diversity is ensured when each litter could be the produce of more than one sire, and bitches mate with different sires at each season.

There is little need for cooperation for food, as this is largely scavenged. Opportunistic scavenging is a relatively low-cost strategy, consisting of waiting around until low-grade food is dumped nearby or handed out. Dogs are therefore reasonably sociable with neighbours, but are prepared to defend resources when these are limited.

The use of aggression is as expensive as it is for wolves and is therefore mostly limited to self-defence or defence of a valued resource. The ancestral posturing and body language, also found in the wolf, continues to serve the domestic dog as a means of avoiding direct confrontation.

The Concept of 'Dominance'

The correct ethological usage of the concept of dominance is where, in a relationship, one individual secures access to resources, or sometimes to a single resource, over the other individual. It is a descriptive term: dominance in the relationship itself is not the goal, but the consequence of a successful competition over a resource. Dominance is conferred when one individual defers; until that point the pair is still in a state of competition. Dominance is an outcome of a competition for a resource, not an aspiration in itself nor a property of an individual.

'Dominance' put forward as a diagnosis for problem domestic dog behaviour was developed from a misinterpretation of wolf behaviour. Studies of unrelated groups of captive wolves showed that they were constantly in conflict with each other. Only the most competitively aggressive wolves secured access to the available resources. Through these observations it was presumed that conflict was the natural state of a wolf pack and that the most competitive wolf sought dominance in order to secure access to all rights.

While many took these assumptions to apply also to relationships between dogs, and in dog–human relationships, it quickly became apparent that using 'dominance' as a cause of behaviour in dogs was unsubstantiated.

There are two main flaws in the explanation. The first is that the captive wolves studied were not exhibiting usual pack behaviour. They were often a conglomeration of unrelated individuals with limited access to food, which was not hunted but provided daily by their keepers and, more importantly, in a restricted space. In the small artificial enclosures, they were constantly in competition with each other. Some won and some lost, but the losers had nowhere to escape to and continued to be locked in a cycle of conflict, but being unable to get out of the way. Winners were constantly trying to drive the losers away, but were not able to do so.

The interpretation of abnormally competitive and violent captive-wolf behaviour as the norm resulted in wolf ethologists assigning the term 'alpha' to the male or female most able to fight off the others; the 'top-wolf' of the artificial pack. However, when the study of wolf behaviour moved away from artificial accumulations of unrelated wolves and began to focus on wolf packs in their natural habitat, there was a swing to a more accurate description of the pair as parents that hold the pack together.

Wolf biologist L. David Mech pointed out in his 2008 article *Whatever Happened to the Term Alpha Wolf?* (Mech, 2008) that in the 2003 book *Wolves: Behavior, Ecology, and Conservation* (Mech and Boitani, 2003), written by 23 authors, the term 'alpha' is only ever mentioned to explain why it has been superseded.

The second flaw is that there is evidence suggesting that the social behaviour of domestic dogs is completely different from that of wolves. Whilst it is unclear exactly when social behaviour diverged from that of the wolf, studies of modern-day feral and group-housed domestic dogs have identified a much more flexible social structure.

The original proposition that a wolf must dominate all others or in turn be dominated filtered down to dog trainers and behaviourists. It became popular belief that badly behaved dogs were trying to exert dominance over their owners, particularly if the behaviour involved aggression. Whilst this is now recognized as incorrect, it should be emphasized that it made sense given the information available at the time.

Many leading animal behaviourists are concerned that this explanation of pet dog behaviour continues to survive, despite the accumulating evidence that it is at best unhelpful and at worst highly detrimental. Labelling a dog as 'dominant' gives a trainer or owner licence to fight back, whereas labelling it 'in need of understanding and guidance' often provides a more thought-out attempt at resolution of inappropriate behaviour.

There is no doubt that using the types of coercive methods that derived from interpretations of behaviour based on 'dominance' can work through subduing a dog, provided that the practitioner is able to compete on the dog's terms. If this is in the use of aggression, the handler has to out-fight the dog, perhaps repeatedly, with all the attendant risk. From the dog's perspective, because the underlying determination of the dog's behaviour is incorrect, it often does not address the cause, but attempts to deal with the symptoms as they arise. This additionally only indicates to the dog the behaviour that is not accepted, with usually no indication of what behaviour is expected.

This type of approach also causes welfare concerns for dogs. This applies to the more combative elements such as leash 'corrections', lifting a dog by the scruff, rolling it over and holding it down by the throat, as these techniques are likely to cause fear, anxiety and pain. However, welfare can also be compromised by the apparently more benign but insidious recommendations such as withholding attention for long periods and social exclusion by banishing the dog to the 'outer edges of the pack'. Suggestions such as eating a biscuit before feeding the dog, in order to show that the 'alpha eats first', probably do not cause the dog any harm, but neither do they empower the owner to take control of their dog's behaviour, or resolve any problems that owners may have with their pet.

Wolf ethologists revised their views of wolf social organization and have published their amendments extensively, as have the proponents of domestic dog behaviour, but some trainers and owners still cling to the outdated and discredited idea that the aim of problem dogs is to dominate their owners.

The Modern Dog

A feral dog, captured from its dam at six or eight weeks old, can become a successful pet. Its needs are few and it is able to form strong social bonds with humans. It is also remarkably flexible in behaviour and in the possibilities of breeding for specific traits.

Breeding for specific traits in dogs was probably one of the first deliberate acts of animal husbandry by humans, for example when they realized that one dog was faster than another, or barked more readily at intruders than others did. Or, even in the early stages of selection, there may have been a preference for prettier pets with cuter faces.

Whatever the reasons, the earliest representations of dogs show different types, intentionally bred for different functions such as hunting by sight and scent, or guarding, and it is these ancient artificial selections that are reflected in our pedigree dogs today. Current pedigrees focus mainly on looks or adherence to a specified breed standard, but this is a relatively recent phenomenon, the UK Kennel Club being founded in only 1873, fewer than 150 years ago and less than 1% of the time period over which dogs have been domesticated. For at least the previous 6000 years, when we have our first proof in the depiction of different breeds on ceramic pottery, dogs were bred for function, with form a secondary consideration.

Individual breeds of dog have a unique suite of behaviours ('breed-specific behaviours') which they are motivated to perform. Breed-specific behaviour is inherited and there is a distribution across each breed as to strength of motivation to show these behaviours; in other words, some individuals have a greater desire compared with others. Hence, although overall Whippets are more likely than Pyrenean Mountain Dogs to chase rabbits, one Whippet may be more so inclined than another.

There are two main issues associated with this when considering pets. The first is that owners often expect dog behaviour to be 'standard' for a breed and the second is that they often fail to understand the impact of breed-specific behaviour. For example, it can be considered normal for most Golden Retrievers to want to hold something in their mouths and for most Border Collies to want to round things up.

It is a strange state of affairs when we penalize dogs for being too good at what we bred them to do, but sometimes it is necessary to explain to owners that a factor in the development of a 'problem' behaviour may be a characteristic which actually makes the dog a good specimen of the breed.

Acknowledging that breed differences exist, we should also acknowledge differences between the sexes. For example, several epidemiological studies have suggested that male dogs have a higher risk of showing behaviours such as aggression as compared with females.

The final major consideration in the way a dog behaves is underlying personality characteristics, or individual differences. One such characteristic is considered to be a continuum from 'bold' to 'shy' (Wilson *et al.*, 1994; Turcsán *et al.*, 2011). Shy dogs are more likely to be anxious in novel situations and will approach life very differently from bolder individuals.

Breed considerations, differences between the sexes and personality traits produce a very individual animal that develops its own behavioural characteristics as it learns and grows, trying to fit into a modern family environment.

The modern dog has gone through several stages of modification; from ancient wolf to feral, from feral to worker, and from worker to 21st century pet. It seems that the last stage, adapting to our 21st century requirements, has proved the most difficult. UK National Health Service (NHS) figures show that the incidence of hospital admissions for 'dog bite or strike' doubled in 20 years, from 6.34 per 100,000 population in 1998 to 14.99 per 100,000 in 2018. Whilst the numbers of children admitted remained relatively static, the increase was driven by a tripling of incidence in adults and is estimated to have cost the NHS over £25 million in the financial year 2017/18. Tullock *et al.* (2021) speculated on several possible causes, including an increase in dog ownership, changes in the dynamics of ownership and failing to meet dogs' exercise needs, but did not pin down any one overriding factor. The overwhelming majority of bites, in the region of three-quarters, are to people known to the dog, often family members and most distressingly children. These statistics suggest a failure in communication between species.

Communication

Dogs are suggested by some researchers to have an almost inherent ability to understand some aspects of human communication. Some of this ability comes from similarities between the species; both dogs and humans find low tones more threatening and higher tones more attractive. Other abilities seem to have emerged from the domestication process; dogs do not appear to need to learn to follow the direction of a pointing finger (other animals, including wolves, at best look at the end of the finger). Research from the Eötvös Lorand University, Budapest, much of it detailed in Adam Miklosi's (2007) book *Dog Behaviour, Evolution, and Cognition*, suggests that dogs are particularly tuned to and attentive towards human communication, for example investigating where we look when we have signalled that it might be pertinent to them (Téglás *et al.*, 2012).

This ability makes dogs uniquely suited to sharing our lives and homes. However, because dogs adapt to our lives relatively 'easily', it can also mean that people may not make sufficient effort to understand the subtleties of their communication. Unfortunately, when owners fail to understand behavioural signals, the result can be undesired behaviours including aggression and the breakdown of the owner–pet bond.

Dog social communication is expressed through body postures and is based on a fairly universal canine template. The same postures, with the same meanings, are identifiable in most other branches of the canid family.

Unfortunately, we have bred domestic dogs of such diverse shapes that many of the original 'template' communications are either not available to them or difficult to read. Bulging eyes, flattened or folded faces, jowls, droopy ears, shaggy coats, floppy fringes and shortened, curly and permanently upright tails hamper their ability to both give and receive visual signalling to such an extent that many breeds are incapable of a full range of meaningful canine communication.

They also restrict owners' ability to recognize and understand when dogs express different feelings. It can be difficult for an expert to tell when a solid-coloured Cocker Spaniel is feeling uncomfortable from its facial expression when it starts out with droopy eyes and ears and a lack of facial mobility. For a 6-year-old child hugging it on the sofa it is almost impossible. This can be an important factor in the occurrence of a disfiguring bite to the face that 'came out of nowhere'.

Whilst dogs appear to be doing their best to communicate with us, many of the ways that we interpret them leave much to be desired, as do many of the ways that we instinctively interact with them. For example, dogs can find close face-to-face contact threatening, as it is not a natural form of dog communication. Similarly, dogs do not hug each other and it restricts their ability to move away – one of the most important alternative behaviours available to a worried dog.

Conversely, humans, and especially children, are extremely attracted to close face-to-face contact and to hugging. Teaching children not to hug dogs is a first step in bite prevention. Teaching dogs that being hugged can be pleasant is the second.

The two other main areas of conflicting communication are the dog greeting and the dog threat-reducing behaviours, although there is some overlap between the two because greeting can also be a form of threat reduction for the dog.

Dogs are often in emotional conflict when greeting people. They want to make contact because people are exciting and rewarding, but they also want to show they mean no harm. When puppies do this to another dog it involves jumping up and licking at their mouth. When dogs try to do it to a human they often get shouted at or otherwise punished. This makes them feel more worried about the contact and they try harder to show they are no threat, which can make them jump up more, or alternatively do the 'ultimate puppy' by rolling on their back and urinating. Of course, it is possible to use each of these threat-reduction behaviours in isolation, so some dogs will just roll over, some will lick and some will urinate, but we should recognize them for what they are – signs that the dog is trying to be non-confrontational.

How owners respond to these behaviours is important in the development of subsequent problems. The natural inclination for most people is to bend over the dog and pat them on the head. Whilst this is meant in the best possible way, it can make the situation more difficult for the dog. It involves looming towards, leaning over, and reaching with the hand to a part of the body the dog cannot see – all of which can be perceived by the dog as threatening. Anxiety in dogs when this happens can result in attempts to avoid contact, including aggression. The best way for owners to greet a worried dog is to crouch down to their height, sideways on and offer themselves for greeting, then wait until they approach. If a dog jumps in greeting, teach it another behaviour out of context ('sit' is popular) then quietly insert it into greeting by asking for it and rewarding with the contact they seek.

When dogs are worried about close social contact, they may attempt to divert the threat through small gestures such as averting their eyes, dropping their ears and turning their head away and down, and if some of these gestures have been tried and have not worked to make the person move away, a dog may lick at the face. If this is interpreted anthropomorphically (and children really cannot be blamed for doing so) it will appear that the dog is 'kissing' the child and they may respond with more affectionate hugging – unfortunately and dangerously, exactly the opposite of what the dog was trying to achieve.

The 'roll-over' too is often misinterpreted as a gesture that is encouraging contact, when it can be the exact opposite. Unless the dog has specifically learned to have its tummy tickled for pleasure, the roll-over is a natural threat-reduction gesture, intended to remove contact, not encourage it (Fig. 2.1).

If any of these attempts to reduce contact fail, the dog's worry increases and it may reach

Fig. 2.1. Collie 'Fox', initially looking worried **(a)**, turns her head to avoid Lucas's play **(b)**, then rolls over **(c)**. This is not submission or an invitation to rub her tummy, but an attempt to avoid contact. (Photos: author's own.)

the point where it feels it necessary to get out of the situation. If it is being held or confined, cuddled, petted or hugged, the next option may be a snap, which will appear to owners to have come out of the blue (Fig. 2.2).

Of course, many dogs are hugged every day and some may learn to enjoy it. So, why do some adapt to this form of human interaction and others do not? As well as the individual differences mentioned above, another important factor is the experience that dogs have from a very young age.

Learning

Lifelong learning is an accepted and well-promoted concept in humans, but often owners think that 'dog training' is something practised in a village hall for an hour a week when the dog is about nine months old.

The laws of learning are like the laws of gravity in that they are never switched off, and anyone who expects a balanced relationship with their pet should be prepared to invest time and effort in preparing them for the challenges of daily life, throughout their lives.

In other species there is good evidence to show that being born in a stressful environment can alter response to stress later in life (Meaney *et al.*, 1991). This is because it is adaptive for offspring born from mothers living in an unpredictable or stressful environment to be more reactive to changing circumstances. Obviously, in a domestic situation, puppies leave their maternal

Fig. 2.2. The close face-to-face contact enjoyed by children can be threatening for dogs. Seen here, Fox is displaying canine threat-reducing gestures of turning her head away with narrowed eyes and dropped ears (a), then lip-licking (b), and finally licking at Lucas's face (c). If this is unsuccessful in getting the child to move away, the next step the dog may take may be to bite. (Photos: author's own.)

environment early in life and increased response to environmental change is not desirable; hence it is important for prospective owners to be sure that their future dog is acquired from an environment that was not overly stressful for the bitch.

Although the importance of what is generally known as the 'socialization period' will be explored in more detail later in the book (see Chapter 8), it is important to note that the neural development necessary for simple associative learning begins in the neonatal puppy within the first few weeks of life. It is therefore possible to train simple responses in very young pups, although they will still be hampered by their lack of development in other aspects. Realistically, owners will begin interacting with their pet when they take them home, at around eight weeks old.

Even at this early age, pups will show differing responses depending upon their inherited predispositions. An awareness that temperament and personality types exist, even in young pups, may help owners.

Training for life is about guiding the dog at every choice it makes. Training should be synonymous with living together. Every minute of every day each pup is making choices. Guiding them towards the most appropriate choice early in life is the most important advice that a veterinary practice can give an owner. Owners need to be aware that it is important to guide puppies towards the right choice from the start, rather than waiting for the pup to make the wrong choice and then correcting it. Although this sounds a little labour intensive, it does not have to be once the owner is in the 'parenting' frame of mind. If a pup has just woken up, eaten, drunk or finished playing, it probably needs to go to the toilet. If owners take them outside at the right time, then no mistakes occur. Toilet training becomes an errorless lesson and a pleasant experience for both pup and owner.

Tuning in to what the puppy needs and making this available when they show appropriate behaviour requires a little twist in how most owners perceive 'training', but can be easily explained in terms of good manners: asking the pup to behave appropriately to earn what it wants helps the owner gain control effectively, without conflict.

Some learning opportunities are less obvious to owners because the value to the puppy is not understood. For example, when a puppy goes out of the door to play, going through the doorway into the garden becomes rewarding in itself and the behaviour the dog offers in order to be let out is reinforced by the opening of the door. The door opening being conditional on calm behaviour being demonstrated by the puppy encourages the dog to be calm when it wants to go out.

Learning most benefits from tiny interactions throughout the day, because everything the dog wants can be guided in the right direction and reinforced, providing the owner with a well-behaved dog without the need for using force.

Dog Sociability

The relative sociability of dogs to other dogs also has a strong learnt component. Although the motor patterns required for social signalling are inherent, each puppy needs to learn how to use them in the right contexts. As with all learning, this is best started at a young age and the sensitive period for learning social skills starts when puppies are able to interact with their litter-mates (see Chapter 8 for further detail).

Because dog breeds are so diverse, it would be too much to expect puppies that have learned the body language skills of their own breed to transpose that understanding to an interpretation of all other breeds' signalling. Dogs need to be appropriately exposed to canine conversation with as many different breeds as is possible, for them to learn that their different styles of interacting are not necessarily threatening.

A shy dog that avoids body contact will need to be persuaded that a confident bouncy dog means no harm. Both dogs benefit from a gentle introduction where neither is allowed to intrude upon the other's sensitivities; the shy dog to learn to tolerate another dog's behaviour and the bouncy dog to learn to moderate its natural exuberance. Well-run puppy classes will help with this (see Chapter 12).

This is also a good time to teach puppies a 'come away' cue. By watching the interaction between two pups it will become apparent when either is starting to be stressed by the proximity or behaviour of the other. That is the time to call the puppy away, breaking the contact and rewarding with the increased distance from the threat (and some praise for making the right decision). Teaching a puppy to do this from a young age establishes a pattern of behaviour where the dog walks away from potential trouble with other dogs, reducing the risk for aggression.

Many books define an age where the sensitive period for social development closes but this can be too proscriptive and misleading. Whilst it is very beneficial to have early social learning, that exposure to other dogs must continue for life in order for social skills to be retained and refined. Withdrawing a dog from dog society even later in life can make it harder for them to reintegrate. It is also believed that a second sensitive period may exist around puberty, when dogs change from puppy behaviours to more adult ones. Keeping up benign exposure to other dogs at this time helps them adjust to adult dog society.

Of course, it is vital that exposure to other dogs *is* benign. The only thing possibly worse than no exposure to other dogs at all is repeated aversive encounters, such as fights. Almost all cases of aggression towards other dogs can be traced either to a lack of social exposure (especially during the sensitive periods) or to specific aversive encounters. In some cases, a single encounter with an aggressive dog can be enough to set up a generalized fear response.

As dogs also need to interact with a range of different people, everything that has been said about dog communication also goes for dog/human communication. Dogs need to be positively introduced to a variety of types of humans and their paraphernalia for them to understand that they are no threat. If dogs have only ever seen quiet adults, a squealing 6-year-old or an unstable toddler, arms outstretched for a cuddle, is like an alien species to them. Puppies need to learn that whatever the type of human, and no matter what they do, it is not a threat to them.

In the same way that aversive encounters with other dogs can set up a fearful aggressive response, aversive encounters with types of people, especially 'non-standard people', can do the same. 'Non-standard' is simply defined by what the dog is not used to and can include anything from men with beards, through ladies with hats, to umbrellas, back-packs, walking-sticks or bicycles. Conversely, once the dog has seen a variety of types, each new type becomes potentially less threatening. Essentially, the more varied a pup's range of experience, the less likely they are to be anxious about new situations.

Frightening encounters lead to anxiety that the next similar encounter will also be negative. The first time the dog is confronted with a screaming toddler and hides under the table sets up a fear response for the future. A puppy hiding under the chair for six weeks of a puppy class is probably learning little other than to hide; one dragged out from the chair and forced to interact is probably learning to be afraid of people and dogs.

Aggression

Aggression is a normal canine response. It is one response from a number of possibilities associated with negative emotional states. The more intense the emotion becomes, the more obvious the aggression appears. It is normally used defensively, through fear, or as a consequence of frustration. However, once an aggressive response has 'worked' for a dog, either by removing a perceived threat or by reducing frustration, it can become a learned response that can be applied when the dog encounters a similar situation in the future.

If a dog learns to use aggression to resolve perceived problems in many contexts, it can become the favoured response to new situations that cause anxiety. Thus if a dog learns to use aggression when frightened by another dog, it can ultimately start to use it when frightened by people. When it becomes a successful strategy, the dog is also likely to use the aggression earlier in future encounters.

Dogs have a limited number of responses with which they are able to deal with a threatening situation. It must be remembered that a 'threatening situation' is one that is judged so by the dog. Whilst we know that a toddler hugging a dog is not an actual threat, if it is perceived to be so by the dog it will deal with it as it feels appropriate. This is where inadequate social learning is so important. Things appear threatening through a lack of understanding that the situation is *not* so, or through having previously learned that it *is*.

Threats can be either avoided or challenged. If a dog that has not previously experienced this situation as safe sees a toddler approaching, arms outstretched, and it bolts under the table, it avoids the perceived threat on that occasion. However, it does not actually deal with the perception for the future (i.e., toddlers are still scary), but the dog has learnt a strategy: it can run away and hide. If the toddler pursues the dog under the table, it now has nowhere to run and hide and no means of avoiding the threat. It will probably try social communication that reflects its anguish, such as dropped ears, lowered body posture, tail clamped under and a hunted expression, but we know that humans are very poor at interpreting dogs, and toddlers more so. It may try snarling or growling as a last resort, but if the toddler presents what it regards to be an immediate danger, it may also bite. Having tried every other means in its repertoire, it has no option (Fig. 2.3).

As well as perceiving threat from the unfamiliar, dogs can develop an aggressive response as a result of aversive experiences, such as being attacked by another dog whilst being walked on a lead in the park. From a single encounter like this, dogs can learn to be wary of unfamiliar dogs because some may bite. It learns that the particular dog that attacked is to be feared, but also that similar dogs may be dangerous (more so the more like the original dog they are); it learns that the park is a potentially dangerous place; it learns that it cannot avoid conflict when it is on a lead; and it learns that its owner does not protect it.

The very next time the dog is taken to the park it will remember the fight and may well be anxious, with elevated stress levels just from being there. If it is approached by another dog whilst on the lead, it knows it is vulnerable and its stress levels will rise some more. It will hold a more rigid body and perhaps a defensive posture, expecting trouble and ready to defend itself. The other dog, which may well have had friendly intentions to start with, will recognize

Fig. 2.3. Fox's worry at Bullie's approach is reflected in her lip-licking (a) and yawning (b). She sits up and growls with a lip-curl when Bullie encroaches too far (c). Note the similarities with the lip-licking shown towards Lucas in Fig. 2.2. Without guidance, her next step may be to bite. (Photos: author's own.)

the defensive attitude and as a result may well become defensive itself. This makes the dog even more anxious, stressed and defensive, which is reflected in the other dog ... and so on until eventually one quite literally snaps and a fight breaks out.

The dog is now more convinced than ever that other dogs are a problem and the park is dangerous, although now stress levels will increase at any time or place when another dog appears. From the dog's point of view, it has to deal with this stress and fear of other dogs, but it cannot avoid them – it keeps being taken for walks on a lead to places where they appear.

The dog must develop a strategy for dealing with this stress. Being faced with a threat, and frustrated at the inability to reduce it by moving away because it is on a lead, aggression often comes to the fore. Of course, temperament dispositions have some bearing as some types slip into aggression more easily than others, but learning has far greater impact.

In anticipation of negative encounters, the dog will start to bark and snarl when it sees another dog in the distance and if the owner turns around and walks away in embarrassment they effectively reinforce the aggressive response by increasing the distance from the feared dog.

After successive encounters the dog rapidly learns that the way to get out of the situation is to be aggressive in advance. Each time it shows aggression the owner drags it away. Once they are removed, the relief they experience is a powerful reinforcer for the aggressive behaviour. The next time the dog feels the stress of a possible encounter it uses the aggression earlier and resolves the situation sooner. If it is not immediately taken away from the other dog, it increases the level of aggression, then the next time it skips immediately to the more extreme level of aggression as soon as it sees another dog.

In some dogs this progression can take place over weeks or months, but with others it may take place in only a few encounters.

When training a dog not to show fear-based aggression it is important to understand that the reward the dog is working for is the feeling of relief and that is what the trainer should be endeavouring to arrange through an alternative behaviour. Any behaviour other than aggression that provides the relief at removal of the fear-inducing stimulus (such as a request to 'come away') can be used as a substitute, but it must be used before any expression of aggression.

Frustration, the other major source of aggression, is caused by preventing a dog reaching what it regards as an important goal, the dog experiencing unmet expectations or the dog experiencing restraint. Again, underlying temperament will play a role, but learning is likely to be an important factor.

Access to a figure to which a dog is strongly attached, most often the human owner, may be very important to some dogs, and loss of that access may cause frustration, resulting in an aggressive outburst in some cases. Such dogs have been known to attack their owners when the telephone rings, as it is a reliable signal that the owner's attention is not on offer for a period of time. Unable to control this loss of attention the dog becomes frustrated, and there is no other target for the aggression but the owner.

Once this happens, there is also a learned component as the owner quickly puts the phone down and talks to the dog instead. The dog is subject to the removal of the owner's attention, causing frustration and aggression, which results in the reward of regaining the very attention sought. This clear source of reinforcement for a successful strategy can have implications in many other areas, such as visitors arriving, the kettle boiling, or putting on the 'shopping coat'. Even worse for the owner, if they are able to resist some of the time, the level of aggression will increase through even more frustration and intermittent reinforcement.

Once an aggressive response becomes a successful strategy for resolving negative states of fear or frustration, dogs will progressively become more confident in displaying the behaviour. For example, imagine a dog reacting to the daily visit of a postman coming to the door and rattling the letterbox. Originally fearing an intruder into the home, the dog barks aggressively. The postman leaves immediately (as they were doing anyway). On some days the postman lingers for longer as more mail is delivered and the dog increases the level of aggression but, from the dog's point of view, it always succeeds in driving the intruder away. The relief felt when the postman leaves reinforces the aggressive strategy and the dog starts to learn how to deal with the challenge. The distress felt reduces to the extent that the event now becomes a situation that the dog has control over; it identifies the cues which predict the visit, reacts by barking, and successfully sends the postman away.

Any regular event the dog can control through the use of aggression is unlikely to cause negative emotional states in the dog, as they can readily identify predictive cues, show an immediate behavioural response and resolve the 'problem' – this immediate response will be reinforced through the relief of avoiding threat and becomes more 'automatic' than considered over time.

It is debatable if aggression used in play can accurately be described as such, as true aggression is regarded as starting from a negative emotional state, whereas play is a positive emotion. However, to the person, dog or other animal bitten in play, the definition may be immaterial; dogs can seriously injure or kill in play.

Many biting accidents are exactly that, an accidental application of the teeth too strongly when playing; a rough-house that goes too far. Dogs need to be able to judge how far to go in play and some simply get it wrong and overstep the mark, resulting in injuries, especially to fragile-skinned children and the elderly. It is therefore vital that owners teach their puppies how to inhibit their play with humans, dogs and other animals (see Chapters 8 and 12).

The original function of many breeds involved using their mouths to grip, for example dogs bred for vermin control, herding and dog-fighting. Although most pets no longer fulfil their original purpose, they still have the potential to become absorbed in breed-specific behaviours. This is why some (but by no means all) Staffordshire bull terriers are difficult to break up once they start to fight; it is why collies nip at heels and terriers may struggle to let go of their ragger. The motivation to show these behaviours can result in injury, even where dogs are given appropriate early experiences.

Usually dogs bound up in breed-specific behaviours do not intend to harm people, but accidents do happen. Owners need to recognize when their pet (irrespective of breed) is going beyond the bounds of acceptable behaviour during play and step in with guidance before any injuries occur.

Separation-related Behaviour

Guiding dogs to show appropriate behaviours is a theme that runs throughout our relationship with dogs and is paradoxically most important when they are left on their own.

As dogs develop within human families, it is not surprising that they attach very strongly to people. In many species there is a strong bond between offspring and parent which in humans we call 'love', and the analogous relationship between owner and dog is often referred to as 'attachment'.

Dogs show an increase in oxytocin, the neuro-hormone released when humans experience love, when they are petted and indulge in gentle play with their human owners (Nagasawa *et al.*, 2015). Although subjective feelings and emotions are difficult to define, some researchers suggest that dogs have the same kind of feelings we have, if not to the same degree (Bekoff, 2012).

Without this capacity to attach so strongly to us, it is doubtful that we could keep dogs as the pets we do. Unfortunately, for this huge benefit there is also a huge disadvantage. If we take it that dogs attach as strongly to us as a child does to their parent, that they experience happiness and generate a supply of beneficial hormones when we are around them, we must also accept that they experience a loss when we are not there; that, alone, they feel at best less happy and at worst miserable.

Bringing home a new puppy will be dealt with in greater depth in Chapter 8, but the time of separation of the pup from its dam is a crucial one for the development of future behaviour. Puppies taken from their mother will be extremely anxious and will rapidly form a relationship with their new owner to feel secure. This is the start of the bond between owner and pet that can last a lifetime.

There are differences in individual puppies because of individual personality characteristics, but it is mainly the way in which the puppy is treated during these very receptive weeks that sets the pattern for their future behaviour.

Our tendency to lavish care and attention on the puppy all day long is in direct contrast with the behaviour of the pup's natural dam, which at this time would be weaning and encouraging independence. Care and attention are important, but we should also be encouraging the puppy to relax quietly on its own. Learning to cope alone when owners leave the room is a good start towards avoiding the development of separation-related problem behaviours.

Separation-related behaviours (defined as undesired behaviours that occur when dogs are left without human company, including toileting, vocalizing and destruction) regularly make up the third biggest portion of a behaviourist's caseload after aggression towards people and other dogs. The distress is exhibited most strongly at the time of departure, although it can continue for many hours afterwards.

Puppies left on their own for the first time will always feel that loss, but it is how they deal with it that shapes their behaviour for the future. If they are left to their own devices, without guidance, they may develop a range of behaviours including vocalizing (a contact call to try to solicit company), chewing items (a comforting 'thumb-sucking' behaviour) and digging at doors (attempting to break out).

Even adult dogs that have been happy to be left alone can develop problems when the circumstances change. A dog previously happy to stay at home on its own may have no concept that the same rules apply, that their owner will return, when the family moves to a new house, and in their own home dogs can face challenges with which they are unable to cope, such as building work in the flat above, that can lead to a fear of being at home alone.

Extraordinary events will be dealt with more easily if the dog has had the opportunity to learn that being left alone in a range of circumstances is a normal part of life.

Conclusions

The threads that run throughout this chapter are worth emphasizing and bringing together.

- A dog is a unique species, with different social structures and behaviours from the wolf. Dogs have evolved to live in harmony with us, not to rule over or 'dominate' us.
- Despite their incredible ability to relate to humans, treating a dog like a human is as much a mistake as treating it like a wolf – it is important for owners to understand the communication signals of their dog so they can adjust their behaviour accordingly.
- Providing a dog with a range of positive experiences, of both physical environments and different social experiences throughout life, is essential for it to grow into a well-balanced pet.
- Breed, temperament and individual characteristics will guide a dog down certain paths and if these paths do not correspond with appropriate behaviour they will need to be identified early and diverted into appropriate responses.
- Experiencing fear or frustration, emotions apparent in many dogs exhibiting problem behaviour, needs to be managed and kept to a minimum.
- All dogs need guidance. Some owners are so good at it that they hardly notice they are doing it, and others need to work at it. Controlling the environment to ensure that dogs learn the desired response is an essential part of being a dog owner.
- Guidance does not have to be heavy-handed. Dogs are best managed by gently pointing them in the right direction for every decision they need to make, until they only ever make the right ones.
- Do not wait for mistakes that have to be corrected, but anticipate and guide before they happen. In the extreme circumstance where that cannot be done, next time help the dog make the right choice as soon as possible.

References

Bekoff, M. (2012) *Mark Bekoff Action4Dogs Seminar*, 9/10th June 2012. Craiglockhart Campus, Napier University, Edinburgh.

Bonanni, R., Cafazzo, S., Valsecchi, P. and Natoli, E. (2010) Effect of affiliative and agonistic relationships on leadership behaviour in free-ranging dogs. *Animal Behaviour* 79(5), 981–991.

Bradshaw, J. (2011) *In Defence of Dogs*. Allen Lane, London.

Meaney, M., Mitchell, J., Aitken, D., Bhatnagar, S., Bodnoff, S. *et al*. (1991) The effects of neonatal handling on the development of the adrenocortical-response to stress: Implications for neuropathology and cognitive deficits in later life. *Psychoneuroendocrinology* 16, 85–10.

Mech, L.D. (2008) Whatever happened to the term ‘alpha wolf’? *International Wolf* 18, 4. Available at https://www.wolf.org/wp-content/uploads/2013/12/winter2008.pdf (accessed February 2022)
Mech, L.D. and Boitani, L. (2003) Wolf social ecology. In: Mech, L.D. and Boitani, L. (eds) *Wolves: Behavior, Ecology and Conservation*. University of Chicago Press, Chicago, Illinois, pp. 1–34.
Miklosi, A. (2007) *Dog Behaviour, Evolution and Cognition*. Oxford University Press, Oxford.
Nagasawa, M., Mitsui, S., En, S., Ohtani, N., Ohta, M. *et al*. (2015) Oxytocin-gaze positive loop and the coevolution of human-dog bonds. *Science* 348 (6232), 333–336.
Pal, S.K. (2011) Mating system of free-ranging dogs (*Canis familiaris*). *International Journal of Zoology* 2011, 314216.
Téglás, E., Gergely, A., Kupán, K., Miklósi, A. and Topál, J. (2012) Dogs' gaze following is tuned to human communicative signals. *Current Biology* 22, 3.
Trut, L.N. (1999) Early canid domestication: the farm-fox experiment. *American Scientist* 87, 160–169.
Tulloch, J.S.P., Owczarczak-Garstecka, S.C., Fleming, K.M., Vivancos, R. and Westgarth, C. (2021) English hospital episode data analysis (1998–2018) reveal that the rise in dog bite hospital admissions is driven by adult cases. *Scientific Reports* 11, 1767. doi: 10.1038/s41598-021-81527-7
Turcsán, B., Kubinyi, E. and Miklósi, A. (2011) Trainability and boldness traits differ between dog breed clusters based on conventional breed categories and genetic relatedness. *Applied Animal Behaviour Science* 132, 61–70.
Wilson, D.S., Clark, A.B., Coleman, K. and Dearstyne, T. (1994) Shyness and boldness in humans and other animals. *Trends in Ecology and Evolution* 9, 442–446.

3 Cat Behaviour

Jon Bowen

Origins and Domestication of the Domestic Cat

The domestic cat is a subspecies of *Felis silvestris*, a species of cat that is native to Europe, Africa, India, China and parts of Asia. Human settlements in these regions provided opportunities for cats to catch rodents that were attracted to food stores. Cats were therefore tolerated and encouraged by humans for their purposes of pest control. These opportunities introduced a selection pressure, which favoured cats that were able to live in close proximity with people and other species, such as the domestic dog, as well as in dense communities of conspecifics. The ability of cats to cope with solitary and group living has enabled this species to achieve an exceptionally large natural geographical distribution. Its association with humans has carried it even further to regions where it is not a native species, such as Australia and the Americas, which it has gone on to colonize successfully. Because cats are relatively small, non-threatening to people and carry out hunting behaviour without any human involvement, there has been no need to domesticate cats to the degree that was necessary with other species such as the dog. As a result, most of the behavioural repertoire of the ancestral species remains intact. This poses a challenge in modern suburban environments where pet cats lack the freedom to perform their normal behaviour and may be forced to conform to human expectations of social and feeding behaviour.

Social Behaviour

We have a long history of misunderstanding the social organization of other species. A prime example is that of the wolf, as discussed in Chapter 2. The tolerance of group size varies considerably within *Felis silvestris*, with some subspecies such as the Scottish wildcat (*Felis silvestris grampia*) being almost completely solitary, and others such as the African wildcat (*Felis silvestris lybica*) being more social and willing to live in groups. This may reflect climate differences in the locations where these subspecies evolved; in temperate and cold climates where human population density is also low, sources of food and shelter may be more sparse and subject to greater competition.

As a result of the depiction of cats in popular literature, and the results of early studies of cat behaviour, cats have been mistakenly characterized as solitary animals. As is the case with the dog, this is a false perception that has persisted in the popular imagination. Whilst they do use distance-maintaining behaviour to avoid direct conflict with each other, the importance of social interaction in this species is clear from

DOI: 10.1079/9781800621312.0003

the presence of specific distance-decreasing and affiliative behaviours.

Observational studies by authors such as Leyhausen (1965), Laundre (1977) and Fagen (1978) expanded knowledge of the social communication and organization of cats and revealed their capacity for group living (Fig. 3.1). The work of authors such as Dards (1978, 1981, 1983), who studied dockyard cats in the UK, and Denny *et al.* (2002) have identified a now well-recognized functional template for domestic cat organization in which related female cats form social groups along with their offspring and juvenile cats.

The function of the groups is to aid in successful kitten rearing. Males are loosely associated with these groups of females and roam a much larger territory that encompasses the home ranges of several groups of females.

These groups of females form in areas where food and shelter resources are abundant, but group members still continue to act independently of one another. Unlike some species, in which females are hostile to the offspring of conspecifics, nursing queens seem to show little discrimination when caregiving to kittens; they will feed and groom any kitten that approaches them (Ohkawa and Hidaka, 1987). This adaptation provides the basis for the formation of female social groups, as it enables kittens to be protected and cared for by members of the group whilst others are hunting. Even so, the only sharing of food between cats is between mothers and their kittens; meals are not shared between adult females.

Until they are completely weaned and able to catch and kill their own prey, kittens are not a source of competition for resources with adult cats, because their nutritional requirements are met by the mother. Once young cats become fully independent, they do become potential competitors for food and shelter resources. If these are not sufficiently abundant, then females may leave the group.

Whilst there are situations in which resources naturally occur in high density, human settlements appear to be a significant driving force behind the formation of more permanent feline groups (Kerby and McDonald, 1988). Where people are absent, and locations of high resource density are therefore absent, cats favour a solitary lifestyle other than when they are raising kittens (Van Aarde, 1978).

Aggression is seen between females that are not members of the same social group, as there is competition between groups for access to food and shelter. However, juvenile cats are more receptive to social contact with unfamiliar

Fig. 3.1. Cats do spontaneously form social groups, which are composed of related adult females with their juvenile offspring. (Photo: iStock.)

individuals, which may favour the formation of new groups by young adult cats.

Male cats generally do not tolerate contact with each other and will compete strongly for control over access to an area that encompasses the home ranges of groups or individual female cats. However, some authors report that well-matched males sharing a neighbourhood will occasionally stop fighting and form loose social relationships that are termed 'brotherhoods' (Leyhausen, 1988).

Neutering has an effect on the social behaviour of male cats, reducing their territory size and the level of inter-male conflict. This is observed in large suburban cat colonies in which neutered males participate equally with females.

Key points

- Cats are a discretionary social species; social groups are formed under certain circumstances, but this is not essential to the survival of the individual.
- The functional basis for the formation of social groups is kitten rearing by groups of related females.
- Groups form when resources are abundant and break up when they dwindle.

Communication

Cats use the three sensory channels for communication: sight, sound and scent. Each of these has certain advantages and disadvantages for communication. Communication in the cat has evolved to enable cats to maintain distance from each other in both time and space, primarily through a system of chemical signals.

Visual communication

The function of the visual system of cats is highly biased in favour of predatory behaviour. The cat's retina has three times the rod density of the human eye, contributing to the light detection threshold in cats being eight times lower than in humans. Although their vision is not monochromatic, cats have sixteen times fewer wavelength-comparing retinal ganglia than primates and are therefore essentially colour blind in terms of behavioural responses. The cat's enhanced visual sensitivity to brightness, patterns and movement blocks learning associated with differences in colour; under normal conditions cats do not appear to learn associations based on colour discrimination. The optic nerve in the cat has a much lower density of nerve fibres than in humans, due to the much smaller amount of information transmitted from the cat's retina. This is because the ratio of ganglionic cells to photoreceptors is very high in the cat; there is a far higher level of integration of information at the level of the ganglionic layer of the retina. The benefit of this is that movement detection is hard-wired into the sensory system and able to directly drive fast-responding reflexive systems; the cat is therefore much more rapidly attentive and responsive to movement than a human. In a sense, cats 'see movement' in the way that we see colour. Overall, vision has evolved in the cat to enable it to see in low light levels, detect camouflaged prey and quickly detect movement. As a result of the wild cat's camouflaged coat, crepuscular and nocturnal activity and stealthy use of cover, visual identification of affiliates is less important than the use of other sensory systems such as olfaction.

Visual communication involves facial and body postures, as well as the visual aspects of certain behaviour such as spray marking and clawing. In general, visual communication operates over a limited range, in line of sight, and is rapidly modified. This gives the signaller a great deal of control over the signal, which can be changed according to the response it elicits. Visual signalling is less effective at night, in adverse weather conditions and in dense undergrowth. The main disadvantage of visual communication is that it must be delivered face to face with a competitor, which increases the risk of conflict and injury.

Visual communication in cats therefore tends to be vivid. Whole-body signals are used to make the signaller appear larger and more threatening, or smaller and less threatening. The most powerful signals are those at the extreme, such as the side-on body posture with arched back and tail erect. In the dog, rolling on to the back is considered to be a submissive behaviour intended to appease an opponent during a conflict. During such an interaction, the opponent may approach, stand close to or investigate the individual that is showing a submissive roll. The submissive individual is in a position of extreme vulnerability. After a conflict, dogs that are part of the same social group will engage in friendly post-conflict resolution behaviour in order to reaffirm a relationship that is essential to proper cooperation within the group.

In cats, however, the intention of all body-posture signalling during a conflict is to maintain distance between individuals, either by intimidating an opponent into a retreat, or by indicating such a low level of threat that an opponent has no need for an approach. Also, unlike dogs, the roll is not a position of complete vulnerability in cats and in this position the cat is able to make very good use of its teeth and all sets of claws. After a conflict, both parties go their separate ways and so there is no need for post-conflict resolution.

Facial expression is extremely important in conveying momentary changes in the emotional state of the cat. Changes in ear position, pupil dilation and the display of teeth are not paralleled by body-posture changes, since these are slower to accomplish, especially where the animal has adopted a fully erect or very low body position.

Cats are not asocial; they will live in large colonies where resources are plentiful and they do develop affiliative relationships with each other. So, distance-reducing behaviours are an essential part of communication that allows cats to approach one another.

This is particularly important when we consider that cats have evolved a system of establishing group identity using odour, through allorubbing and allogrooming. It would not be possible to establish a group odour if cats did not have some communicative mechanism for reducing distance.

In the typical affiliative behaviour, a cat will approach and then move its tail into a vertical position (tail-up). A quiet trill or chirrup is issued, and after getting to within around 0.5 metres the approaching cat will sit down. Slow eye blinking is also used to indicate that this is a friendly approach. Physical contact only occurs between cats in about 30% of these approaches. Two cats may approach each other with tail-up, in which case they will often sit down in close proximity, but still without physical contact. When cats do make contact during a greeting, it is often only briefly; one or two rubs or a brief period of grooming. Cats that are younger than 2 years of age are more sociable and also more likely to engage in closer contact after a greeting, possibly even playing.

Vocal communication

Vocal communication includes a wide range of sounds, from quiet purrs and trills to loud distress vocalizations. Vocal signals can vary from short to long range, but are always omnidirectional and carry a risk of being detected by cats other than the intended recipient. Vocal communication can be used at a distance in any weather or lighting conditions.

In general, sounds that are produced with the mouth closed, such as purrs, trills and chirrups, are distance-reducing signals. The same is true of vocalizations that are produced with the mouth initially open but closing whilst the sound is produced, such as the miaow. Strained intensity vocalizations that are made with the mouth open throughout the sound are generally distance-increasing signals or signs of distress, such as growls, hisses, spits, snarls and shrieks (Kiley-Worthington, 1984).

Olfactory communication

Olfactory communication involves the deposition of scent marks in the environment to convey a signal indirectly to another individual (Fig. 3.2). Scent signals are persistent, localized and indirect. They cannot be removed by the signaller once they have been deposited and they can therefore convey information about the presence of the signaller that may be to its detriment. For example, rodents have been shown to exhibit unconditioned fear and avoidance behaviour in response to feline odours (Dielenberg *et al.*, 2001). In a further twist, *Toxoplasma gondii* infection in rodents has been shown to have a highly specific effect on this aversive response, producing not just a loss of fear of feline urine pheromone odours, but also an attraction to them (Vyas *et al.*, 2007). This enables the parasite to continue its life cycle by infection of its feline definitive host. The behavioural changes seen are completely specific to feline odours,

Fig. 3.2. Cats use olfaction to investigate their environment in search of odour signals deposited by other cats. (Photo: iStock.)

with other fear and conditioning responses remaining unaffected. Scent signals can also carry a considerable distance, as is the case in urinary scent marks by females during the reproductive season.

Scent signals carry both information and meaning. The information relates to the signaller – its identity, health and reproductive state. The personal identification element of the signal has common features between claw marks, spray marks and facial/flank marks, so that the recipient of any of these signals will be aware of which individual is responsible.

Previous experience with that individual adds value to this information. Meaning is conveyed in the position and type of signal used. Facial and flank marks are signals largely used in a cat's core territory, where it only expects to encounter other familiar cats. Spray marks are thought to be used to maintain temporal separation between cats, such that neighbouring cats can indicate information about usage of space, as well as when a particular cat visits that location. Claw marking is used to stretch back muscles after resting, as well as to maintain the condition of claws, but is also a form of communication (Fig. 3.3). When scent is released from pedal glands during clawing, this creates a signal that indicates territorial residency.

Fig. 3.3. Cats leave claw marks as an indicator of territorial ownership. (Photo: iStock.)

The meaning of the chemical signal is conveyed using pheromones that form part of the spray, claw or facial/flank mark. Unlike the releaser pheromones that are found in insects, which have a direct and immediate effect on behaviour, mammalian chemical signals have a more subtle effect on mood, emotion and behavioural priorities. For example, in the presence of female sex pheromone a male cat will continue with its current behaviour but its priorities will shift towards seeking out a mate.

In cats, pheromone signals are composed of combinations of fatty acids that are deposited together on a scent-marked site. Cat urine also contains the strong-smelling compound

3-methyl-3-methylbutan-1-ol (MMB), which has a chemical signalling function. This is produced by the action of the enzyme cauxin, which hydrolyses the chemical felinine that is synthesized in the liver and excreted in urine.

The recipient of the signal will sniff the location and may perform a Flehmen behaviour in order to draw a larger quantity of pheromone through ducts opening from the roof of the mouth and floor of the nasal cavity into a specialized chemosensory organ called the vomeronasal organ (VNO). The VNO consists of a pair of tubular organs 1–2 cm long in the hard palate. Once in the VNO, pheromone chemicals bind to a pheromone-binding protein before being presented to cell-surface receptors. Signals from the VNO are directed towards primitive brain structures, such as the amygdala, that are involved in subconscious changes in mood, emotion and motivation.

Claw, urine and facial/flank marks are generally deposited on vertical surfaces, often on objects that are close to an entry point to a particular space, so that other cats will notice them.

The rate of spray marking by male and female cats is increased during the period when females are in oestrus. As a result, the rate of spray marking in males and females decreases after neutering, as does the pungency of the male spray mark.

Both spray marking and claw marking also include an element of visual display.

During spray marking, the cat will sniff the site to be marked, using a Flehmen behaviour to collect more of the pheromone signal. Then it will back up to the site to be sprayed, with its tail raised. During the act of spraying the cat's tail twitches and moves, the cat has a glazed look in its eye and it will often make treading movements with its hind feet. This pattern is distinct from urination, which is usually on horizontal surfaces (unless the cat has urinary tract disease) and is not accompanied by tail movements or treading.

When cats rub against each other and groom each other, they transfer scent to create a common group odour. In wild or feral cat groups, individuals may already share strong odour similarities since in most cases groups are composed of related females. Allorubbing and allogrooming of this kind are essential for social bonding but do not contribute to self-maintenance; cats can groom themselves successfully without assistance from others.

Key points:

- The visual system of cats is predominantly adapted to suit predatory behaviour.
- Movement detection is greatly enhanced in the cat's visual system, to the detriment of colour vision and visual acuity.
- Recognition of the identity of conspecifics is therefore more likely to be based on other sensory information, such as odour.
- Distance-increasing visual and vocal signals between cats tend to be vivid, so that they can be delivered at sufficient distance to avoid fighting.

Territory

The size of the territorial range of cats (home range) varies between feral and pet cats, male and female cats and neutered and entire cats. Although evidence regarding the absolute size of home ranges is inconsistent between studies performed in different countries and habitats, intact-male home ranges are, on average, three times larger than intact-female home ranges (Liberg *et al.*, 2000). The same author found that pet females in Sweden had home ranges of around 30–40 ha and rarely roamed further than 600 m from their homes (Liberg, 1980), and that feral females had home ranges that were about four times larger than those of pet females (Liberg, 1984). It appears that male home range size is determined by the availability of reproductive opportunities, whilst that of females is governed by the availability of food. Range size has been found to vary between 0.1 ha in a Japanese fishing village to 170 ha in the Australian bush (Bradshaw, 1992), indicating that cats do not have a specific need to maintain a large territory, only to hold sufficient territory to satisfy specific survival needs. A larger territory will also be needed if suitable resting sites are distant from hunting locations.

The home ranges of neutered male and female domestic pet cats are likely to be smaller than those of their feral counterparts, and in one UK study were found to be 0.45 and 0.27 ha, respectively (Bradshaw, 1992). This reflects the high-density food resources, the close proximity of hunting, resting and latrine locations, and competition for space between the numerous cats in a typical suburban area. Home range size is suggested to be inversely proportional to population density (Bradshaw, 1992) but there is likely to be considerable variation with local circumstances.

Within the home range, the cat will also have a smaller core territory that is primarily used as a secure site for resting, feeding and self-maintenance behaviour. In wild cats this core territory is typically around 100 m in diameter and makes up 0.2–4% of the total home range (Turner and Bateson, 1986). This core territory is ideally away from direct view or intrusion by other cats that are not part of the social group.

The extent to which female cats will defend their home range relates to the abundance of food resources. Studies by Foley *et al.* (2005) and Driscoll *et al.* (2009) indicated that cats defended their territories, whilst a study by Corbett (1979) identified a lack of competition between cats due to the reliable excess of food in the abandoned farms on the Scottish islands where the study took place.

Conflict is avoided by the use of scent marks and cats following strict timetables for their movements around their territories. Urine spray marks provide a signal between cats that a particular area is likely to be in use by different individuals at certain times of day. The degradation of a urine mark over time provides other cats with an indication of when the depositor was last in that location. With regular re-marking, this provides cats with a geographical and temporal map that enables them to pass through an area without encountering one another. Claw marks are used to indicate firmer territorial boundaries that discourage the presence of non-resident cats.

Apart from providing other cats with information about ownership and usage of territory, scent marks also convey information about the identity of the depositor, its sex, health and reproductive status. This enables more complex social organization than merely avoidance, with cats choosing to avoid an area or attempt to gain control of it, depending on the threat posed by the current resident and the need to gain access to the resources included in that area.

Facial and flank marks are deposited in core territory where a cat does not anticipate meeting unfamiliar cats. In areas where facial and flank marks are left, other forms of marking (spraying and clawing) are reduced or absent.

Scent marks are not only a signal to other cats, but also carry meaning for the depositor. In the core territory, cats are more likely to be in a parasympathetic state of arousal, in preparation for activities such as resting, feeding and grooming. Outside the core territory, some degree of sympathetic arousal is always required, as the individual is vulnerable and may be engaged in hunting. Facial and flank marks may provide a chemical signal to the cat that indicates a more relaxed state in all individuals that are present, reducing the likelihood of conflict in colonies where several cats share a common core territory.

The natural organization of territory in cats poses some problems for domestic pet cats. In a domestic setting, the difference in area between the home range and core territory may be minimal; and for indoor-only cats that have a view of a garden, the boundaries of the indoor space are the absolute limits of territory, leaving no opportunity for conventional territorial organization or distance maintenance from neighbourhood cats that are easily visible from indoors.

In urban areas the density of cat populations may be high, exceeding 50 cats per square kilometre. In a study by the author (Bowen and Fatjo, 2013, unpublished data), 81% of 734 UK cat owners whose cats were allowed outdoor access indicated that their neighbours also had at least one cat that was allowed outside, and 66% reported seeing a neighbour's cat in their garden at least once a week. Owner reports of the number of different cats regularly seen in the garden and the frequency of cats visiting the garden were both correlated significantly with the frequency of injuries due to conflict with non-resident cats.

Of those households, 41% reported some level of home entry by neighbourhood cats, with 18.7% reporting that cats came in to fight with their cat and 25.6% reporting that cats came in to steal food. For households that had a plain cat door without security features (as opposed to a selective-entry cat door that only allows resident cats in and out) the figures were significantly higher: 24.8% of cat-owning households reported that neighbourhood cats came into their home to fight with their cats, and 39.4% reported that cats came in to steal food.

Given that the territorial marking behaviour of cats has evolved to reduce direct conflict and maintain distance, it is worrying that so many domestic pet cats regularly experience home invasions and injuries from fights with non-resident cats. This ought to be a significant source of stress for domestic pet cats.

Cats that had experienced injuries due to conflict with other cats showed 3.9 times the rate of indoor spray marking compared with cats that had not experienced injuries. They also waited by the cat door before going out, and became agitated and afraid when they saw another cat in the garden.

Cats who had experienced home invasion of any kind showed 24% higher levels of scratching behaviour compared with cats that had not experienced home invasion. They also waited by the cat door before going out, became agitated and fearful when they saw other cats in the garden, and showed significantly higher rates of hair loss and skin disease.

There was a strong, and significant, correlation between cats entering the home for food and fighting with resident cats in the home. Only 45% of owners in the survey fed their cats *ad libitum*, indicating that a lack of readily available food may be a strong motivation for many cats to enter homes seeking sources of food, and thereby ending up in conflict with residents.

Domestic gardens also often lack suitable landmarks for scent marking, such as posts for scratching. As a result, the boundaries between the necessarily small territories of domestic cats become blurred. There are few clear indicators of territory ownership to deter a cat from approaching another cat's home.

Forty-one per cent of cat owners that gave their cats outdoor access did not provide a cat door, meaning that these cats are reliant on a person to let them in and out of the house. Cats follow a strict timetable of movement around their territories so that they can avoid contact with other cats. By being tied to the owner's routine, the cat is unable to follow this kind of pattern.

However, even though outdoor access is fraught with problems due to a lack of owner understanding of the actual needs of cats, there is evidence that a lack of outdoor access contributes to problem behaviour. Heidenberger (1997) found that cats that were allowed to go outside when they wanted, or at least 2–3 times weekly, were less problematical to their owners.

Key points:

- The size and layout of a cat's territory is adapted to the availability of resources, such as food.
- Cats can choose to occupy small territories if food resources are sufficient.
- In the UK, suburban environments often have a high cat population density and a lack of suitable marking locations, so that territorial boundaries become blurred and unclear.
- Since many cats are not fed *ad lib*, their need for food may drive them to investigate and invade the core territories of other cats.

Timing of Activity

Cats are now commonly described as crepuscular and this has been confirmed in radio-telemetry studies (Konecny, 1987). This is also implied by the structure and function of the cat's eye, which has a number of adaptations that improve its function in low light conditions. However, cats have previously been described as nocturnal and diurnal, based on findings from different studies. Under laboratory conditions in isolation from human contact, cats showed random patterns of activity in continuous light, and free-running circadian rhythms in constant darkness (Randall *et al.*, 1985). Food intake was stimulated by simulated starlight and the presence of people. There was also a significant level of idiosyncratic variation in entrainment to different light–dark cycles, indicating that some individuals were more capable than others of adapting to different day lengths.

The activity of domestic cats is also significantly influenced by human activity. In conditions that more closely resemble a normal domestic environment, cats that were more closely involved with their owners showed greater levels of activity during daylight (Piccione *et al.*, 2013). Given that artificial lighting, including light pollution, maintains the outdoor environment in towns and cities at a constant level of illumination that resembles the crepuscular period, many cats are exposed to a distorted photoperiod that does not favour proper entrainment and is likely to favour more idiosyncratic patterns of behaviour. For many generations domestic cats have been exposed to a different set of selection pressures related to photoperiodicity of behaviour and the availability of prey, such that entrainment would have a much lower survival value.

Key points:

- Cats often adapt their activity patterns to the presence of humans.
- Experimentally cats show variation in activity pattern and ability to adapt to light conditions, which may be why some individuals exhibit problematic nocturnal activity.

Predatory Behaviour

Cats most commonly catch prey that is around 1% of their own body weight (Pearre and Maas, 1998), with 80% of reported prey caught being < 50 g in weight according to one study (Brio *et al.*, 2005). In an observational study of feline predatory behaviour in a suburban nature preserve in New York, mammals and birds made up 52% and 23% of kills, respectively, with an overall success rate for killing of 13% (Kays and DeWan, 2004). Cats show preferences in terms of their predatory targets and are tolerant of high failure rates in predatory attacks.

The selection of prey target is also influenced by hunger. Although hunger does not appear to affect the rate of hunting behaviour, larger and potentially more dangerous prey will be targeted when cats are hungry (Biben, 1979). In the same study, hunger was shown to increase the speed of killing behaviour, with satiated cats toying with prey for longer.

The motivation to hunt, or at least engage with potential prey, is separate from hunger. Adamec (1976) showed that cats would stop feeding in order to catch and kill a live rat that was introduced into the environment. They would then move the dead prey close to the feeding area and return to eating the meal. Despite having killed the rat, cats showed almost no interest in eating it. The indication is that predatory behaviour is activated by stimuli, such as odour, prey size and movement, that are unrelated to stimuli connected with feeding or hunger.

The amount of time spent searching for food and hunting does vary between pet and feral cats, with feral cats spending twice as much time per day (12 versus 6 hours) (Turner and Meister, 1988).

Cats employ a variety of strategies when hunting, which are refined through experience. Two approaches predominate: stalking and sit-and-wait. Sit-and-wait is the primary strategy for hunting rodents. Both strategies share the common features that they employ stealth and enable the cat to get close enough to its prey that it can make a final brief attack from close range.

Cats appear to identify specific hunting locations where they anticipate a higher probability of encountering vulnerable prey. They move speedily between these locations, rarely deviating to take up unanticipated hunting opportunities. This implies that during an active period of hunting, cats prioritize the catching of prey, rather than the finding of new hunting sites. This would make sense, since actual hunting opportunities are time sensitive due to the activity patterns of prey, whereas information about potential future hunting sites is available all the time.

At a hunting location the cat will survey the area systematically searching for evidence of the presence of prey. The cat's hearing extends into the ultrasound, enabling the cat not only to detect scratching sounds and vocalizations of rodents but also to precisely locate prey using hearing alone. When such sounds are heard, active prey seeking is activated.

Once the cat finds an area where there are signs of prey, such as the scent of rodent urine close to a hole, it will withdraw to a distance or area of cover and wait. An attack occurs when prey moves into an open space away from a position of safety and the cat judges that it will be able to mount a successful attack. Attack is activated by movement of anything of around the correct prey size within the cat's field of vision. During the last phase of a waiting period before an attack, the cat may perform paddling behaviour with its rear feet, which may function to prepare hind leg muscles to deliver maximum output during a dash and pounce.

If the attack is unsuccessful, the cat will return to cover or explore the hunting area for another opportunity. If unsuccessful on several occasions, or no further opportunities are detected, the cat will move on to another hunting area.

Stalking behaviour uses the cover of plants and bushes to move closer to prey, such as birds, that are more mobile and not tied to a burrow or shelter location. Cats will also predate young birds in the nest, dig moles out of their burrows and ambush young rabbits as they leave their burrows. However, the rate of predation of larger species, such as rabbits and squirrels, is low in domesticated pet cats compared with the rate of predation of rodents and birds.

Apart from small mammals, birds and rodents, cats will also hunt and consume insects. Evidence for the specific nutritional contribution of insects to the diet of the cat is limited. However, insects (particularly spiders) are rich in taurine (Weisenborn, 2012), an essential amino acid in the cat.

Feeding and Elimination

Feral and wild cats use latrine sites that are located in their home range, but not their core territory. Latrine locations are chosen on the basis of location and substrate preference, both of which are learned during development. The natural preference is for a location that is private, where the cat will not be disturbed whilst eliminating, and a substrate that is easy to dig, as cats tend to bury urine and faeces unless they are using them for marking. Cats will tend to choose separate locations to eliminate urine and faeces.

Latrine locations are away from hunting and resting areas, but close enough to those places for the cat to be able to use them without a significant interruption of their other hunting and territory marking routines.

Cats are obligate carnivores and exhibit a preference for prey that is different from what they have eaten in the past. This is termed a 'monotony effect' and is argued to be a way of ensuring a balanced diet (Bradshaw, 1992). Since cats hunt independently and catch only small prey that provide a single meal for an individual, they do not share food. Feeding is a solitary activity that carries no social meaning. Cats eat 10–20 small meals each day, according to the availability of prey.

The core territory of domestic cats contains an incompatible arrangement of resources that would not be present in a wild situation. For example, in a study by the author, 31% of households provided a food bowl near the litter tray and the same percentage located a water bowl near the litter tray. From the cat's perspective, having a latrine location inside the core territory is undesirable. Twenty-seven per cent of households placed the litter tray next to the cat door, which might seem an ethologically appropriate choice given that this is the closest possible indoor location to the garden. However, it also means that in multi-cat households, or homes with a plain cat door, the latrine lacks privacy.

In the same study, one-third of cat owners fed their cats at specific mealtimes and 65% indicated that food was not available for the cat all day. This forces cats into close proximity with each other for an activity that is usually solitary, as the cat is forced to consume as much food as it can during the periods of food availability.

The Impact of Predation

Felis silvestris is a native species in most of Africa, Europe, central Asia, India, China and Mongolia. In these regions it forms a natural part of local ecology. Due to mutual tolerance between cats and humans, the density of cat populations can be much higher within and around human communities than in rural areas.

Felis silvestris is not a native species in the Americas or Australia and so the impact it might have on wildlife in those regions may be considered much more serious and intrusive to local ecology. In the USA, it is estimated that free-ranging domestic cats kill 1.4–3.7 billion birds and 6.9–20.7 billion small mammals each year (Loss *et al.*, 2013).

This raises the issue of the impact of pet versus feral populations. No current estimate of the UK feral cat population could be found, but the total cat population was estimated at 6 million in 1980, 1.2 million of which were feral (Tabor, 1981; Harris *et al.*, 1995). The current estimate of the UK pet cat population is 8 million, according to the Pet Food Manufacturers Association. This situation can be contrasted with Australia and the USA:

- Australia: 3 million pet cats, 10–20 million feral cats (Jongman and Karlen, 1996; Dickman and Denny, 2010).
- USA: 86.4 million pet cats, 60–100 million feral cats (APPA, 2011; Dauphine and Cooper, 2009).

So, whilst in the UK the feral cat population is perhaps 15% the size of the pet cat population, in the USA these populations are similarly sized and in Australia feral cats outnumber pet cats by a factor of 10–20 to one.

Trap–neuter–return (TNR) programmes have been shown to be effective in a number of studies. Levy *et al.* (2003) showed a 66% reduction in population size over an 11-year period in one free-roaming population. In the UK, nationwide trap-and-neuter programmes of this type have been in operation for more than a decade. In the UK and Germany, levels of neutering are generally high, which helps to limit population growth (Heidenberger, 1997)

However, the most significant factors in the maintenance of a large feral population are climate and the availability of prey. The temperate climate of Northern Europe limits kitten survival, compared with the warmer climate of Australia. Australia also has a diverse population of prey species that have evolved in the absence of an equivalent predator to *Felis silvestris*.

In a study of domestic cat predation in the UK, the estimated mean predation rate was estimated at 18.3 per cat per year, with 65% of households reporting no prey brought back in a given year, falling to 22% when averaged over several seasons (Thomas *et al.*, 2012). The same study found that only 20% of cats returned four or more dead prey annually. A retrospective study by the author found an average prey return rate of 3.3 birds and 12 rodents per cat per year for households where cats had outdoor access; 44.6% never returned a bird and 39.6% never returned a rodent.

The number of birds observed in a garden correlated significantly with the amount of environmental enrichment provided in a garden (bird feeders, scattered food and nesting boxes), as well as the amount of natural features present (long grass, trees, wild flowers, bushes). Of particular note was a correlation between bird numbers and indicators of active gardening, such as the presence of a greenhouse, vegetable patch and compost heap. However, the number of birds caught did not correlate with scores for the number of birds observed in the owner's garden. This is supported by findings from Thomas *et al.* (2012,) who reported that numbers of five bird species were not correlated with level of their predation.

The predation of robins was just significantly correlated with observed numbers ($p = 0.046$). This may relate to some aspect of the vulnerability of this species when feeding and nesting, but the result would also become insignificant when measures are applied to counteract the problem of multiple comparisons.

The study by Thomas also showed a significantly negative correlation between mean annual predation rate, cat population and housing density; the number of prey returned was lower in high residential and cat population density.

In the author's study, 62% of cat owners indicated an interest in encouraging more birds in the garden, with 48.1% of those giving a positive response also indicating that owning a cat prevented them from doing so. Those who reported that having a cat restricted their ability to attract birds to the garden provided similar numbers of bird tables and hanging feeders, but significantly fewer bird boxes and scatter-feeding opportunities for ground-feeding birds.

It appears that improving a garden to attract and support bird numbers does not increase the general risk of predation, and cat owners should be encouraged to do so (Fig. 3.4).

Key points:

- *Felis silvestris* is native to Europe and would exist here without human intervention.
- The impact of this species on wild bird and mammal species must be recognized.
- Increasing bird numbers by feeding birds and providing an insect- and bird-friendly garden does not appear to increase predation risk.

Fig. 3.4. Cats will hunt small birds and mammals but the cat is a native species in the UK and Europe, and evidence is inconclusive regarding the impact of predation by pet cats on wild bird populations. (Photo: iStock.)

References

Adamec, R. E. (1976) The interaction of hunger and preying in the domestic cat (Felis catus): An adaptive hierarchy? *Behavioral Biology*. 18, 2, 263–272.

APPA (2011) *2011–2012 APPA National Pet Owners Survey*. American Pet Products Association, Greenwich, Connecticut.

Biben, M. (1979) Predation and predatory play behaviour of domestic cats. *Animal Behaviour* 27, 81–94.

Bowen, J.E. and Fatjo, J. (2013) unpublished data.

Bradshaw, J.W.S. (1992) *The Behaviour of the Domestic Cat*. CAB International, Wallingford, UK.

Brio, Z., Lanszki, J., Szemethy, L., Heltai, M. and Randi, E. (2005) Feeding habits of feral domestic cats (*Felis catus*), wild cats (*Felis silvestris*) and their hybrids: trophic niche overlap among cat groups in Hungary. *Journal of Zoology* 266, 187–196.

Corbett, L.K. (1979) Feeding ecology and social organization of wild cats *(Felis silvestris)* and domestic cats *(Felis catus)* in Scotland. PhD thesis, University of Aberdeen, Scotland.

Dards, J.L. (1978) Home ranges of feral cats in Portsmouth. *Carnivore Genetics Newsletter* 3(7), 242–255.

Dards, J.L. (1981) Habitat utilization by feral cats in Portsmouth dockyard. In: *The Ecology and Control of Feral Cats.* Universities Federation for Animal Welfare, Potters Bar, UK, pp. 30–49.

Dards, J.L. (1983) The behaviour of dockyard cats: interactions of adult males. *Applied Animal Ethology* 10, 133–153.

Dauphine, N. and Cooper, R.J. (2009) Impacts of free-ranging domestic cats (*Felis catus*) on birds in the United States: a review of recent research with conservation and management recommendations. In: Rich, T.D., Arizmendi, C., Desmarest, D.W. and Thompson, C. (eds) *Proceedings of the Fourth International Partners in Flight Conference: Tundra to Tropics*, pp. 205–219.

Denny, E., Yakovlevich, P., Eldridge, M.D.B. and Dickman, D. (2002) Social and genetic analysis of a population of free-living cats (*Felis catus* L.) exploiting a resource-rich habitat. *Wildlife Research* 45(4), 405–413.

Dickman, C. and Denny, E. (2010) Strategies to reduce conflict: managing feral and stray cats. In: Tensen, M. and Jones, B. (eds) *Proceedings of the RSPCA Scientific Seminar*. RSPCA Australia, Deakin West ACT, pp. 41–45.

Dielenberg, R.A., Hunt, G.E. and McGregor, I.S. (2001) 'When a rat smells a cat': the distribution of Fos immunoreactivity in rat brain following exposure to a predatory odor. *Neuroscience* 104(4), 1085–1097.

Driscoll, C.A., Macdonald, D.W. and O'Brien, S.J. (2009) From wild animals to domestic pets, an evolutionary view of domestication. *Proceedings of the National Academy of Sciences* 106(1), 9971–9978.

Fagen, R.M. (1978) Population structure and social behavior in the domestic cat (*Felis catus*). *Carnivore Genetics Newsletter* 3(8), 276–281.

Foley, P., Foley, J.E., Levy, J K. and Paik, T. (2005) Analysis of the impact of trap-neuter-return programs on populations of feral cats. *Journal of the American Veterinary Medical Association* 227(11), 1775–1781.

Harris, S., Morris, P., Wray, S. and Yalden, D. (1995) *A Review of British Mammals: Population Estimates and Conservation Status of British Mammals Other Than Cetaceans*. Joint Nature Conservation Committee, Peterbourgh, UK.

Heidenberger, E. (1997) Housing conditions and behavioural problems of indoor cats as assessed by their owners. *Applied Animal Behaviour Science* 52, 345–364.

Jongman, E.C. and Karlen, G.A. (1996) Trap, neuter and release programs for cats: a literature review on an alternative control method of feral cats in urban areas. In: Hassett, S. (ed.) *Proceedings of 5th National Conference on Urban Animal Management in Australia.* Australian Veterinary Association, Brunswick, Victoria, pp. 81–84.

Kays, R.W. and DeWan, A.A. (2004) Ecological impact of inside/outside house cats around a suburban nature preserve. *Animal Conservation* 7, 273–283.

Kerby, G. and McDonald, D.W. (1988) Cat society and the consequences of colony size. In: Turner, D.C. and Bateson, P. (eds) *The Domestic Cat: The Biology of its Behaviour.* Cambridge University Press, Cambridge.

Kiley-Worthingon, M. (1984) Animal language? Vocal communication of some ungulates, canids and felids. *Acta Zoologica Fennica* 171, 83–88.

Konecny, M.J. (1987) Home range and activity patterns of feral house cats in the Galapagos Islands. *Oikos* 50, 17–23.

Laundre, J. (1977) The daytime behaviour of domestic cats in a free-roaming population. *Animal Behaviour* 25, 990–998.

Leyhausen, P. (1965) The communal organization of solitary mammals. *Symposia of the Zoological Society of London* 14, 249–63.

Levy, J.K., Gale, D.W. and Gale, L.A. (2003) Evaluation of the effect of a long-term trap-neuter-return and adoption program on a free-roaming cat population. *JAVMA* 222, 42–46.

Leyhausen, P. (1988) The tame and the wild – another Just-So-Story? In: Turner, D.C. and Bateson, P. (eds) *The Domestic Cat: The Biology of its Behaviour*. Cambridge University Press, Cambridge.

Liberg, O. (1980) Spacing patterns in a population of rural free roaming domestic cats. *Oikos* 38, 336–349.

Liberg, O. (1984) Home range and territoriality in free ranging house cats. *Acta Zoologica Fennica* 171, 283–285.

Liberg, O., Sandell, M., Pontier, D. and Natoli, E. (2000) Density, spatial organization and reproductive tactics in the domestic cat and other felids. In: D.C. Turner and P. Bateson (eds). *The Domestic Cat: The Biology of its Behaviour*, 2nd edn. Cambridge University Press, Cambridge, pp. 119–148.

Loss, S.R., Will, T. and Marra, P.P. (2013) The impact of free-ranging domestic cats on wildlife of the United States. *Nature Communications* 4 (1396).

Ohkawa, N. and Hidaka, T. (1987) Communal nursing in the domestic cat, *Felis catus. Journal of Ethology* 5(2), 173–183.

Pearre, S. and Maass, R. (1998) Trends in the prey size-based trophic niches of feral House Cats *Felis catus* L. *Mammal Review* 28(3), 125–139.

Piccione, G., Marafioti, S., Giannetto, C., Panzera, M. and Fazio, F. (2013) Daily rhythm of total activity pattern in domestic cats (*Felis silvestris catus*) maintained in two different housing conditions. *Journal of Veterinary Behavior* 8, 189–194.

Randall, W., Johnson, R.F., Randall, S. and Cunningham, J.T. (1985) Circadian rhythms in food intake and activity in domestic cats. *Behavioral Neuroscience* 99(6), 1162–1175.

Tabor, R. (1981) General biology of feral cats. In: *The Ecology and Control of Feral Cats*. The Universities Federation for Animal Welfare, Potters Bar, UK, pp. 5–11.

Thomas, R.L., Fellowes, M.D.E and Baker, P.J. (2012) Spatio-temporal variation in predation by urban domestic cats (*Felis catus*) and the acceptability of possible management actions in the UK. *PLoS ONE* 7(11), 1–13.

Turner, D.C. and Bateson, P. (1986) *The Domestic Cat: The Biology of its Behaviour*. Cambridge University Press, Cambridge.

Turner, D.C. and Meister, O. (1988) Hunting behaviour of the domestic cat. In: Turner, D.C. and Bateson, P. (eds) *The Domestic Cat: The Biology of its Behaviour*, 1st edn. Cambridge University Press, Cambridge.

van Aarde, R.J. (1978) Reproduction and population ecology in the house cat, *Felis catus,* on Marion Island. *Carnivore Genetics Newsletter* 3(8), 288–316.

Vyas, A., Kim, S., Giacomini, N., Boothroyd, J.C. and Sapolsky, R.M. (2007) Behavioral changes induced by Toxoplasma infection of rodents are highly specific to aversion of cat odors. *PNAS* 104(5), 6442–6447.

Weisenborn, W.D. (2012) Sulfur contents of spiders and insects in desert riparian habitat. *Florida Entomologist* 95(4), 952–960.

4 Understanding the Behaviour of Small Animals: Rabbit and Rodent Behaviour

Anne McBride and Emma Magnus

Small mammals are increasingly popular companion animals. This has escalated demand for specialized veterinary services to meet the challenges of maintaining their physical health and understanding their behavioural needs (Yeates, 2018).

Commonly kept species are the rabbit, rat, mouse, hamster (Golden and Dwarf), guinea pig, gerbil, degu and chinchilla. Other species seen in the pet market include chipmunk, African (giant) pouch rats and African pigmy hedgehogs (Meredith and Delaney, 2010). Due to space limitations of this book, it is only possible to provide an overview of general principles of the natural behaviour and behaviour welfare provision of these small prey animals. Information is drawn from a range of academic fields. Readers are encouraged to explore this wider literature to understand this fascinating group of creatures more fully.

All species are the product of evolutionary adaptation to the pressures of the environment in which they originated. This adaptation moulded physical, behavioural and social characteristics (Wolff and Sherman, 2007), enabling species survival. At the individual level, these characteristics facilitate biological success, where your genes are represented in future generations. That entails getting enough to eat and not getting killed before you find a mate and rear your offspring so they too can continue this cycle of life.

Such basic biology is often disregarded in how we keep animals. This is compounded by humans having altered the physical appearance of many through artificial selection via selective breeding. Rabbits and some rodent species were primarily kept for their meat, for fur or for use in laboratories and were selectively bred for size or coat colour and texture. The human ideals of aesthetics in the show and pet markets have driven selection for more extreme change, including of body and head shape. As seen in other species, from cows to dogs, selective breeding and altering the natural form carries great potential for inherited and acquired welfare problems. Regrettably such problems, for example those related to brachycephalia, are increasingly documented in rabbits and rodents (Broom and Fraser, 2015; UFAW, n.d.).

These physical changes also reinforce the view that these animals are 'something different' from their wild cousins. Indeed, simply calling or portraying an individual animal as pet, farm, laboratory or wild alters our perceptions of its emotional life, intelligence and character (Herzog, 2010; González-Redondo and Contreras-Chacón, 2012; Leighty *et al.*, 2015; Wilkins *et al.*, 2015).

Such misperceptions can lead to serious physical and mental welfare issues, through inappropriate housing, diet and misinterpretation of behaviour. Prevention of these issues

 Companion Animal Behaviour Problems: Prevention and Management of Behaviour Problems in Veterinary Practice (eds. R. Casey, S. Heath and H. Zulch)
DOI: 10.1079/9781800621312.0004

requires knowledge of the species' natural ecology, ethology, sensory abilities, behaviour and communication.

Species-specific phenotypes (physical appearance and behaviour) have evolved to aid survival in their different natural environments, be they specialist desert dwellers like the hamster, or generalists able to live in a variety of habitats like the rat and mouse. Most of these natural characteristics persist in our pet species despite centuries of domestication (McBride, 2000; Berdoy, 2002; Brandão and Mayer, 2011; Yeates, 2018). For example, domesticated rats are smaller and less aggressive than their wild relatives, but otherwise show the original wild-type behaviour repertoire with respect to finding food, building homes, avoiding predators, and their social and parental behaviour (Berdoy, 2002).

Like humans, animals are individuals. Increasingly, research indicates that personality is another phenotype that was selected for as part of the natural evolution of a species (Dingemanse and Réale, 2005; Nettle, 2006). More confident (optimistic) animals have better cognitive abilities (Hasleton and Nettle, 2006), health and reproductive success compared with less confident individuals, even when the basic needs for survival are met. These differences in turn affect the offspring (Czerwinski *et al.*, 2016). We therefore need to consider the personality of the animals we keep and breed (Smith and Blumstein, 2008).

Survival means that animals have to find food; construct safe homes that are warm in winter and cool in summer; interact with each other as peacefully as possible; and avoid predators. For rabbits and rodents, predation is a major challenge, as they provide the bulk of the food for terrestrial and aerial predators, including snakes, lynx, wolves, foxes, stoats, weasels, badgers, eagles, hawks and owls. They are also hunted by humans and their cats, dogs and ferrets. But life is not all about danger and fear.

Animals are motivated to find suitable mates, rear young and partake in pleasurable activities including playing, socializing, exploring, grooming, resting and sleeping. Ensuring good welfare is more than just providing enough to avoid pain, illness, fear and distress. We need to provide opportunities for pleasurable and cognitive experiences (Balcombe, 2007; Wynne and Udell, 2013). Each individual animal should have a life worth living (Mellor, 2016) regardless of why we keep them (McBride and Baugh, 2022).

Table 4.1 summarizes some natural features of a few species. This chapter provides a short overview of rabbit and rodent adaptations to original habitats in terms of physiology and behaviour and begins with a brief introduction to some commonly kept species.

Rabbit

Rabbits living wild in the UK, across Europe and Australia are European rabbits, *Oryctolagus cuniculus*. The original home of this species is the dry scrubland of the Iberian Peninsula (Southern Spain), an area not known for its prolific supply of green, lush grazing, but rather for dry, tough grasses.

The domesticated rabbit is a direct descendant and has been selectively bred for 1500 years. It was originally, and still is, kept primarily as a source of food and fur. As laboratory animals, rabbits also contribute to scientific research. Increasing popularity as a pet and for show over the past 150 years has led to selective breeding for more extreme appearances, such as lop ears, brachycephalic heads and miniature size, resulting in the creation of physically diverse breeds. Approximately 65 rabbit breeds are recognized by the British Rabbit Council and the rabbit is the UK's third most popular mammalian pet (PDSA, 2015, 2020) (Fig. 4.1).

Guinea Pig

Guinea pigs, *Cavia porcellus*, belong to the family Caviidae and originate from the relatively dry, tall grasslands and lower slopes of the Andes in South America. Domesticated for some 5000 years, they were and are kept as an important food source for many South American peoples. Introduced to Europe during the 16th century, they continue to be popular as pets due to their size, longevity and docile nature. Guinea pigs are used in research and for show, resulting in various breeds (McBride, 2010).

Chinchilla

The long-tailed chinchilla (*Chinchilla lanigera*) originates from the high, rocky, arid slopes of the

Table 4.1. Comparison of general features of rabbits and rodent species.

	Rabbit	Guinea pig	Chinchilla	Degu	Gerbil	Golden hamster	Dwarf hamster	Rat	Mouse
Approximate hearing range (Hz)[a]	360–42,000	54–50,000	20–30,000	?	100–60,000	80–60,000	80–60,000	200–76,000	1000–91,000
Activity time	Crepuscular / nocturnal	Crepuscular/ diurnal	Crepuscular / diurnal	Crepuscular/ diurnal	Crepuscular / diurnal	Crepuscular / nocturnal	Crepuscular / nocturnal	Nocturnal	Nocturnal
Diet	Herbivore	Herbivore	Herbivore	Herbivore	Omnivore	Omnivore	Omnivore	Omnivore	Omnivore
Coprophagic	Yes	Yes	Yes	Yes	No	No	No	No	No
Hoard food	No	No	No	Yes	Yes	Yes	Yes	Yes	Yes
Continuously growing teeth	Yes	Yes	Yes	Yes	Yes	Yes	Yes	Yes	Yes
Social	Yes	Yes	Yes	Yes	Yes	No	Yes	Yes	Yes
Best social groupings for owners Group housed animals are best bought as animals known to each other	Neutered pair	Neutered male and one or more females	Neutered pair	Harem groups 1 neutered male 2–4 females	Harem groups 1 neutered male 2–4 females	One – solitary	Winter Whites, Roborovski and Campbells same-sex groups Chinese: solitary	Same-sex groups or neutered male plus female(s)	Same-sex groups or neutered male plus female(s)
Home	Complex burrows	Tall grasses / rock crevices	Simple burrows / rock crevices	Complex burrows	Complex burrows	Complex burrows	Simple burrows	Complex burrows	Complex burrows
Altricial or precocial young	Altricial	Precocial	Precocial	Precocial	Altricial	Altricial	Altricial	Altricial	Altricial
Age at sexual maturity: female	14 weeks	8–12 weeks	6–8 months	4–6 months	10 weeks	3–4 weeks	3–4 weeks	5–6 weeks	5–6 weeks
Age at sexual maturity: male	16 weeks	12–16 weeks	5–6 months	4–6 months	10 weeks	3–4 weeks	3–4 weeks	5–6 weeks	5–6 weeks
Lifespan in captivity	10 years	5–6 years	15–20 years	5–8 years	3–5 years	2 years	2 years	3–5 years	18 months–2 years

[a]Data for hearing ranges taken from Fay, R.R. (1988) *Hearing in Vertebrates: a Psychophysics Databook*. Hill-Fay Associates, Winnetka, Illinois; and Warfield, D. (1973) The study of hearing in animals. In: Gay, W. (ed.) *Methods of Animal Experimentation, IV.* Academic Press, London, pp. 43–143; and Albright, J. and De Matos, R. (2010) Hamsters. In: Tynes, V.V. (ed.) *Behaviour of Exotic Pets*. Wiley-Blackwell, Chichester, UK, pp. 127–137.

Fig. 4.1. Wild rabbit **(a)** versus Lop rabbit **(b)**, illustrating the effects of domestication on body conformation. (Photo: iStock.)

Chilean Andes. Whilst there are still wild populations in existence, the species is considered to be endangered, due to habitat loss and illegal hunting for their fur. Originally it was kept for its fur, for which it is still farmed. Using chinchilla fur in clothing became less fashionable during the late 20th century and animals were sold as pets. The colour and plush, soft feel of the fur contributes to its attractiveness as a pet. Like other small species, chinchillas are also used in research (Fig. 4.2).

Fig. 4.2. Chinchilla. (Photo: iStock.)

Degu

The degu (*Octodon degus*) originates from central Chile, along the western side of the Andes. Though this area has semi-desert conditions in summer, the winters are cold and wet. Degus have been kept as laboratory animals since the 1950s, primarily for diabetes research. They are physiologically unable to regulate glucose and, if fed incorrectly, are prone to developing diet-related diabetes and cataracts.

Gerbil

The Mongolian gerbil (*Meriones unguiculatus*) comes from the arid steppes and semi-desert conditions of Mongolia (Fig. 4.3). They were brought to Europe in the 19th century and became popular as pets and as laboratory animals (Fig. 4.4).

Unlike most other species discussed in this chapter, gerbils, degus, chinchillas and guinea pigs are active both in the day and by night, though they return to the shelter of burrows and rock crevices when it is too hot or cold.

Hamster

There are two types of hamster kept as pets. First is the golden or Syrian hamster (*Mesocricetus auratus*)

Fig. 4.3. Wild gerbil habitat. (Photo: iStock.)

Fig. 4.4. Domesticated gerbils. (Photo: iStock.)

Fig. 4.5. Golden or Syrian hamster (*Mesocricetus auratus*). (Photo: iStock.)

(Fig. 4.5). Originally from the dry, rocky, scrubby slopes of north-west Syria, this species is extremely rare in the wild. Second are the four species of dwarf hamsters. The two Russian and the Roborovski are from the *Phodopus* (meaning short-tailed) genus, and the Chinese dwarf hamster is from the *Cricetulus* species, the long-tailed hamsters. All four come from the Central Asian steppes, a land of mountains and semi-dry deserts. Each has special adaptations for its own particular type of habitat.

Mice

Pet mice are direct descendants of the house mouse *Mus musculus*. They were kept as pets 3000 years ago in ancient China and Egypt (Cucci *et al.*, 2005). During the 17th century, mice became popular in Europe, with selection for different coat colours or textures leading to the development of many breeds. Due to their size and ease of breeding, they are the most frequently used animal in laboratory research.

Rats

Pet rats are direct descendants of the wild brown rat *Rattus norvegicus* that we see around our towns and cities. They have been kept as pets since the early 19th century. Rat-catchers caught wild rats for the popular blood sport of rat baiting with terriers. Wild animals occasionally have young that are colour mutants. In rats these include albinos and the 'hooded' black-and-white type. People are attracted to unusual colours and tend to keep and selectively breed such animals. By 1828 rats were being used in research and both colours continue to be widely used in laboratories (www.ratbehavior.org, accessed 29 April 2022). In the 20th century they became popular as pets and, by the 1970s, as show animals (www.nfrs.org, accessed 29 April 2022) with further selective breeding for different colours and markings.

Physical Adaptations

Physical adaptations enable animals to find food, avoid predation and survive, and indeed to live comfortably, in their original habitats. The rabbit and many rodents have evolved to live in environments that are arid for at least part of the year, some being hot and dry, others cold and dry. In dry conditions it is important to conserve body fluids and avoid dehydration; relevant adaptations include hard dry faecal droppings, concentrated urine and an inability to produce sweat. This means that these animals are very susceptible to overheating when in conditions above their ambient temperature range.

Body fur provides insulation from the cold and also reduces dehydration risk. Several species have fur-covered toes insulating them from hot or cold ground. The rabbit also has fur on the soles of its feet, which acts as a shock absorber when hopping and provides extra grip on sand and snow.

The natural colour of animals is the result of the need for camouflage. Blending in with the background is important for predators trying not to be seen by their prey, and for prey to avoid predators. Prey animals that stand out are unlikely to survive long.

As Table 4.1 shows, rabbits and rodents are typically most active at night (nocturnal) and in the early evening and early morning (crepuscular), when light is low and shadows are long. Some are diurnal (active in the day) but still seek safety in the shadows of rocks and vegetation. The natural fur pattern is 'agouti', a darker colour on the top and lighter underneath (Fig. 4.6).

This reduces detection in low light. Rabbits, mice, rats, guinea pigs and degus are brown-grey on top, which imitates the colour of the vegetation, soil and leaf litter. Chinchillas are famous for their natural colouring of pearly blue-greys and browns that enables them to blend with the rocks of their home. Gerbils and golden and Campbell's Russian hamsters are more yellow-brown, mimicking the sands of their semi-desert habitats. The summer top-coat of the white winter Russian hamster is grey-brown, with a wide dark stripe running along the centre. As the name suggests, the winter coat is white, giving good camouflage in snow. This is not seen in captivity, as artificial light does not trigger the coat change.

Predator avoidance also requires being agile and swift. Rabbits and rodents have fragile, light skeletons, reducing overall bodyweight. They have elongated, highly muscled back legs, particularly obvious in rabbits, chinchillas and degus. This enables them to move quickly: even the tiny golden hamster can reach speeds of 5 miles (8 km) an hour. The long hind legs also are springboards for jumping: chinchillas and rabbits can jump over 3 ft (1 m). Some species have long tails that aid balance, as when

Fig. 4.6. Wild gerbil showing the natural agouti colouring. (Photo: iStock.)

Fig. 4.7. Golden hamster using a lookout site and its eyes, ears and nose to gather distance information about its environment: for example, if there is danger, other hamsters, or food about. (Photo: iStock.)

chinchillas leap from rock to rock, or rats and mice climb bushes and trees. Long back legs enable animals to sit bolt upright, the extra height increasing chances of early detection of any threat, be that by sight, smell or by hearing wing beats in the air, foot falls or scales slithering across the ground.

Senses

Our senses gather information from our surroundings and our body. For each species, senses evolve in response to the physical environment and associated pressures. They are integral to species-specific behaviours, including recognizing conspecifics, communication, exploration, foraging and detecting predators (e.g. Finlay and Sengelaub, 1981) (Figs 4.7 and 4.8).

Understanding and accounting for animal senses is essential in optimizing welfare (Olsson *et al.*, 2003; McBride, 2017a,b). Though rabbits and rodents have well-developed senses, including taste, pain and proprioception, only some general points are given here.

Vision

The eyes of small prey mammals are located to the side of the head and above the midline. This gives them panoramic vision, seeing behind and above them. Being crepuscular or nocturnal, they are adapted to very low light levels. What we might think of as normal room lighting levels is for them bright and aversive and can cause physical damage (Castelhano-Carlos and Baumans, 2009). Even the relatively diurnal species, gerbils and guinea pigs (Van den Broek *et al.*, 1995), avoid bright sunlight, retreating to the dark of burrows and rock crevices, or the dappled shade of tall grasses and bushes. The truly nocturnal species have limited visual acuity and are rather short-sighted, compared with crepuscular and diurnal species. All species are very responsive to movement, as it may herald a predator's approach. They can see in the ultraviolet (UV) range (Honkavaara *et al.*, 2002), which may help in finding food from UV-reflecting plants and UV is reflected by light-coloured underbellies, acting as an unmistakeable 'Watch out!' signal when an individual, suspecting possible danger, stands upright. The urine of some rodents reflects UV, 'lighting' up pathways.

Fig. 4.8. Chinchilla in the wild on a lookout post, showing the long tail it uses to balance. (Photo: Adobe stock.)

Whiskers

Whiskers are an important and underrated sense organ, informing animals about their world. They enable small animals to judge if their body will go through a hole: if whiskers go through, shoulders and hips will too. Whiskers also provide animals with a picture of objects. Those near the end of the nose are shorter than those further back. Each length resonates at a different frequency of air movement, either as it 'bounces' off stationary objects, or as it ripples off a moving object, like the ripples left in water by a passing boat. A rat's whiskers are far more sensitive than human fingertips (Carvell and Simons, 1990).

Hearing

The human hearing range is 20–20,000 Hz. Rabbits and rodents hear higher (ultrasonic) frequencies (Table 4.1). The exception is the chinchilla, whose hearing is similar to ours, hence their use in research into human hearing loss.

The rabbit is perhaps most famous for its very large, concave, upright ears. Large ears capture more sound and thus hearing is more acute. Rabbits can move each ear independently, allowing them to pinpoint the location of the noise accurately. Rodent ears are also relatively large and more or less mobile. Ears are sparsely covered in fur and have a network of surface veins which help regulate body temperature. When an individual is warm, more blood flows through the ears, where it cools before being transported back around the body.

Hearing and vocal communication are linked, and most rodent vocalization is in the ultrasonic range (Sales and Pye, 1974). Even for animals we can hear, we know little about the complexity of sounds they make, let alone what they mean. We know even less about those we can only hear using technology. Recent work on rodent communication shows an intriguing complexity to vocalizations and their meanings (see for example Panksepp and Burgdorf, 2000; Berdoy, 2002; Long, 2007).

Olfaction

Smell is very important, especially for animals who spend most of their time in low light or, when underground, in absolute dark. Smell helps distinguish friends from strangers, warn of predators, find food and investigate objects. It is also used to communicate information, through depositing secretions from various glands around the body, including the submandibular (chin), flank, inguinal (groin) and anal glands. Deposits from flank and chin glands are rubbed on objects, on other individuals and on oneself. Urine and faecal deposits act as territorial markers (you may have seen such marker mounds of rabbit droppings (latrines) when on a countryside walk). Scent-laden urine has other uses too. Male rabbits accurately spray urine when running past a rival male in a territorial dispute, or at a female they are courting. Chinchillas, when frightened and cornered, will spray urine to try to deter the threat.

As scent detection is so important, rabbits and rodents are obligate nasal breathers, meaning it is difficult to breathe through the mouth. This evolutionary adaptation allows them to keep smelling for danger while eating. They only breathe through their mouth in extremis, as when suffering from a severe respiratory infection.

Diet and Digestion

Smell, sight and taste are important in finding and identifying food and indicating its freshness.

Rabbits, guinea pigs, chinchillas and degus are true herbivores, known as fibrevores, a term reflecting the importance of fibre in their diet (McBride and Meredith, 2018). They spend several hours every day (or night) walking and nibbling, using their senses to graze the tastiest grasses and herbs selectively (Fig. 4.9). The other rodents are omnivorous: the main part of their diet may comprise seeds, grasses, nuts and fruit, but insects, birds' eggs and meat, be that baby birds found in a nest or an animal's carcass, are welcome additions.

Rabbits and rodents cannot vomit so must be careful about what they eat. Because there is little in their natural environment that is poisonous, fibrevores tend to eat novel plants. This is a problem in captivity, as many garden and house plants may be toxic.

In contrast, omnivores take a very cautious approach to new foods and initially will only eat a small portion. If it makes them feel ill, even several hours later, then they will not return to it, a phenomenon known as 'learned taste aversion'. Information about new food is passed to others in the group who associate the individual's behaviour (well or ill) with the scent of the food they ate on their breath (Galef and Giraldeau, 2001).

Digestion starts with chewing. Unlike those of dogs, cats and humans, the teeth of rabbits and rodents grow surprisingly quickly throughout life, at a rate of 2–2.5 mm a week, approximately 0.3 mm every day (Addison and Appleton, 1915; Meredith and Lord, 2014). This evolutionary adaptation is imperative where food is hard, like nuts and seeds, or abrasive, like grasses. The natural diet ensures the teeth are worn down appropriately and do not become misaligned. Misaligned teeth mean the animal cannot chew and would slowly starve to death. In captivity, inappropriate diet causes misaligned teeth, resulting in pain, abscesses, weight loss and, often, euthanasia.

An animal's digestive system is designed for its species' specific diet. The natural diet of fibrevores is grasses and herbs. These are high in fibre, much of which is indigestible but essential to the healthy working of the gut (Meredith and Redrobe, 2002). Fibrevores have evolved special adaptations to process their food more than once, in order to extract as much of the nutrients as possible. Cattle and sheep do this by burping up partly digested food boluses, a process known as 'chewing the cud'. Rabbits, guinea pigs, chinchillas

Fig. 4.9. Wild rabbit habitat. Note the arid landscape and the variety of green and dry vegetation (natural hay). (Photo: iStock.)

and degus use caecotrophy. In the stomach the larger indigestible fibre portions and smaller digestible particles are separated. The large particles are excreted as hard, dry pellets. The small particles pass into the caecum, are fermented by good bacteria and formed into soft, smelly pellets, known as caecotrophs. As these pass out of the anus, the animal pops its head down to catch them before they drop to the ground (a behaviour known as refection). The pellets are chewed, swallowed and digested again to extract all the remaining goodness. For the crepuscular and nocturnal species this mostly happens in the daytime when they are safe in their hideaways. The hard droppings of waste are excreted when the animal is out on the surface foraging. These contain a lot of nitrogen and effectively fertilize the ground, helping to grow more grass and herbs in the future.

Home Environment

Most of these species are social, except the golden hamster (and the gerbil *Psammomys obesus* known as fat sand rat) (Sørensen *et al.*, 2005). All are active on the surface in the cooler parts of the day and night. They avoid the heat and cold by spending time in the relatively stable temperature of rock crevices, deep vegetation or below ground in tunnels (burrows) (Fig. 4.10). Tunnels act as refuges from the weather and from predators. They are nurseries for young, places to socialize and rest and may also be larders for those that store food, like degus, rats, mice and hamsters.

Burrows may simply be a single tunnel with a wider resting area at one end or they may be extensive and complex with branching side tunnels. Even the home of the solitary golden hamster is complex and the tunnel can be over 6 ft (180 cm) dug at a depth of 30 in (65 cm) below ground.

Complex burrows have interconnecting tunnels with many junctions and several entrances and exits. They are common where predators such as snakes, weasels, stoats and ferrets are likely to invade. The complexity reduces the chance of being caught: an individual has a chance to evade the predator by changing direction, rather than getting trapped in a dead end.

Digging and maintaining tunnels, whether simple or complex, are feats of engineering, requiring time and physical effort. The depth and complexity of the tunnel system is partly determined by the soil structure. For example, rabbit burrows dug in sand dunes are smaller than those on chalky soil, as sand is more liable to collapse.

The solitary golden hamster has to do all its own digging, but degus and rats work cooperatively (Flannelly and Lore, 1977; Ebensperger and Bozinovic, 2000). For rabbits, tunnel digging is the prerogative of the females whilst the males often simply watch, or scrape small areas of ground. This does not mean that males do not or cannot dig, but they prefer to let their mate do the 'housework'.

Communication

Whether a species is highly social or mainly solitary, communication with others is essential. Communication lets others know how you are feeling and what you intend to do and enables the transfer of knowledge, such as your territory boundary, social standing (status), where tasty food may be, or your prowess at keeping rivals at bay. A nice example of the latter is the male degu's way of marking his territory. At the burrow entrance he makes a mound out of twigs, pebbles, dung or anything else, and adds to its height each time he drives off a rival male (Fulk, 1976).

Olfaction and vocalization are important communication channels for animals that spend most of their time in dark or low-light conditions (Fig. 4.10). Small prey species wish to go unnoticed by predators. Consequently, their visual signals, both body and facial movements, are subtle. For example, unlike the exuberant greeting behaviours of dogs (tail wagging, circling and sniffing), rabbit greetings are low-key, decorous affairs; a gentle nose-to-nose sniff, or T-junction approach (McBride, 1998).

Fig. 4.10. Wild chinchilla resting near the safety and coolness of a rock crevice. (Photo: Adobe stock.)

Social species frequently groom each other (allogrooming). This relaxing, comforting behaviour helps strengthen relationships. It is important in establishing bonds between animals and their owners. Animals have preferences for where and how they are touched. For example, rabbits seem to prefer being stroked around the head, which corresponds with normal rabbit–rabbit allogrooming, while rats enjoy gentle tickling, reminiscent of rat–rat play, and respond with laughter (Panksepp and Burgdorf, 2000).

Most aggression occurs between members of the same sex and is related to getting and keeping a territory, access to mates and resources such as good places to have babies. It is always more prevalent during the breeding season, as providing the best for the current or soon-to-be born offspring is a driving concern (Myers and Poole, 1961).

Direct fighting may mean injury, so most disputes are resolved by non-violent confrontations. This is done through escalated signalling, such as growling, chasing, leaping, bouncing and paw scraping. For example, male rabbits run in parallel to size each other up and decide who is likely to win if they fight, and whether the more sensible option would be to avoid a fight and retreat. This 'avoid being hurt' approach is adaptive and common across species. In captive situations, where retreat is not possible, fights can escalate (DiVincenti and Rehrig, 2016) even to the point of death.

Increasing fear is also indicated by a sequence of escalating signals, used particularly if the animal cannot escape the threat. The escalation sequence conveys the individual's increasing intention to defend itself by using any means available, including biting and kicking. With respect to dogs (Shepherd, 2007) and cats (UK Cat Behaviour Working Group, 1995; Braastad *et al.*, 2022), signals indicating anxiety, fear, or pleasure are specific to each species (McBride and Hinde, 2022).

Humans frequently miss or misinterpret animal communication signals. It is imperative that owners and others learn about the communication of species in their care, and watch the animals, if they are to detect physical or behavioural problems in the early stages (McBride, 2017a). For example, loud grinding of teeth by rabbits indicates pain, while thumping of back legs is a signal of fear (Stone, 2011). A degu warbling and beating its tail on the ground is excited and happy, whilst a frightened one will bark and drum its back legs (Long, 2007).

Different body and ear postures indicate confidence, pleasure, pain, fear and behavioural intent (e.g. Berdoy, 2002; Stone, 2011; McBride and Hinde, 2022). In general, animals that are feeling relaxed and are not in pain will be curious about their environment, move freely when active, interact socially, play and lay outstretched when resting (Fig. 4.11). As seen in Fig. 4.12, the difference between a relaxed rabbit and a tense rabbit is subtle but clear. A scared rabbit will crouch down to make itself small and as inconspicuous as it can. Its muscles tense and its eyes bulge as it tries to see as much as possible. A lower-ranking rabbit in a stable, compatible relationship will also crouch down in greeting, but will not have bulging eyes and will have relaxed muscles.

Fig. 4.11. Wild rabbit relaxing on bed of sandy soil. (Photo: iStock.)

Animals that are ill or in pain will be less active than normal, often sitting in a hunched position. Just like humans, they may be grumpy and more aggressive when not well. Recent work on Facial Grimace Scales show that these small species clearly indicate when they are in pain (Hawkins *et al.*, 2016; posters available from NC3Rs, n.d.). Indeed, we know that rats can recognize pain in other rats simply from looking at photographs (Nakashima *et al.*, 2015).

Misinterpreting natural behaviour can lead to people unwittingly causing their animal serious stress, even when their intention is to make the animal feel happy, relaxed and safe. The phenomenon of 'trancing' is an illustrative example.

Animals can be put into what appears to be a relaxed trance by placing them on their backs, or holding them firmly (McBride, 2015). But appearances are deceptive, and this is neither a relaxing experience nor a 'trance' but a state known as tonic immobility – or the 'death feint' (Darwin, 1839). Tonic immobility occurs across the animal kingdom, being shown by reptiles, birds and mammals, including humans (Marx *et al.*, 2008; McBride, 2015). Initiated by the animal being terrified for its life, 'playing dead' is a behaviour that has evolved as a last resort to limit injury and enable escape from a predator or other extreme threat. Predators often relax their jaws or lose interest if the prey goes completely still and appears dead. Indeed, this may save the life of some small pets caught by a dog or cat.

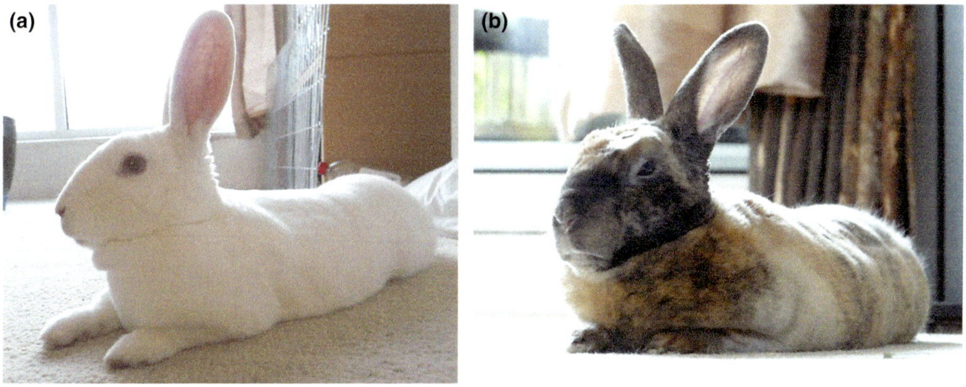

Fig. 4.12. (a) This rabbit is alert and anxious. Note the wide eyes and tense, elongated neck muscles. **(b)** This rabbit is relaxed. Note the semi-closed eyes and head resting back on the shoulders. (Reproduced from *BSAVA Manual of Rabbit Medicine* (2014) with permission from BSAVA.)

Research using various species shows that tonic immobility is associated with physiological and behavioural indicators of stress and fear. These include elevated respiration and heart rate, wide eyes, flattened ears, struggling and increased muscle tension. This state of stress continues after the animal has exited the 'trance-like' state (Carli, 1979; McBride *et al.*, 2006).

Putting animals into tonic immobility is widely used for health checks and minor procedures such as nail trimming. This can be appropriate, but there are more welfare-friendly alternative methods (Hawkins *et al.*, 2008; Oxley *et al.*, 2018). Tonic immobility is no way to bond with a pet.

Social Living

Social groups vary (Table 4.1). Groups may comprise a mated pair, a male with several females (a harem), or several males and several females. Members may be of a similar age or be of different generations (Wolff and Sherman, 2007). These differences are primarily determined by the species but will be influenced locally by the availability of resources, especially food and places to live. For example, where digging is hard and food plentiful, as on the chalk downlands of England, rabbits may live in family groups of 2–14 individuals, with several groups sharing a large tunnel system. However, where the ground is soft and prone to collapse, group size will be smaller, often a single pair in sandy areas (Cowan and Garson, 1985; Cowan, 1987).

Overall, group living has benefits, not least, enhanced predator detection. Stable groups have some form of hierarchical structure that is maintained peaceably by clear communication. Two individuals in an established group reinforce their relative status by the subordinate appropriately signalling their acknowledgement of the status quo. In this way animal societies remain relaxed and fighting is a rarity. Aggression only increases when resources become scarce or particularly important, as during the breeding season.

Successful reproduction is obviously important for species survival. As rabbits and rodents are a major food source there is a good chance that an individual's life will be short. Indeed, offspring mortality is high, with most dying before they reach adulthood (McBride, 1998). Sexual maturity is reached when a few weeks old for the smaller species and a few months for rabbits and chinchillas. This means that those born early in the breeding season may be mothers themselves the same year (McBride, 2014a,b).

Some species, including degu, guinea pig and chinchilla, invest a lot of energy into pregnancy, giving birth to a few precocial babies in each litter. The young are born fully furred and able to move swiftly within minutes. Mice, rats, hamsters and others have evolved a different approach. They have shorter pregnancies and altricial young, who are born in a fetal-like state, unfurred, relatively immobile and with undeveloped eyes and ears. This strategy means that they can have a greater number of babies per litter and more litters per year. Compare the four to six precocial babies a chinchilla will have in a year, with the 30 a rabbit or the 100 a mouse or rat can produce. Hence the expression 'breed like rabbits'.

Mental Abilities

Free-living animals do not show the abnormal, repetitive behaviours seen in captive animals. These include over-grooming, repetitive biting of objects such as cage bars, back flipping or running in a wheel for long periods of time. Arising from anxiety or frustration, these abnormal behaviours are a loud cry for help – a strong indication that we have not provided for the animal's well-being in terms of feeling safe or having enough to do.

We tend to consider smaller animals as lacking in intelligence. Consequently, we frequently keep them in rather barren environments that change little and provide few opportunities for either physical or mental stimulation. Yet the real world is neither barren nor boring, and often provides challenging problems to solve. For example, individuals may need to navigate rapidly back to their home if chased by a predator, or social pressures may force an individual to leave and seek its fortunes elsewhere. A rainstorm can cause runways to be altered, tunnels to flood and collapse and sources of food to become inaccessible. These species have evolved to be curious about what is going on in the environment and have excellent spatial abilities and cognitive maps. Much of our understanding of how people navigate comes from research using rats, because they are so good at learning mazes.

Research shows that mice, rats and other rodents are excellent problem solvers and have a range of cognitive abilities (Wynne and Udell, 2013). For example, degus can use a degu-sized rake to get food placed out of reach (Okanoya *et al.*, 2008); and rats and mice show empathy and altruistic behaviour (Langford *et al.*, 2006; Chen *et al.*, 2009; Ben-Ami Bartel *et al.*, 2014).

Animals have to recognize other individuals and recall important information about them, including how they are likely to behave in different circumstances. Researchers are just starting to investigate differences in animal personality in these smaller species (Dellu *et al.*, 1993; Czerwinski *et al.*, 2016) and beginning to confirm what many already knew: that these animals are individuals with their own personalities and preferences in life.

To provide good welfare we need to be aware of an animal's needs at species and individual level. While their attributes are very different from our own, they are no less intricate and amazing. The veterinary team and other professionals have an important role in helping owners understand and provide for the physical and behavioural needs of these small animals to ensure they live healthy *and* fulfilling lives.

Sources of Further Information

Rats and mice:

American Fancy Rat and Mice Association: http://www.afrma.org/rminfo4b.htm
The National Fancy Rat Society: www.nfrs.org
The National Mouse Club: www.thenationalmouseclub.co.uk
www.ratbehavior.org
www.ratlife.org

Rabbits:

The Rabbit Welfare Association: www.rabbitwelfare.co.uk

Degus:

www.degutopia.co.uk

Hamsters:

The National Hamster Council: www.hamsters-uk.org

Chinchillas:

Chinchilla Chronicles: https://chinchillachronicles.com

References

Addison, W.H.F. and Appleton, J.L. Jr (1915) The structure and growth of the incisor teeth of the albino rat. *Journal of Morphology*, 26, 43–96. doi: 10.1002/jmor.1050260103
Albright, J. and De Matos, R. (2010) Hamsters. In: Tynes, V.V. (ed.) *Behaviour of Exotic Pets*. Wiley-Blackwell, Chichester, UK, pp. 127–137.
Balcombe, J. (2007) *The Pleasurable Kingdom: Animals and the Nature of Feeling Good*. Macmillan, London.
Baugh, S. and McBride, E.A. (2022) Animal welfare in context: historical, scientific, ethical, moral and One Welfare perspectives. In: Vitale, A. and Pollo, S. (eds) *Human/Animal Relationships in Transformation - Scientific, Moral and Legal Perspectives*. Palgrave Macmillan, Cham, Switzerland, pp. 119–147.
Ben-Ami Bartel, I., Rodgers, D.A., Sol Bernardez Sarria, M. *et al.* (2014) Pro-social behaviour in rats is modulated by social experience. *eLife* 3, e01385.
Berdoy, M. (2002) *The Laboratory Rat – A Natural History*. Video (27 minutes). Available at: youtube.com/watch?v=giu5WjUt2GA (accessed 29 April 2022) [This video is strongly recommended. The principles apply across species, namely that despite generations of selective breeding animals retain behavioural features in common with the wild relatives that are expressed when given the opportunity.]
Brandão, J. and Mayer, J. (2011) Behavior of rodents with an emphasis on enrichment. *Journal of Exotic Pet Medicine* 20, 256–269.
Broom, D.M. and Fraser, A.F. (2015) *Domestic Animal Behaviour and Welfare*. CABI, Wallingford, UK.
Carli, G. (1979) Plasma corticosterone and its relation to susceptibility to animal hypnosis in rabbits. *Neuroscience Letters* 11, 271–274.
Carvell, G.E. and Simons, D.J. (1990) Biometric analyses of vibrissal tactile discrimination in the rat. *Journal of Neuroscience* 10, 2638–2648.
Castelhano-Carlos, M.J. and Baumans, V. (2009) The impact of light, noise, cage-cleaning and in-house transport on welfare and stress of laboratory rats. *Laboratory Animals* 43, 311–327.
Chen, Q., Panksepp, J.B. and Lahvis, G.P. (2009) Empathy is moderated by genetic background in mice. *PLoS ONE* 4(2), e4387. doi: 10.1371/journal.pone.0004387
Cowan, D.P. (1987) Group living in the European rabbit (*Oryctolagus cuniculus*): mutual benefit or resource localization? *The Journal of Animal Ecology* 56, 779–795.
Cowan, D.P. and Garson, P.J. (1985) Variations in the social structure of rabbit populations: causes and demographic consequences. In: Sibly, R.M. and Smith, R.H. (eds) *Behavioural Ecology: Ecological Consequences of Adaptive Behaviour*. Blackwell Scientific Publishers, Oxford, pp. 537–555.
Cucci, T., Vigne, J.-D. and Auffray J.-C. (2005) First occurrence of the house mouse (*Mus musculus domesticus* Schwarz & Schwarz, 1943) in the Western Mediterranean: a zooarchaeological revision of subfossil occurrences. *Biological Journal of the Linnean Society* 84, 429–445.
Czerwinski, V.H., Smith, B.P., Hynd, P.I. and Hazel, S.J. (2016) The influence of maternal care on stress-related behaviors in domestic dogs: what can we learn from the rodent literature? *Journal of Veterinary Behavior: Clinical Applications and Research* 14, 52–59.
Darwin, C. (1839) *Voyage of the Beagle*. Published (2007) by Forgotten Books, Charleston, South Carolina.
Dellu, F., Mayo, W., Piazza, P.V., Le Moal, M. and Simon, H. (1993) Individual differences in behavioral responses to novelty in rats. Possible relationship with the sensation-seeking trait in man. *Personality and Individual Difference* 15, 411–418.
Dingemanse, N.J. and Réale, D. (2005) Natural selection and animal personality. *Behaviour* 142(9-10), 1159–1184.
DiVincenti, L. Jr and Rehrig, A. (2016) Social behaviour of adult male New Zealand White rabbits housed in groups or pairs in the laboratory. *Journal of Applied Animal Welfare Science* 20, 86–94.
Ebensperger, L.A. and Bozinovic, F. (2000) Communal burrowing in the hystricognath rodent, *Octodon degus*: a benefit of sociality? *Behavioral Ecology and Sociobiology* 47, 365–369.
Fay, R.R. (1988) *Hearing in Vertebrates: a Psychophysics Databook*. Hill-Fay Associates, Winnetka, Illinois.
Finlay, B.L. and Sengelaub, D.R. (1981) Toward a neuroethology of mammalian vision: ecology and anatomy of rodent visuomotor behavior. *Behavioural Brain Research* 3(2), 133–149.
Flannelly, K. and Lore, R. (1977) Observations of the subterranean activity of domesticated and wild rats (*Rattus norvegicus*): a descriptive study. *Psychological Record* 2, 315–329.
Fulk, G.W. (1976) Notes on the activity, reproduction and social behaviour of *Octodon degus*. *Journal of Mammology* 57(3), 495–505.
Galef, B.G. and Giraldeau, L.A. (2001) Social influences on foraging in vertebrates: causal mechanisms and adaptive functions. *Animal Behaviour* 61(1), 3–15.

González-Redondo, P. and Contreras-Chacón, G.M. (2012) Perceptions among university students in Seville (Spain) of the rabbit as livestock and as a companion animal. *World Rabbit Science* 20(3), 155–162. doi: 10.4995/wrs.2012.1147

Haselton, M.G. and Nettle, D. (2006) The paranoid optimist: an integrative evolutionary model of cognitive biases. *Personality and Social Psychology Review* 10(1), 47–66.

Hawkins, P., Hubrecht, R., Buckwell, A., Cubitt, A.S., Howard, B.A. *et al.* (2008) *Refining Rabbit Care: A Resource for Those Working with Rabbits in Research.* UFAW/RSPCA, Southwater, UK.

Hawkins, P., Atkinson, J., Birt, R., Cruden, J., Duran, A. *et al.* (2016) Report of the 2015 RSPCA/UFAW rodent and rabbit welfare group meeting. *Animal Technology and Welfare* 15(1), 9–22.

Herzog, H. (2010) *Some We Love, Some We Hate, Some We Eat: Why it is So Hard to Think Straight About Animals.* Harper Collins, New York.

Honkavaara, J., Koivula, M., Korpimäki, E., Siitari, H. and Viitala, J. (2002) Ultraviolet vision and foraging in terrestrial vertebrates. *Oikos* 98(3), 505–511.

Langford, D.J., Crager, S.E., Shehzad, Z., Smith, S.B., Sotocinal, S.G. *et al.* (2006) Social modulation of pain as evidence for empathy in mice. *Science* 312, 1967–1970.

Leighty, K.A., Valuska, A., Grand, A.P., Bettinger, T.L., Mellen, J.D. *et al.* (2015) Impact of visual context of public perceptions of non-human primate performers. *PLoS ONE* 10(2), e0118487. doi: 10.1371/journal.pone.0118487

Long, C. (2007) Vocalisations of the degu (*Octodon degus*), a social caviomorph rodent. *Bioacoustics* 16, 223–244.

Marx, B.P., Forsyth, J.P., Gallup, G.G., Fusé, T. and Lexington, J.M. (2008) Tonic immobility as an evolved predator defense: implications for sexual assault survivors. *Clinical Psychology: Science and Practice* 15, 74–90.

McBride, A. (1998) *Rabbits and Hares.* Whittet Books, Stansted, UK.

McBride, A. (2000) *Why Does My Rabbit ...?* Souvenir Press, London.

McBride, A. (2010) *Guinea-pigs: Understanding and Caring for Your Pet.* Magnet & Steele, Llandow, UK.

McBride, A. (2014a) Normal behaviour and behaviour problems. In: Meredith, A. and Lord, B. (eds) *BSAVA Manual of Rabbit Medicine.* BSAVA Publications, Gloucester, UK, pp. 45–58.

McBride, A. (2014b) The Year of the Rabbit: Spring. *Rabbiting On*, Spring 2014, 10–11.

McBride, A. (2014c) The Year of the Rabbit: Summer. *Rabbiting On*, Summer 2014, 11–12.

McBride, A. (2015) Animals in trances: peace of mind or panic? *Rabbiting On*, Winter 2015, 10–12.

McBride, E.A. (2017a) Small prey species' behaviour and welfare: implications for veterinary professionals. *Journal of Small Animal Practice* 58, 423–436. doi: 10.1111/jsap.12681

McBride, E.A. (2017b) Stress in the veterinary surgery – small mammals. *The Veterinary Nurse* 8, 376–381.

McBride, E.A. and Baugh, S. (2022) *Animal welfare in context: historical, scientific, ethical, moral and One Welfare perspectives.* In: Vitale, A. and Pollo, S. (eds) *Human/Animal Relationships in Transformation –Scientific, Moral and Legal Perspectives.* Palgrave Macmillan, Cham, Switzerland, pp. 119–147.

McBride, E.A. and Hinde, J. (2022) Behaviour. In: Rendle, M. and Hinde, J. (eds) *BSAVA Manual of Practical Veterinary Welfare.* BSAVA Publications, Gloucester, UK.

McBride, E.A. and Meredith, A. (2018) Guinea pigs, chinchillas and degus (caviomorphs). In: Yeates, J. (ed.) *Companion Animal Care and Welfare: The UFAW Companion Animal Handbook.* Wiley, Oxford.

McBride, A., Day, S., McAdie. T., Meredith, A., Hickman, J. and Lawes, L. (2006) Trancing rabbits: relaxed hypnosis or a state of fear? In: *Proceedings of the VDWE International Congress on Companion Animal Behaviour and Welfare.* Vlaamse Dierenartsenvereniging v.z.w., Sint-Niklaas, Belgium, pp. 135–137.

Mellor, D.J. (2016) Updating animal welfare thinking: moving beyond the 'Five Freedoms' towards 'A Life Worth Living'. *Animals* 6(3), 21. doi: 10.3390/ani6030021

Meredith, A. and Delany, C.J. (2010) *BSAVA Manual of Exotic Pets*, 5th edn. Wiley, Oxford.

Meredith, A. and Lord, B. (eds) (2014) *BSAVA Manual of Rabbit Medicine.* BSAVA, Gloucester, UK.

Meredith, A. and Redrobe, S. (2002) *BSAVA Manual of Exotic Pets.* BSAVA, Gloucester, UK.

Myers, K. and Poole, W.E. (1961) A study of the biology of the wild rabbit in confined populations. 11. The effects of season and population increase on behaviour. *CSIRO Wildlife Research* 4, 14–26.

Nakashima, S.F., Ukezono, M., Nishida, H., Sudo, R. and Takano, Y. (2015) Receiving of emotional signal of pain from conspecifics in laboratory rats. *Royal Society Open Science* 2(4), 140381. doi: 10.1098/rsos.140381

NC3Rs (n.d.) *Grimace Scales.* National Centre for the Replacement Refinement & Reduction of Animals in Research, London. Available at: https://nc3rs.org.uk/grimacescales (accessed April 2021).

Nettle, D. (2006) The evolution of personality variation in humans and other animals. *American Psychologist* 61(6), 622.

Okanoya, K., Tokimoto, N., Kumazawa, N., Hihara, S. and Iriki, A. (2008) Tool-use training in a species of rodent: the emergence of an optimal motor strategy and functional understanding. *PLoS ONE* 3(3), e18860.
Olsson, I.A.S., Nevison, C.M., Patterson-Kane, E.G., Sherwin, C.M., Van de Weerd, H.A. and Würbel, H. (2003) Understanding behaviour: the relevance of ethological approaches in laboratory animal science. *Applied Animal Behaviour Science* 81, 245–264.
Oxley, J.A., Ellis, C.F., McBride, E.A. and McCormick, W.D. (2018) A survey of rabbit handling methods within the UK and the Republic of Ireland. *Journal of Applied Animal Welfare Science* 22(3), 207–218. doi: 10.1080/10888705.2018.1459192
Panksepp, J. and Burgdorf, J. (2000) 50-kHz chirping (laughter?) in response to conditioned and unconditioned tickle-induced reward in rats: effects of social housing and genetic variables. *Behavioural Brain Sciences* 115, 25–38.
PDSA (2015) PDSA Animal Wellbeing (PAW) Report 2015. Available at: https://www.pdsa.org.uk/what-we-do/pdsa-animal-wellbeing-report/past-reports (accessed 22 June 2022).
PDSA (2020) PDSA Animal Wellbeing (PAW) Report 2020. Available at: https://www.pdsa.org.uk/what-we-do/pdsa-animal-wellbeing-report/past-reports (accessed 22 June 2022).
Sales, J. and Pye, D. (1974) *Ultrasonic Communication by Animals*. Springer, Dordrecht, Netherlands. doi: 10.1007/978-94-011-6901-1
Shepherd, K. (2007) *The Canine Commandments*. Broadcast Books, Bristol, UK.
Smith, B.R. and Blumstein, D.T. (2008) Fitness consequences of personality: a meta-analysis. *Behavioral Ecology* 19, 448–455.
Sørensen, D.B., Krohm, T., Hansen, H.N., Ottesen, J.L. and Hansen, A.K. (2005) An ethological approach to housing requirements of golden hamster, Mongolian gerbils and fat sand rats in the laboratory – a review. *Applied Animal Behaviour Science* 94, 181–195.
Stone, T. (2011) *Understanding Your Rabbit's Habits*. T-squared Publications, Linslade, UK.
UFAW (n.d.) *Genetic Welfare Problems of Companion Animals*. Universities Federation for Animal Welfare, Wheathamstead, UK. Available at: https://www.ufaw.org.uk/genetic-welfare-problems-intro/genetic-welfare-problems-of-companion-animals-intro (accessed 16 July 2018).
UK Cat Behaviour Working Group (1995) *An Ethogram for Behavioural Studies of the Domestic Cat* (Felis silvestris catus L.). UFAW, Wheathamstead, UK.
Van den Broek, F.A.R., Klompmaker, H., Nakker, R. and Beynen, A.C. (1995) Gerbils prefer partially darkened cages. *Animal Welfare* 4, 119–123.
Warfield, D. (1973) The study of hearing in animals. In: Gay, W. (ed.) *Methods of Animal Experimentation, IV*. Academic Press, London, pp. 43–143.
Wilkins, A.M., McCrae, L.S. and McBride, E.A. (2015) Factors affecting the human attribution of emotions toward animals. *Anthrozoös* 28, 357–369.
Wolff, J.O. and Sherman, P.W. (2007) *Rodent Societies: An Ecological and Evolutionary Perspective*. University of Chicago Press, London.
Wynne, C.D.L. and Udell, M.A.R. (2013) *Animal Cognition: Evolution, Behaviour and Cognition*, 2nd edn. Palgrave MacMillan, New York.
Yeates, J. (ed.) (2018) *Companion Animal Care and Welfare: The UFAW Companion Animal Handbook* Wiley, Oxford.

5 Parrot Behaviour

Clare Wilson

Introduction

Understanding the natural behaviour of wild parrots is an essential basis for advising on the welfare issues and behavioural problems that develop in our captive parrot population. The vast majority of behavioural problems observed in captive parrots arise due to misunderstanding of the communication strategies and emotional or physical needs of these incredible creatures. Therefore, education plays a highly significant role in both preventing and treating these problems. The most common problems seen in captive parrots stem from instinctive natural behaviours that are either insufficiently catered for by their carers, for example foraging behaviour, or are not tolerated due to a lack of understanding of normal parrot behaviour, such as messiness and vocalization. The aim of this chapter is to highlight the natural behaviour of wild parrots as a basis for understanding why behaviour problems develop in the captive situation, and how to address those problems when they arise.

Parrots are an extremely numerous and diverse group of birds and it is not possible to discuss each species in detail within this text. The focus across all three chapters on parrots will therefore be on the most common species that are kept as companions in the home. Rather than going into specific details about individual species, this text will discuss the main groups – the macaws, cockatoos, true parrots (Greys, Amazons, Poicephalus, etc.), parakeets (budgerigars, cockatiels, ring-necked, conures, etc.) and lories – as within these groups there is considerable overlap in behavioural tendencies and requirements. However, it is vital that owners and professionals advising on behavioural problems look into the details of the individual species as, although there are broad similarities within the groups, there are key differences in their natural ethology which may be relevant to the ideal manner in which the individual species is kept. Within parrot species, there is also variation in the environments that these birds inhabit in the wild and this can be linked to the types of behaviour that they show. Parrots occupy habitats from arid savannahs, to temperate forests to tropical rainforests, but the vast majority of parrot species are from tropical or subtropical equatorial regions. It is preferable for people to research all of this information before acquiring a pet parrot to ensure that they choose a species that is most suited to their lifestyle and the expectations they have from their companion.

It is absolutely crucial that owners are aware that parrots are not domesticated creatures (Fig. 5.1). Domestication is the process whereby selective breeding (either through deliberate human intervention or by environmental pressures) results in a change in genetic profile such that the animals are better able to

DOI: 10.1079/9781800621312.0005

Fig. 5.1. This Peach-faced lovebird, Parsnip, although very tame, is still genetically a wild creature. (Photo credit: Clare Wilson.)

cope with living with humans than were their wild ancestors (Meehan and Mench, 2006).

Although parrots have historically been kept as pets and aviary birds for thousands of years, it is only relatively recently that they have been bred in captivity and the vast majority of pet birds one encounters are very few generations from their wild counterparts. Sadly, some are still wild-caught birds. Owners should therefore perceive their companion parrots as wild animals that can be tamed to cope with living in the domestic setting but be fully aware that they continue to be strongly motivated by their wild innate behaviours in ways that domesticated animals are not. This is especially significant in terms of reproduction-related behaviours, as parrots cannot be routinely neutered to avoid undesirable behaviours driven by sex hormones.

It is therefore essential to study their behaviour in the wild carefully if we are to understand their true needs and engineer appropriate captive circumstances for them to ensure the highest level of welfare. Only a few parrot species have been bred sufficiently in captivity to have altered behaviour and be considered domesticated to a degree. However, even these species have mainly been bred for appearance and conformation rather than for temperament, so their behaviour has been altered very little. Most affected are the little parakeets, the budgerigars and cockatiels, but there are also ring-necked parakeets and lovebirds which have been bred for a considerable time in captivity. Budgies and cockatiels are far more commonly kept as companion animals in the house and so, in terms of breeding for temperament and compatibility with the domestic environment, these species are perhaps better adapted than others which have been primarily kept as aviary birds. Budgerigars have been captive bred for showing since the mid-19th century; they have been bred in a variety of different colour mutations and are larger than their wild counterparts, unlike the larger parrots who all still occur in their natural plumage and natural size. Cockatiels have also been bred in a variety of colour mutations. Breeding success in captivity may have selected for those individuals that are best able to cope with their captive context.

Behaviour problems are rarely encountered in pet budgies and cockatiels but it will be interesting to observe whether more problems start to arise as hand rearing of these small birds becomes increasingly common. As will be discussed later, hand rearing is a significant risk factor in the development of behaviour problems in larger parrots. It is likely that the lower intellectual capacity of these small parakeets compared with the astounding cognitive abilities of medium-size and large parrots makes them less vulnerable to developing behavioural disorders in captivity. Although it is unusual for budgies and cockatiels to present with behaviour problems, this does not mean that their emotional welfare in the domestic setting is ideal. Rather, they do not generally manifest such stress with behaviour problems that are sufficiently significant for their owners to seek help, but stereotypies and other behavioural indicators of stress are common and often related to lack of space and social stimulation (Polverino *et al.*, 2015; Phillips *et al.*, 2018).

Captive parrots are unable to show key aspects of their normal behavioural repertoire, such as flight, particularly with the aim of finding feeding grounds, and socializing with conspecifics, as flight is commonly restricted either by size of cage or by wing clipping. Parrots are very commonly kept alone, or sometimes with a bird of a different species, so social interaction is extremely limited. Different species appear to cope with these restrictions to different degrees and the extent to which owners will be able to provide opportunities for parrots to show these behaviours is an important factor when choosing an avian companion. The significant behavioural activities of wild parrots and how these relate to the domestic setting are covered in more detail in the following sections.

Behavioural Effects of Parrots Being Prey Species

All species of parrot are vulnerable to predation in the wild. They are most vulnerable to attack by birds of prey, which catch them in the air or from trees. They are also susceptible to being hunted by snakes, which climb trees and either catch parrots as they rest or enter nest cavities. Monkeys will also take eggs and nestlings from nest cavities. Humans are responsible for hunting and trapping parrots and have had a major effect on wild populations of parrots. Wild parrots therefore prioritize keeping themselves out of harm's way and use alarm calls to alert the rest of the flock to potential danger. If they feel threatened, they will usually take flight. They are less vulnerable to attack if they are higher from the ground and also less vulnerable from aerial predators such as hawks if they fly up high, as birds of prey generally attack from above.

Parrots therefore feel safest if they are perched up high and will also keep a close eye on their environment for potential threats (Fig. 5.2).

In the domestic setting it is therefore crucial that parrots are provided with a safe and secure environment where they do not feel at risk of predation. This requires special consideration in birds that have clipped wings and are unable to fly to a safe location. These birds are especially vulnerable to stress. Wing clipping is discussed in more detail in Chapter 11. Parrots are often inadvertently stressed by normal human day-to-day activities in the domestic environment, so owners must attempt to see the environment from their bird's perspective. For example, placing a cage in the middle of a room or close to a doorway within the house or by a window looking out onto a street can make a parrot feel vulnerable. In contrast, if the cage is placed well away from the doorway, the bird is able to assess people or other pets that enter the room from a relatively safe distance. Contrary to advice given in many texts, it is crucial to provide parrots with high perches to enable them to feel safe and keep a lookout for potential danger. Providing perches

Fig. 5.2. These captive hybrid macaws are very relaxed preening high in the trees, well away from ground predators. (Photo credit: Clare Wilson.)

does not make handling difficult where birds are trained to recall (see Chapter 11).

Social grouping and structure

Sociality is one of the most significant behaviours of wild parrots: all parrot species are social to some degree or another. However, many birds spend significant proportions of time in solitary conditions in captivity. Lack of social interaction is a major welfare consideration and is a high-risk factor for the development of behaviour problems such as excessive vocalization and self-damage to feathers. Most parrots are generally not territorial with other parrots in the wild, both of their own species and other species, due to their gregarious nature and the flexible nature of natural flock formation. However, there are marked differences between species and there is also variation depending on the time of year, particularly during the breeding season, and the age of the birds (see below in reproductive behaviour). This non-territorial nature might suggest that it would be easy to introduce a new parrot to a household, but this is not always the case and it is not uncommon to see territorial responses in companion parrots, probably due to differences in space, resources and social structure in the domestic situation. This will be discussed in more detail in later chapters.

Their natural sociability is a major factor in the appeal of parrots as companion birds, just as it is with dogs. As long as they are appropriately socialized and tamed, parrots can form meaningful and significant relationships with humans as well as with their own species (Fig. 5.3).

Sociality is distinct from flocking behaviour. Parrots do congregate in flocks, but this has different motivations to social interaction between individuals or small groups. Types of social bonds vary considerably between species, but the most common situation found is male–female pairs living within a larger group containing juveniles and sometimes other bonded pairs. There may also be older individuals that are not pair-bonded, either due to a 'divorce', which is most commonly the result of poor reproductive success, or because of the death of one partner.

Fig. 5.3. This Congo African Grey provides invaluable companionship and pleasure to her owner. (Photo credit: A.G. Turner.)

These older individuals may pair up again in the future when they find a suitable mate, but in the meantime would usually interact playfully and socially with birds of the same sex as well as the opposite sex. There is a tendency for the larger social groups to include the same bonded pairs over a period of time, but these groups are flexible and generally parrots would not be hostile to unfamiliar birds entering the flock.

Aggressive behaviour does occur between individuals in the wild but, as with all social species, parrots use a ritualized set of communication behaviours to avoid conflict. It is extremely rare to see severe injuries occurring in the wild. Research has shown that when there is a higher risk of damage (i.e. larger, more powerful beak) parrots have more strategies for avoiding serious confrontation (Serpell, 1982). Although parrots are considered to form long-term stable social relationships, there is considerable debate about whether or not parrots within flocks form hierarchical relationships, due to limited research on wild populations and the difficulties of observing these active creatures in the wild. As with all social groups, interactions between individuals appear to be flexible and will vary according to both the context and resource involved in any competitive encounter (Friedman, 2001; Friedman *et al.*, 2006). Research has failed to identify 'dominant' pairs within flocks and has also failed to find predictable relationships between individual parrots whereby the outcome of a competitive encounter can be predicted by previous experience or characteristics of the individual birds, such as age, gender and size. So, although wild parrots do communicate with each other using assertive and submissive gestures, they do not generally appear to form stable hierarchical systems. Some research on captive populations, including Quaker parrots (also known as monk parakeets) (Hobson *et al.*, 2014) and cockatiels (Seibert and Crowell-Davies 2001), has suggested the formation of linear hierarchies and de Souza Matos *et al.* (2017) reported non-linear hierarchies in blue-fronted Amazons, but this is very limited in terms of the species and populations observed, so does not as yet provide sufficient usable evidence. Monk parakeets are also unique amongst parrots in their reproductive behaviour, using communal nests, and therefore identification of hierarchical structure must at this time be interpreted as a unique feature of captive monk parakeets and not extrapolated to other species or to the wild situation.

Reproductive behaviour

Many species of wild parrots form very strong pair bonds and many species will pair for life unless a partner dies or they experience reproductive failure. These pairs remain together throughout the year and, although they will be part of a much larger flock out of the breeding season, they still remain in close proximity and closely bonded to each other. Some species are monogamous during the breeding season and form strong pair bonds for the purpose of producing young but return as individuals to the main flock out of the breeding season and form new pairs the following season. The initial courtship behaviour of parrots is a ritualized affair and varies between species. It involves vocalizations and body posturing, such as raising and lowering the crest feathers, strutting back and forth and head bobbing. The pair bond is strengthened and maintained by continued courtship and other affiliative behaviours, including allopreening, allofeeding and close physical contact such as flying wing tip to wing tip and sitting very closely together when perching. Wild parrots would not usually show these affiliative behaviours, other than allopreening, to other members of their flock other than their own chicks. Affiliative behaviours towards chicks declines once they become independent, the age of which varies considerably with species. It is very common for parrots, both males and females, to show strong defence of the pair bond relationship if challenged by a rival.

The tendency to make strong bonds with one person (or one bird) varies with species. This strong desire to pair bond, and defend that relationship, can have serious consequences in the pet home and this is discussed in Chapters 11 and 20. Unfortunately there is not as yet a routine method for surgical neutering of pet parrots and therefore reproductive urges can cause seasonal disturbance in their behaviour. This will again be discussed in the later chapters, as owners may need to alter management practices at particular times of year when their birds come into breeding condition.

Given their intellectual capacity, it is not surprising that parrots can be choosy about companions and mates. This is less of an issue in the parakeets and small parrots, but the larger parrots clearly assess other parrots as to whether they are appropriate mates and rarely pair up with the first partner offered to them. They can show similar preferences to their human flock members and often show clear preferences for certain family members. It is not uncommon for parrots to take a dislike to, or favour, a particular individual.

Chicks are altricial when they hatch (Fig. 5.4) and in most species there is cooperative parental care between the male and female. Again, depending on species, nesting strategies vary. Some species are territorial around their nest sites, such as blue-and-gold macaws, African grey parrots, cockatoos and Amazons. These species form flocks out of the breeding season but during the breeding season will live as a pair and defend their own individual nest site from other flock members. The distance around the nest site that is protected varies between species and also with circumstances, such as predation risk and food availability, and this must be borne in mind when considering breeding in captivity and whether birds can share or need individual aviaries. Others, such as cockatiels and budgerigars, breed in colonies with several nests in close proximity to each other within a shared territory. The Quaker parrot (monk parakeet) commonly builds really large nests with several entrances so that several pairs share the same nest. Care of fledglings is also variable between species, with

Fig. 5.4. Newly hatched lovebird chicks. (Photo credit: Clare Wilson.)

some living in small family groups whilst others use communal nursery areas when adults travel to feeding sites. Nest defence is shown by males and females but tends to be more strongly observed in the male in most species. Some species are female dominant, such as the eclectus parrot, and female lovebirds can become particularly aggressive to other females in the breeding season. Such territorial behaviour can cause significant problems in the home.

There is considerable variation in the time it takes for young birds to become independent of their parents. Small parrots such as lovebirds are independent, although still immature, by 8–10 weeks of age. Many of the large parrots have a much longer dependent stage and this is especially significant in many cockatoo species. Larger parrots will often live with their parents for the first year of their life or more. Depending on the species, sexual and social maturity is not reached until a few to several years old. During the dependent stage parrots learn vital skills from their parents, siblings and other flock members. This long period of learning in the larger parrots is crucial for appropriate behavioural development. In Chapter 11 the issue of inadequate learning due to inappropriate hand-rearing practices will be discussed, as this can have severe consequences for parrots living in a domestic setting.

In relation to social maturation, it is again important to emphasize that parrots are not domesticated. Neotonization of dogs as a consequence of domestication (see Chapter 2) generally makes them compliant and compatible with being nurtured and cared for as if they were still dependent young. As with cats, no such process has occurred with parrots, so when they reach social and sexual maturity they are independent creatures that are ready to live an adult life, caring for themselves and for any young they produce. This means that in the domestic setting, where they are forced to be dependent on their carers, there is risk of conflict between the desires and needs of the parrot and those of their owner.

Flocking

Flocking occurs at times in all parrot species. However, this is a flexible behaviour and may vary according to predation risk, season and food supply as well as between species. It is clearly impractical for owners to create a large flock for their pet parrots, but flocking behaviour is relevant to the domestic setting in terms of how the parrot might view other family members, extended family members, other household pets and new parrots that are brought into the household. It is also relevant to the importance of security with regard to seeing birds of prey (or aeroplanes) out of windows, which could potentially cause far more stress to an individual bird than to a bird who is a member of a large flock. Birds flock for several reasons. The capacity for predator detection and avoidance is far greater. For example, parrots are more likely to congregate in larger flocks for risky activities such as coming down to the ground to drink, as they are more vulnerable to predation in this context. In locations where there is a higher risk of aerial threat, flock sizes tend to be larger which is thought to give greater protection to the individuals. There is greater efficiency in searching for food and water sources and increased opportunity for pair bonding. A flock is better able than smaller groups to defend an area of territory that contains good-quality nest sites or food sources.

As an example of the variation in flocking behaviour, orange-fronted conures in Costa Rica will fly off in flocks of anywhere between four and 20 to forage for food during the day. However, several of these smaller flocks may well congregate to form a far larger flock – hundreds of birds – at plentiful food supplies or good roosting sites for a midday rest or overnight. This pattern is similar for many other parrot species and it is not uncommon to find mixed flocks of different parrot species, for example at clay licks. Although it is uncommon for individuals of different species to interact with each other, observations at clay licks have shown a greater tendency for parrots to show aggression to their own species and mainly to ignore other species. In the captive setting birds can sometimes learn to interact with parrots of other species but they tend to show a preference for their own species.

Foraging

Wild parrots spend many hours a day, around 40–60% of their waking hours, foraging for food and also visiting drinking locations. One might interpret a caged bird's life as easy and trouble free with food being provided easily. However, foraging behaviour is not just important physical and mental stimulation; it also provides a social experience for wild parrots. All parrot species travel considerable distances, many kilometres, when foraging. Some, such as African greys and Amazons, tend to return to regular roosting sites overnight, whereas others, such as budgerigars and some cockatoo species, are nomadic and will move depending on food availability. Parrots may attempt to increase flock size at plentiful food resources to reduce predation risk or during searching activities to help increase the chances of finding food. They do this by using contact calls.

Foraging behaviour varies between species depending on the type of diet. There is huge variation between parrot species in terms of the contents of the diet, but they are all opportunistic foragers and by the nature of this feeding strategy they spend a considerable portion of their time budget on this activity. Some parrots have quite a general diet; for example, the Amazons and macaws eat a mixture of nuts, seeds, leaves, fruits from a wide variety of plant species, along with occasional insects. Others have very specific requirements, such as the greyheaded parrot, which seeks the kernels of unripe fruits, and the lories and lorikeets of Australasia, which mainly feed on nectar (Fig. 5.5). There will also be seasonal variations depending on availability of food types. In the domestic setting, owners commonly provide a bowl of food *ad libitum* to their birds. This not only risks over-eating and obesity but also seriously limits the physical and mental stimulation that foraging activities could provide.

Fig. 5.5. This dusky lory feeds on a very specific diet of nectar. (Photo credit: Clare Wilson.)

Feeding behaviour

One of the most common reasons why parrots are rehomed or relinquished is due to their messy nature. In the wild, parrots will often feed in sites that are plentiful with food and therefore they do not mind wasting some. It is important that owners understand the reasons behind their parrot's apparently messy behaviours, so that they can be realistic in their expectations and accept that this is an aspect of parrot owning that cannot be avoided. Fruit-eating parrots will often be seeking out the energy-rich seeds in the centre of the fruit rather than the soft pulp of the fruit body, or vice versa in fruits where seeds contain toxins. Owners may find this wasteful or even think that the parrot did not enjoy its grape, whereas this is natural parrot behaviour. Parrots also thoroughly enjoy the activity of eating, just as we do. They have highly developed touch sensors in their beaks and feet for handling food and often enjoy chewing the cases from nuts as well as eating the nuts themselves. Feeding a purely pelleted diet can restrict a parrot's enjoyment of manipulating food items. It is therefore very important to emphasize to owners the nature of a parrot's natural diet in the wild when providing advice on feeding them in the captive situation.

Another aspect of feeding behaviour that is important in understanding the natural tendencies of parrots is how much exploration is required during the foraging process. Parrots eat many different plant parts, including flowers (or unopened buds), nectar, pollen, fruits, grains (such as grass seeds or agricultural crops), nuts and leaves. Some also eat insects that they encounter whilst foraging in plants, such as those feeding in flowers or hiding within cracks in the bark of trees. The level of exploration required by different species in their foraging activities in the wild may influence their capacity for coping with an environment of limited complexity in the domestic setting. Some food items are far easier to find than others; for example, leaves are abundant and within easy view, whereas flower buds and insects may require greater searching amongst leaves and branches to locate them. The natural diet, complexity of environment and extent of exploration to find desired food items of different species in the wild will affect their activity levels and desire to carry out exploratory behaviour in captivity (Mettke-Hoffman *et al.*, 2002, 2005). This is another aspect that owners should research prior to selecting a suitable species for their circumstances. It is also key when providing advice to owners about provision of appropriately complex environments and foraging opportunities for their chosen species.

Language: vocalization and visual communication

Interactions between parrots within a flock are complex and poorly understood: this is an area where much more research is required. Postural signs and vocalizations are both very important methods of communication in parrots. Parrots vocalize in many situations, just as people do, and are able to identify individuals by the sounds that they make. They communicate with each other via vocalizations when they are playing, trying to find each other, defending their territory, alerting others to a predator or when they find an exciting plentiful food source. Being noisy is a normal part of parrot life. Parrots usually have regular times of day when they tend to be noisier. Most species have a very active noisy period when the sun first rises and the flock is gathering and preparing to set off foraging. This is especially notable in Amazons but occurs in all parrot species. Alarm calls are another important vocalization that wild parrots use. These can also be very loud.

Vocalizations are especially important in forest-dwelling species. It is these birds that tend to have the greater volume of contact call in order to be able to find each other in dense vegetation. However, all parrots can have loud contact calls, as these are used to recruit perching parrots whilst flying at a distance overhead and alerting them to good food sources and thus need to be heard over considerable distances. In contexts and species where territorial defence occurs, this may also involve loud vocalizations. Noise is a common reason why parrots are relinquished and it is therefore crucial for owners to understand the reasons why they vocalize. This makes it easier for owners to reduce the frequency through appropriate management, but also to learn to accept what is normal behaviour for parrots that cannot be altered by training. Conures and macaws are renowned for being particularly loud. Prospective owners often fail to consider this with conures, which are deceptively small in size for the amount of noise they are capable of producing.

Parrots regularly use contact calls to keep in touch with members of their social group or the wider flock and owners may misinterpret these calls and respond inappropriately. Inappropriate responses to parrots making contact calls to their owners can result in the very commonly encountered problem of screaming behaviour.

Communication via body language involves alterations in body posture and in feather position. This is very variable between species but there are some common themes that indicate relaxation, anxiety and fear which will be discussed in detail in Chapter 11.

Parrots also appear to communicate with each other via the appearance of their plumage and gain information using the ultraviolet spectrum. Their ability to see a greater range of light than humans is important to bear in mind when considering whether their environment is suitable.

Roosting behaviour

Most parrots roost in flocks on branches of trees and this may be close to the feeding area in nomadic species or close to nesting areas in territorial species. Breeding birds who are not yet incubating eggs would usually sleep close to their nest site. Although many parrots live in tropical climates that can reach high temperatures during the day, most species are capable of surviving at far colder temperatures, provided that they have protection from wind and rain.

Bathing and preening

Keeping the plumage in pristine condition is a vital part of a parrot's daily activity. Grooming and bathing are essential for feather maintenance to ensure insulation against heat and cold, and a good condition for flying long distances during foraging trips. Feathers on the head cannot be reached by the parrot's own beak, so allopreening often involves the birds grooming each other's heads. One bird may approach another with a lowered head to request this activity. Allopreening is an essential part of maintaining the pair bond and caring for young but is also seen between other flock members. This can be a very rewarding behaviour for birds and head scratching is often a useful reward to use in training pet birds and in affirming the owner–bird bond.

Water is essential to maintain healthy plumage in birds and wild parrots bathe or shower on a daily basis (Murphy *et al.*, 2011). Many parrots prefer to shower whilst it is raining, or rub themselves on recently rained-on leaves in the trees rather than to come down to the vulnerable ground level position for a bath.

Daily routine

Despite the variation between species in many aspects of their lives, one thing that parrots have in common is a regular routine to their daily activities. As the morning light appears flocks start to gather and this is generally a vocal affair. This involves contact calls, greetings and general excitement about the imminent departure for foraging. The flock will then head off on a foraging expedition and may travel long distances before finding a suitable site. If the destination provides fragmented feeding areas, then the flock may break up into smaller groups. As most parrots are in tropical or subtropical climates, it is common for the heat of the day to become too great to allow high levels of activity, so they will usually have a resting period in the middle of the day (Gilardi and Munn, 1998). As the weather cools down and the sun starts to set, the vocal flock gathering occurs again as the group flies back home to the roosting site. This pattern of activity is linked to circadian rhythms and therefore parrots in captivity will show a similar timetable. This may conflict with the domestic setting in many ways. Owners or neighbours may find the morning and evening vocalization sessions difficult to live with; they may be out of the house during the parrot's most active periods, which can compromise the opportunity to provide social, mental and physical stimulation. In addition, the hours of dawn and dusk in the summer months of temperate climates may not be compatible with the parrot's natural requirements for 10–12 hours of sleep or with the owner's requirement for peace and quiet at 5am.

Chewing

Although parrots use their strong beaks for accessing food, such as breaking nut shells, they also use them for chewing objects during exploratory investigation and play. In addition, they use their beaks to enlarge natural tree cavities by chewing in order to improve nest sites. This destructive behaviour can often cause problems in the home and so owners must provide ample alternative chewing opportunities for their birds (Fig. 5.6).

Learning

As with all social species, young parrots do not hatch with all the knowledge they need to be successful individuals within a social structure and a complex environment. Learning occurs continually through interactions with parents, siblings, other flock members and their physical environment, enabling young birds to become socially competent and learn the physical and cognitive skills appropriate to their needs.

Fig. 5.6. This hybrid macaw is thoroughly enjoying destroying this wooden fence, showing the importance of owners providing appropriate chewing outlets. (Photo credit: Clare Wilson.)

Neonatal parrots respond to touch, food and vocalizations and even at this early stage of development they are learning through moving about in the nest cavity, manipulating food and interacting with siblings and parents. Once their eyes are open and their legs strong enough to support their weight, they will peep out of the nest box to start learning about the world around them. Initially they are only capable of making sounds to request feeding but even whilst they are still in the nest they learn different vocal signals through communicating with their siblings, parents and neighbours. Once the young birds have fledged, they have a steep learning curve in terms of learning physical skills such as climbing, flying and food manipulation and also a rapid expansion of social interaction to learn communication skills. All this experience provides stimulation for neurogenesis, resulting in a more complex neural network in the brain, and has long-term consequences on behavioural flexibility. Environments in which there is limited stimulation during this time can result in a poorly adapted parrot that finds it difficult to cope with the challenges it faces for survival. This crucial early learning, which occurs naturally in the wild setting, is difficult to replicate in the human world where eggs are hatched in incubators and chicks reared in sparse plastic tanks, often with very limited opportunities for social and environmental interaction. Just as dogs learn to inhibit biting during development (see Chapters 8 and 12), this is also a vital skill for young parrots to learn. In the wild they have the opportunity to learn from parents and siblings through normal day-to-day interactions. As with other aspects of appropriate social interaction, this is a more difficult behaviour for human 'parents' to teach than the natural parents and siblings and is the most likely explanation for such a high frequency of biting in hand-reared parrots. These potential pitfalls in companion parrot rearing are discussed in more detail in Chapter 11, where preventative advice to rear parrots that are better able to cope with captivity in domestic situations is covered in more detail.

Parrots are amongst a few groups of animals who are capable of lifelong vocal learning. Many birds learn species-specific vocalizations at a very young age and have no capacity to learn or create new vocal sounds as they mature. Parrots, humans, hummingbirds, some songbirds such as starlings and some mammals, including cetaceans, have all been shown to have the ability to learn throughout life from their conspecifics and their environment to produce new sounds. This is the basis for parrots learning to mimic human language and domestic sounds such as doorbells and phone ringtones. Some parrot species have been shown not only to mimic but also to use and respond to human language in context. For those owners who have the knowledge and dedication to educate their parrots this can be enormously beneficial to intra-species communication and welfare.

Summary

In order to optimize the pet bird's welfare, owners need to provide an environment that is as close to the natural environment as possible. They also need to consider ways to ensure that these prey animals feel secure. Cages should be as large as space allows, preferably with birds also having access to an outdoor aviary for flight exercise and to be exposed to natural daylight. Foraging opportunities should be provided rather than simply presenting food in a bowl. Enrichment should consider the natural behaviour for each individual species, be imaginative and be changed on a regular basis to make the environment complex, interesting and novel. The environment should also allow the birds to carry out natural behaviours such as chewing, manipulation of objects and climbing. Parrots need a regular routine to their day, with their natural rhythm of high levels of activity in the morning and evening being catered for. Social needs are also important, to avoid distress: for the vast majority of owners the ideal solution for their bird is for him to have an avian companion. If owners are well educated as to the natural behaviour of their tame wild companion, parrots can make hugely rewarding family members. However, owners need to consider the natural behaviour of each species before deciding on a parrot as a pet and be sure that natural behaviours such as messy eating and loud vocalizations fit with their own lifestyle. Parrots require an enormous time and emotional commitment from owners and must be respected for the intelligent, social creatures that they are.

References

de Souza Matos, L.S., Palme, R. and Vasconcellos, A.S. (2017) Behavioural and hormonal effects of member replacement in captive groups of blue-fronted amazon parrots (*Amazona aestiva*). *Behavioural Processes* 138, 160–169.

Friedman, S.G. (2001) The struggle for dominance: fact or fiction? A bird's eye view. *Original Flying Machine*, Issue 6: May/June.

Friedman, S.G., Martin, S. and Brinker, B. (2006) Parrots and dominance, behaviour analysis and parrot learning. In: Luescher, A.U. (ed.) *Manual of Parrot Behaviour*. Blackwell Publishing, Oxford, pp. 147–163.

Gilardi, J.D. and Munn, A.C. (1998) Patterns of activity, flocking, and habitat use in parrots of the Peruvian Amazon. *Condor* 100, 641–653.

Hobson, E.A., Avery, M.L. and Wright, T.F. (2014) The socioecology of monk parakeets: insights into parrot social complexity. *The Auk Ornithological Advances* 131, 756–775.

Meehan, C. and Mench, J. (2006) Captive parrot welfare. In: Luescher, A.U. (ed.) *Manual of Parrot Behaviour*, Blackwell Publishing, Oxford, pp. 301–317.

Mettke-Hofmann, C., Winkler, H. and Leisler, B. (2002) The significance of ecological factors for exploration and neophobia in parrots. *Ethology* 108, 249–272.

Mettke-Hofmann, C., Wink, M., Winkler, H. and Leisler, B. (2005) Exploration of environmental changes relates to lifestyle. *Behavioral Ecology* 16(1), 257–254.

Murphy, S.M., Bruan, J.V. and Millam, J.R. (2011) Bathing behaviour of captive orange-winged Amazon parrots (*Amazona amazonica*). *Applied Animal Behaviour Science* 132, 200–210.

Phillips, C.J.C., Farrugia, C., Lin, C., Mancera, K. and Doneley, B. (2018) The effect of providing space in excess of standards on the behaviour of budgerigars in aviaries. *Applied Animal Behaviour Science* 199, 89–93.

Polverino, G., Manciocco, A., Vitale, A. and Alleva, E. (2015) Stereotypic behaviours in *Melopsittacus* undulates: behavioural consequences of social and spatial limitations. *Applied Animal Behaviour Science* 165, 143–155.

Seibert, L.M. and Crowell-Davies, S.L. (2001) Gender effects on aggression, dominance rank, and affiliative behaviours in a flock of captive adult cockatiels (*Nymphicus hollandicus*). *Applied Animal Behaviour Science* 71, 155–170.

Serpell, J.A. (1982) Factors influencing fighting and threat in the parrot genus *Trichoglossus*. *Animal Behaviour* 30, 1244–1251.

Part II

The Veterinary Practice Role in Preventing Behaviour Problems

6 The Role of the Practice Environment

Caroline Bower

It is essential for the veterinary team to consider the physical practice environment from the pets' point of view in order to protect the mental welfare of their patients. Looking at facilities through the eyes of a dog, cat, bird or small prey animal pet can often reveal features that are not comfortable or relaxing. This can lead to welfare concerns and also render patients more anxious, stressed, less compliant, potentially aggressive and generally more difficult to examine and treat. We should strive to create an environment that helps pets and their owners to feel relaxed and not under threat and, better still, even enjoy a visit to the practice. If this can be achieved, the whole experience will be more positive for all concerned, staff time will be optimized and clients will be happy to visit the practice more frequently.

The design of each area of the practice, and how it is used, can make a significant difference to the behaviour of our patients. However, it is also important to realize that even the best design features alone are not enough; the knowledge and behaviour of staff within the practice and the way in which they approach every patient are equally, if not more, important.

The whole ethos, with regard to the behaviour of patients, should be to protect good behaviour whenever it is offered and to provide an environment and a practice team that helps patients to remain calm. Mostly this means preventing them from feeling fearful and manifesting behavioural signs of stress, although some dogs can be problematic because of over-excitement and boisterous behaviour; this could occur, for example, if the dog has attended poorly supervised puppy classes in the practice premises.

Stress is an adaptive response to a fear-inducing event and the neuroendocrine changes that occur are associated with the 'fight or flight' response. The behavioural manifestations of stress therefore may involve the pet trying to escape from the room, basket or handler, or showing aggression in its various forms, but some pets will 'freeze', i.e. become immobile, and others may become over-active, which may be misinterpreted as excitability.

To many animals, the practice environment may seem threatening, thus inducing anxiety or fear; there are noises such as phones ringing and dogs barking; there can be noxious smells such as disinfectants and fear pheromones; there may be overcrowding in the waiting room; a prey species such as a rabbit or parrot may find itself near a predator such as a dog or cat. The acute stress response is important to the survival of the animal, but in the veterinary context it generally makes the job of clinicians much more difficult and can cause the experience for both pet and owner to be very unpleasant, even during a simple examination. Add to that the fact that the stressed animal is

DOI: 10.1079/9781800621312.0006

more likely to urinate, defecate, bite, scratch, behave unpredictably and cause damage to itself, the handlers and even the facilities, then it becomes clear why we want pets to remain calm and compliant in our hands. An acute stress response will affect the behaviour of many pets in the waiting room, consulting room and treatment areas, and chronic stress can adversely affect the emotional well-being and clinical recovery of hospitalized patients.

The neuroendocrine response to stress involves increased activity of the sympathetic branch of the autonomic nervous system and increased release of adrenaline and noradrenaline coupled with elevated circulating cortisol (Casey, 2002). Immediate effects include enhanced cardiovascular output, an increase in blood glucose and raised blood pressure. These physiological changes must be considered when assessing animals clinically, interpreting blood sample results (especially with reference to stress hyperglycaemia), or performing procedures such as blood sampling on cats with pre-existing hypertension, when further elevation of blood pressure could potentially cause a serious problem, such as retinal detachment.

Considering the many challenges that face the veterinary team in managing pets and their owners on a daily basis, it is remarkable that so many pets do cooperate and remain positive about visiting veterinary practices and indeed do recover smoothly despite being kept in hospital wards. This may be promoted by a combination of a calm and confident temperament in the pet and client, a practice environment that is welfare-friendly for patients and a practice team who make it a priority to pay attention to the behavioural needs of each and every individual pet that enters the practice.

There are a number of underlying factors that directly affect the way a pet visiting the clinic will behave, including the following.

- The intrinsic temperament and personality of the pet. Calm and confident pets should be expected to cope better, although we need to be aware that behavioural inhibition associated with stress can be mistaken for calmness.
- The degree of socialization that the pet has experienced. Comprehensively socialized animals should experience less stress when encountering their own and other species in a restricted environment (see Chapter 8).
- The amount of habituation the pet has experienced in novel environments.
- Whether or not the pet has been through a process of latent inhibition with regard to the veterinary surgery, i.e. whether or not a positive association has been made with the premises and staff, related to praise and food rewards, thus reducing the likelihood of developing a negative association later.
- Previous experiences in this or another veterinary clinic. Both good and bad experiences will be remembered and can influence the association. Painful procedures and fear-inducing events are typical reasons for pets to develop anxiety about veterinary visits. Anxiety is an emotional response where the pet anticipates a negative outcome
- The behaviour of the owner or person in charge of the pet. The anxious or tense owner can adversely affect the behaviour of their pet.
- The physical condition of the pet. For example, pain can make the pet behave with defensive aggression when approached or handled.

The following sections consider the factors affecting the behaviour and emotional state of dogs, cats, birds and small prey animal pets in key areas of the practice: waiting room, consulting room and ward. The less alien the practice environment looks, smells, feels and sounds, the less anxious or fearful the pet will be, and the calmer and more cooperative it is likely to be. The more pet friendly and like a home environment it can be, the better (Yin, 2009). Staff need to consider all the sights, smells, sounds and contact materials that may upset the patients. All patients ideally need to feel as though they are visiting a friend's house, not only in the way that staff behave, but also in terms of the environment they experience. Of course, there are constraints related to hygiene, cleaning, safety, observation and other practical issues, but thinking about veterinary practices from the pet's point of view and seeing the environment through their eyes helps identify things that the average practice can do to make the environment more pet friendly.

The Waiting Room

The waiting room environment sets the tone for the whole visit. Crowded, multi-species waiting rooms have the potential to cause anxiety or aggression in patients and concerns for owners (Fig. 6.1). The design of the room can make a huge difference to the ease of management of pets within this area but, in addition, a behaviourally aware, attentive receptionist can manage pets and their owners in such a way as to minimize adverse conditions (see Chapter 7).

How to create a pet-friendly waiting room

Even where practices do not have the luxury of a large amount of space, or separate waiting rooms for different species, what they do have can be used wisely. Here are some suggestions for practices with limited space.

- Avoid bottlenecks around exit and entry points and subdivide the waiting room into areas for different species if possible. In an ideal world one would separate dogs and cats and keep predators and prey species apart, for example separate rabbits from dogs, and cats from birds of prey. This does not mean that five separate waiting rooms are needed (although it would be lovely), as even clever use of recessed areas and visual barriers can be very effective. At the very least, try to provide some screening between small pets in baskets and dogs.
- A receptionist who is behaviourally aware can manage waiting space well. For example, they can suggest that a nervous or boisterous dog waits outside or in the car, or they can move a nervous dog, cat or rabbit into a consulting room or to another quiet area to wait for their appointment (see Chapter 7).
- In general, give each pet as much personal space as possible. Overcrowding and close proximity to other strange pets is likely to increase arousal and anxiety (Fig. 6.2).
- Display a notice requesting that clients keep dogs on leads, away from other animals and pet carriers.
- Keep noise levels down as much as possible by taking phones out of reception or have the receptionists wear ear phones to take the calls. Remove noisy dogs to another

Fig. 6.1. The dogs in this small waiting room are being well controlled by their owners, but there is not enough space between them, particularly if one was to become noisy or aggressive. (Author's own image.)

area if possible. Think about the materials used on floors and walls and whether they will absorb or amplify sound.

- Smells that may upset pets include surgical spirit or other chemicals that they may associate with a veterinary visit, fear pheromones from other pets, faeces and urine. Apart from keeping the waiting room clean and removing unpleasant odours, the use of pheromone ('Feliway') diffusers in veterinary waiting rooms can have a calming effect.

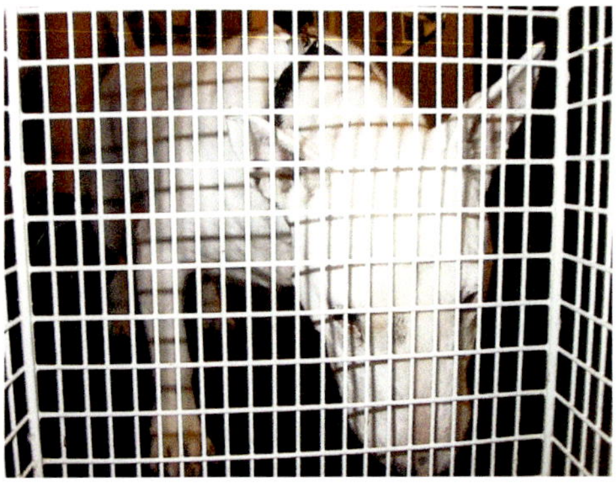

Fig. 6.2. This is a cat's view of an inquisitive bull terrier, taken from within the basket; imagine how threatened this poor cat must feel. (Author's own image.)

- Cats prefer to be raised above floor level in their baskets, so provide secure surfaces such as stools or shelves to allow for this. In addition, encourage owners to cover pet carriers containing cats, birds or small prey animal pets with towels, to provide visual protection from other animals (Fig. 6.3).
- It is best for cats and timid pets to wait for their appointment in an area of minimal human and animal 'traffic'.
- Staff should also try to prevent noises from consulting rooms and treatment areas reaching the waiting room, as shrieks or growls will upset both pets and their owners.
- It is helpful for owners to be given a definite appointment time, and to run on time. If at times there is a build-up of owners and pets in the waiting room, the receptionist needs to be aware of the potential effect this may have on the waiting patients and move them accordingly into a quieter area or even back to the car park, to avoid increased arousal or anxiety.
- It is worth considering special surgery times for birds and exotics, or for cats, keeping them separated from dogs. Obviously, the

Fig. 6.3. The owner of this cockatoo had followed our receptionist's request to cover the basket with a towel and raise it above floor level. The towel may also be used during bird handling in the consulting room. (Author's own image.)

purest form of separation is the species-specific practice, such as feline or avian only.
- To help avoid congestion it is best to avoid situating weigh scales in a busy waiting room.
- Keep a varied supply of treats on reception ready to offer to pets, to encourage a positive association with this area of the practice (Fig. 6.4).

The Consulting Room

The pet who is relaxed in the waiting room is more likely to enter the consulting room in a calm and compliant state, will be easier to examine and is less likely to become fearful and exhibit behaviours such as aggression, urination or defecation, which are unpleasant for all involved. If the patient is physically and mentally relaxed it will be easier to gain an accurate assessment of health or disease, painful foci, valid reaction to palpation, etc. The relaxed pet is also far less likely to require extra staff for restraint and therefore will not 'waste' valuable time; and is far less likely to need any form of chemical restraint, which involves the additional negative stimulus of an injection, more time, greater risk to the pet and more expense for the client.

Various essential features in a consulting or examination room are needed to achieve a relaxed consultation. These include good lighting and a suitable surface for examination of various species of pet. Most vets and nurses like to have a computer and hand-washing facilities available in this room too. Consulting rooms can be uninviting to pets and their owners, the pet may associate the area with previous negative experiences, for example painful examination or injection, and may already be in an anxious state if conditions within the waiting room have

Fig. 6.4. This receptionist is making good use of treats provided on the reception desk, appreciated by both dog and owner. (Author's own image.)

been less than ideal. The behaviour of the vet or veterinary nurse in this room will also influence the way each pet perceives the examination room and consequently how it behaves. Consider whether the pet will be more relaxed if examined on the floor or on the table, and use treats and praise whenever you can to reward calm, relaxed and cooperative behaviour. It may be possible and appropriate to examine some pets outside if there is a quiet and safe place to do so.

Appropriate pet-friendly handling is dealt with further in Chapter 14.

How to create a pet-friendly consulting room

- Minimize noise, such as phones and loud voices, and offensive smells, such as surgical spirit, blood, or anal glands, in or within the vicinity of the consulting room.
- Do everything you can to make the room comfortable for pet and owner alike. Consider offering chairs for the vet or nurse and owner, as nervous dogs tend to feel less threatened if we are not standing over them and the people involved may feel more relaxed sitting down, rather than standing facing each other over a tabletop.
- Consider positioning the examination table so that there is more space in the room and the nervous dog does not immediately feel he is being crowded.
- A washable rug on the floor reduces echo and makes the room more like home.
- Consider placing a towel or blanket over the examination table. Cats in particular tend to relax if they are allowed to stay in contact with a familiar towel or blanket which they have in their travel box. Nervous rabbits stay calmer and can be more safely examined if they are gently wrapped in a towel or provided with a temporary 'burrow' (Fig. 6.5). Caged birds should also be handled using a towel during examination so that they do not associate the hand with this procedure. It helps to have a good supply of clean towels and blankets in the practice.
- Many pets show signs of anxiety if they are forced to stand on metal weigh scales, so it helps if the surface is anti-slip and more inviting through use of rubber matting, a towel or rug. Use paediatric scales for weighing cats, birds and small prey animal pets.
- The use of treats and a calm and confident voice when the pet enters, goes on the examination table or scales, is being examined or having an injection, will help to create positive associations with this area of the practice. Ensure palatable treats for all species can be readily accessed in both dry and wet forms.
- Ensure any equipment that may be needed is to hand before removing a pet from the basket or cage, as minimal handling and restraint are preferable. Abide by the doctrine 'less is more', especially when handling anxious animals.
- With specific reference to birds, ensure the window is closed, blinds are drawn and fans are off.
- Use scissors or quiet clippers to clip hair, especially on cats.
- Allow longer appointment times for pets who are known to be anxious or aggressive,

Fig. 6.5. A burrow created simply using a vet bed helps keep this rabbit calm before examination. (Author's own image.)

as it will not help if the clinician is hurrying due to time pressure. These pets could also have elective appointments at quieter times of the day.

The Ward

Animals in the ward may be there as day cases or for longer periods if they are having a series of investigative procedures or treatments. They may be hospitalized due to requirements for care that cannot be delivered at home, such as oxygen therapy or nebulization, intravenous fluids or analgesia, and other medication; or as post-surgical cases, for example orthopaedic and spinal cases where cage rest as well as effective analgesia are necessary, after gastric dilatation and volvulus surgery, or when a feeding tube is in place.

Whatever the reason, these patients need more than good medical care. For each individual animal, it is essential to consider their behavioural needs and do our best to provide the ideal environment for recovery. The consequences of stress, and the behaviours that may be manifest by animals as a result of acute stress in the practice, are discussed in this chapter. The stress response is adaptive, in that it allows animals to cope with short-term stressors and is involved in the initiation of rapid escape or defence behaviours. Although these behaviours can make animals more difficult to handle, they enable the individual to show a coping response. It is important to consider the potential for acute stress becoming chronic when pets are hospitalized for days or even weeks and the significance of this with regard to welfare and recovery.

Hospitalized pets are unable to escape or control their situation by using a behavioural response and this can result in adverse effects on their physiological and emotional health. If the animal is unable to escape from a source of stress, such as noise, the physiological neuroendocrine response continues and may become maladaptive. Prolonged release of high levels of glucocorticoids, and other neural changes, can have wide-ranging negative effects on the body; deleterious physiological effects include increase in blood pressure, inhibition of the normal inflammatory response and changes in immune function (Carlson, 1994). For example, it was found in a study with rats that the stress of inescapable shock decreased the number of circulating lymphocytes (Keller *et al.*, 1983). Animals in wards that induce stress may therefore be less likely to be able to fight infectious disease agents. This is obviously of particular concern if infection is under treatment, such as a contaminated wound, viral gastro-enteritis, or cat flu.

Older animals, even if they are healthy, do not tolerate stress as well as younger ones (Carlson, 1994) and this should be considered when making decisions about whether or not an older pet really does need to be hospitalized.

Potential sources of stress in a hospital ward include the following.

- Separation from the owner and family.
- Separation from other familiar companion animals.
- Removal from a familiar environment and routine.
- Changes of diet, bedding and litter.
- Being housed with unfamiliar species or other unfamiliar pets of the same species.
- Prey and predator species in the same ward.
- Loss of normal diurnal rhythm due to constant lighting required for observation.
- Inappropriate temperature and humidity.
- Noises such as barking, phones, clattering food bowls.
- Kennel surfaces that have a high level of reflection of noise and light (Fig. 6.6).
- Offensive smells, including fear pheromones from other animals.
- Close proximity to aggressive or vocal animals.
- The presence of many unfamiliar handlers.
- Loss of choice of food, resting surface, litter type (for pets using a litter tray), or access to preferred substrate for going to the toilet at a suitable distance from their bed.
- Loss of ability to withdraw or hide, and lack of opportunity to show normal avoidance responses.
- An unpredictable routine.
- Intrusive handling.

Fig. 6.6. Stainless-steel surrounds and lack of a hiding place or raised surface is not cat-friendly. (Author's own image.)

How to create a pet-friendly ward

All species

- Ensure the five freedoms are observed, i.e. freedom from hunger and thirst, discomfort and pain, fear and distress, and that they are able to behave normally within the constraints of their hospital accommodation.
- Keep noise levels to a minimum by using noise-absorbent materials on surfaces where practical, and avoiding phones ringing loudly in the ward. Barking dogs may settle with sedation, or you may invest in a sound-proofed kennel. There is limited evidence that pheromone diffusers (e.g. Adaptil™, Ceva Animal Health) may have some benefit in keeping dogs calm, to the benefit of other patients.
- Try to provide diurnal rhythm by dimming lights and avoiding any unnecessary disturbance overnight. Consider the use of webcams to facilitate observation without disturbance.
- Avoid unfamiliar hospitalized animals having direct eye contact with each other.
- Avoid overcrowding with staff or visiting owners. Visitors are generally happier to be in a separate, private room with their sick pet, if the pet can be moved.
- Avoid housing predators and prey species together.
- Ensure good air quality, temperature and humidity control. Bear in mind that different species have differing requirements and some will need to have their own individual accommodation, for example vivarium for reptiles, increased heat and oxygen or nebulization for the bird with respiratory disease. Many modern wards have an intensive care unit for such purposes, which may also be mobile.
- Ask the owner about the pet's usual or favourite food and litter type and use it unless it is contra-indicated.
- It may be appropriate for some pets to have a companion hospitalized with them to reduce stress, for example guinea pigs, rabbits and pairs of birds.
- Provide soft, warm, comfortable and washable bedding.
- Avoid any unnecessary disturbance or handling of the pet.

Dogs

Dogs stressed by the hospital environment may show a range of behaviours, including lack of willingness to interact with people, anorexia, behavioural signs of appeasement, aggression when approached or handled, hyperventilation, hypervigilance, withdrawal and hiding. Some dogs become extremely vocal. There is physical tension on examination, which should not be confused with pain but which makes pain assessment challenging.

- All dogs should be in kennels appropriate to their size and must be able to move around freely.
- Dogs should not be able to see other dogs directly from their kennels.
- Avoid placing timid or anxious dogs in a busy part of the ward.
- Use blankets or towels to cover kennels for dogs who are easily aroused by movement, or for very sick and stressed individuals, provided that the ward staff are still able to supervise care sufficiently.
- A vocal dog may be stressed and will cause noise stress in others, including the ward staff. A sound-proofed kennel is a great asset for reducing noise stress to others, but always consider underlying reasons why an individual is barking. If these cannot not be addressed easily, consider chemical anxiolysis or sedation. In some cases, Adaptil diffusers, lavender or chamomile oil on the bedding may have a calming effect.
- Avoid bright lights, noise and external movement stimulation for dogs with seizure potential, and particularly in dogs with status epilepticus or seizures related to other causes, such as toxicity.
- Provide toys or puzzle feeders for young or playful dogs.
- Ensure appropriate opportunities to toilet away from their kennel areas at sufficiently frequent intervals.

Cats

Cats stressed by the hospital environment show a range of behaviours, including lack of interaction with people, lack of investigative behaviour, constant hiding, anorexia, aggression when approached or handled. Fearful body language includes crouching, leaning back, low head position, pupil dilation, flattened ears and tail twitching.

- House cats away from dogs, ideally in a cat-only ward.
- Try to maintain a calm and quiet atmosphere by careful selection of construction materials for surfaces, cat kennels and door catches. For example, providing a plastic cover for metal door latches allows quiet opening and closing of doors (Cat Friendly Clinic, 2012).
- House in kennels above floor level.
- Kennels should be wide and not too deep, so that cats can be easily accessed and it is not necessary to 'drag' them out.
- Provide a place to hide. It is possible to purchase purpose-made cage furniture which gives the cat the choice of either hiding underneath or sitting on top, such as the Hide & Sleep® from Cats Protection (Fig. 6.7). Simple and cheap alternatives include a disposable cardboard box on its side for hiding or sleeping in, or an inverted plastic box with a hole cut in one side and some bedding on the top and inside, so that the cat can choose to hide inside the box or sit on top.
- Consider cat igloos, which encourage the cat to rest and give some privacy. These can also be handy when it comes to lifting the cat out for examination, because the whole bed can be removed, helping to keep the cat relaxed.
- Stock a variety of litters, as cats become accustomed to one type in their home environment and may be reluctant to use the litter

Fig. 6.7. Hide & Sleep® (Photo courtesy of Cats Protection).

tray if the litter itself is unfamiliar or aversive to the cat. Failing that, offer fine gravel litter, which is acceptable to most cats.

- Older cats, who may have restricted range of movement, benefit from a low-sided litter tray to allow easier access.
- Some cats will only toilet in a covered tray, which can easily be provided by the use of a cardboard box on its side.
- Keep a selection of toys for younger cats and kittens to play with.
- House timid and very sick cats in the quietest area of the ward, or if you have a mobile kennel move it to a quiet room, provided that adequate supervision is still possible.
- Cover the kennel front with a towel or blanket if the cat needs more privacy.
- Ensure that the examination surface is non-slip and situated in a quiet area, not in the middle of a busy treatment area. Blood sampling, induction of anaesthesia and blood pressure monitoring should also be carried out in a quiet location.
- Minimize frequency of handling.
- Use scissors or quiet clippers for clipping hair.
- Use water containers that encourage drinking, especially in cats with renal or lower urinary tract disease. Avoid small plastic drinking bowls. Many cats enjoy drinking from a water fountain.

For further information, see Cat Friendly Clinic guidelines, 2012.

Rabbits

Stressed rabbits may show hiding, immobility, anorexia, lack of willingness to interact and aggression when approached or handled. They may also develop gut stasis, which is serious and can be fatal.

- It is very important that the rabbit has adequate space to move around and sit upright.
- Rabbits should be housed at floor level, away from predators, especially vocal dogs.
- The presence of a companion can reduce stress and aid recovery.
- Try to supply the rabbit's familiar diet, even if it is not ideal.
- Provide a place to hide, such as an inverted box with a hole cut in the side, or a piece of carpet tube (Fig. 6.8).
- Provide safe, suitable toys such as puzzle feeders if the rabbit is accustomed to these.
- Ensure that the rabbit has access to water in a bowl, not just in a sipper bottle, as rabbits who are sick or in discomfort often stop using their drinking bottles, especially if they have oral pain.

Birds

Stressed birds may show anorexia, hiding and escape behaviour, unwillingness to interact and aggression when approached or handled.

- As previously mentioned, paired birds can be hospitalized together.

Fig. 6.8. This rabbit's cage is of adequate size and there are materials available for nesting and hiding, but the stainless-steel surface is highly reflective of sound and light. (Author's own image.)

- The requirements for temperature and humidity are likely to vary from those needed by hospitalized dogs and cats, so ideally birds should be in a separate ward facility, or at least in an ICU with controlled temperature, humidity and provision of oxygen or nebulization therapy as required.
- Birds should not be in a noisy environment, but during daylight hours they need some background sound such as music or radio voices; their instinct tells them that when it is silent in the wild it means there is a predator about.
- Sociable birds need plenty of friendly, non-intrusive contact with ward staff, so do not leave them on their own for prolonged periods.
- Aviary-kept, timid and wild birds should be disturbed as little as possible.

Summary

This chapter has provided an overview of some practical ways in which we can modify the veterinary environment to help the mental well-being of our patients (and their owners). There will always be constraints due to the design of premises that prevent us from creating the perfect set-up for all species, but by looking closely and thinking carefully about the surroundings through the eyes of our patients, there will be improvements that can be made. These improvements, coupled with the actions of a behaviourally aware veterinary team, can make a huge difference not only to the welfare of animals in our care, but also to the pleasure the practice team derives from working in a veterinary practice.

References

Carlson, N.R. (1994) Emotions and stress. In: Carlson, N.R. (ed.) *The Physiology of Behaviour*, 5th edn. Allyn and Bacon, Boston, Massachusetts, pp. 359–369.

Casey, R. (2002) Fear and stress. In: Horwitz, D.F. and Mills, D.S. (eds) *BSAVA Manual of Canine and Feline Behavioural Medicine*, 2nd edn. BSAVA, Gloucester, UK, p. 145.

Cat Friendly Clinic (2012) Guidelines. Cat Friendly Clinic, Tisbury, Wiltshire, UK. Available at https://catfriendlyclinic.org (accessed 29 July 2021).

Keller, S.E., Weiss, J.M., Schleifer, S.J., Miller, N.E. and Stein, M. (1983) Stress-induced suppression of immunity in adrenalectomised rats. *Science* 221 (4617), 1301–1304.

Yin, S. (2009) *Low Stress Handling, Restraint and Behaviour Modification of Dogs and Cats*. Cattledog Publishing, Davis, California.

7 The Role of the Practice Staff

Francesca Riccomini and Claire Hargrave

Introduction

A veterinary practice's reputation depends upon the knowledge, abilities, enthusiasm and skills of its staff. The attitude of personnel at all levels towards the well-being as well as the health of the animals in their care is also important. Veterinarians, nursing and support staff obviously need to be well trained and competent, but clients also have an expectation that their pets will be treated holistically and that veterinary professionals will be an interested and reliable source of information and advice on all aspects of an animal's welfare and care.

Awareness of the importance of companion animal behaviour should not therefore be seen as optional. This includes the principles and practicalities that underpin acceptable interactions between any pet, the environment and the other individuals of whatever species it encounters across all life stages. Everyone in the veterinary team should possess an up-to-date working knowledge of behaviour, with the prevention of problematic behaviour being central to practice policy and culture.

Although individual veterinary staff may have limited influence over the design, layout and general running of the premises, they can have considerable impact on animal welfare through the sensitive handling of each patient. This is achieved through awareness of the differing requirements different companion species have (predators versus prey, for example); how each animal's history, world view and perceptions of specific situations will affect its behaviour (see Chapters 2–5); and especially awareness of how challenging the clinic environment can be for pets and owners alike (see Chapters 6 and 14). As well as recognizing the importance of prevention of problematic behaviour, staff should be encouraged to identify high-risk situations, offer valid and appropriate pre-emptive advice and suggest potential solutions. They may additionally be involved in setting up and running related initiatives, such as puppy socialization groups and kitten information evenings (see Chapter 12).

The Role of Management

Integrating behavioural advice is an essential aspect of practice activity and managers have a crucial role in ensuring that staff acquire the necessary relevant education and communication skills in this area. In addition, informed, carefully considered protocols relating to important behavioural issues should be developed. Easily available and regularly checked and updated written resources are essential, detailing, for example, practice policy on continuing socialization during primary vaccination courses, neutering in relation to behavioural problems, sensitive handling of 'special needs' patients and aggressive animals and organizing behaviour

DOI: 10.1079/9781800621312.0007

referrals. These help to inform new staff and serve as reminders when needed. They should take the form of evidence-based but accessible guidelines, which leave scope for individual responsibility and decision making in the face of a specific situation. Management practices should also instil confidence in all members of the clinic's staff, so that they are not inhibited from raising concerns and highlighting behavioural issues with colleagues and tactfully with owners.

Staff education is key to achieving these objectives. It cannot be assumed that veterinarians, nurses and receptionists, who provide crucial front-line contact with owners and their pets, will have enough behavioural knowledge to 'get by' or know instinctively how to react appropriately, for example when dealing with an aggressive or distressed animal. Even recently qualified vets and vet nurses may not have benefited from specific behavioural education, because of the pressure to cover multiple specialist areas in professional courses.

To ensure a good level of knowledge, time and financial commitment for relevant continuing professional development (CPD) are essential. Here personnel on reception who may not generally physically handle patients should also be included. They need to be familiar with the natural behaviour of the main companion species the clinic deals with (see Chapters 2–5) so that they can recognize signs of stress in patients, understand when environmental conditions are likely to cause anxiety and how and when to act to relieve distress or avoid situations that may compromise welfare (see Chapters 6 and 14). In addition, many owners seek advice or raise behavioural concerns during telephone calls or conversations in the waiting area. Reception staff, even those working part time, therefore need to have sufficient knowledge to answer such queries properly or to understand when to refer the client to a more qualified colleague.

The in-house expertise available will determine the degree of external education required to ensure adequate levels of staff knowledge. For example, a veterinary surgeon or nurse who has attained behavioural qualifications could organize regular lectures, seminars and group discussions based around a particular topic, publication or individual behaviour case. Where such expertise does not exist within the practice, a relationship could be established with a local veterinary or non-veterinary behaviourist who holds an appropriate level of recognized qualification (Box 7.1). Most behaviourists are willing to assist with practice training, for example by organizing events aimed at improving the general behavioural education of the staff or addressing pertinent specific topics.

In addition, when a member of the practice attends a behaviour CPD event (Box 7.2) or course, the newly acquired information could be disseminated to the rest of the practice team. This might include making proceedings or lecture notes available to all personnel or giving a talk or leading a discussion centred around the CPD event attended. However, care is needed. Although CPD courses may enhance a staff member's practical competence, they do not confer sufficient knowledge and experience for delegates to claim expertise. Therefore, no one should take on the responsibility of advising clients unless they have attended a course that provides the appropriate level of expertise that qualifies them to do so (Box 7.1).

Establishing and Assessing External Contacts

It is important that practices foster good relationships with other relevant professionals, such as trainers, non-veterinary behaviourists, veterinary behaviourists and kennels/catteries that can offer appropriate help to clients. Practices are responsible for any recommendation or apparent endorsement they make or appear to make, even by displaying advertising literature on waiting room notice boards. There may also be associated liability issues if poor advice is given as a result of recommendation. Therefore, knowledge of each individual's educational background and qualifications plus a clear understanding of what post-nominals mean is essential, especially where referrals are made (Box 7.1). It is also unwise, for instance, to take on trust a trainer or behaviourist's estimation of their own attitudes, skills and handling of pets and people. It is important to ensure that someone with the background necessary to make an educated judgement investigates the validity of specific qualifications and claimed educational attainment. They should also observe anyone the clinic

Box 7.1. Minimum qualifications for identified roles in behaviour and training

The **Animal Behaviour and Training Council** (ABTC) (https://abtc.org.uk) is a regulatory body that sets and maintains the minimum standards of knowledge and practical skills needed to be an animal trainer or animal behaviour therapist. The ABTC has established minimum levels of academic competence for those working in the fields of behaviour and training.

This enables universities and other formally accredited course providers to ensure that their qualifications achieve consistency. Consequently, practice staff wishing to be formally acknowledged as holding competencies in behaviour or training should seek accreditation through attending approved courses.

Minimum UK Level	Role
Level 3 – equivalent to A levels	Animal Trainer
Level 4 – equivalent to foundation degree	Animal Training Instructor (including Puppy Class Instructor)
Level 5 – equivalent to 2nd year of relevant undergraduate course	Animal Behaviour Technician
Level 6 – equivalent to relevant Honours degree	Clinical Animal Behaviourist or Veterinary Behaviourist
Postgraduate (minimum 3-year approved residency programme)	European or RCVS Veterinary Specialist

In addition, the ABTC provides descriptors for knowledge and skills required for each role, enabling veterinary professionals to ensure that staff are not providing advice beyond their level of competency.

A guide to UK Companion Animal Behaviour Training Courses:

- **ABTC** (https://abtc.org.uk/practitioners-info/career-information/)
- **ASAB** (https://www.asab.org/ccab-courses-and-course-validation)

Accreditation schemes meeting ABTC requirements:

- Independent accreditation of Clinical Animal Behaviourists and Veterinary Behaviourists, that is external to specific membership organizations and specific to species specialism, is available through the **Association for the Study of Animal Behaviour** (https://www.asab.org/ccab).
- Independent accreditation of veterinary surgeons as clinical behaviourists is also available through the **European College of Animal Welfare and Behavioural Medicine** (www.ecawbm.com). The ECAWBM approves veterinary surgeons who have reached specialist status through completion of a 3-year full-time equivalent clinical training programme approved by the College, and passed board examinations.
- Behaviour clinicians assessed through the above independent schemes are automatically eligible for membership as Clinical Animal Behaviourists of relevant ABTC practitioner organizations. In addition, the majority of ABTC practitioner organizations have their own internal assessment programmes for animal trainers and animal behaviourists.
- Veterinary surgeons may also wish to become accredited through the **Royal College of Veterinary Surgeons** specialist status (https://www.rcvs.org.uk/education/specialist-status/rcvs-list-of-specialists/). The RCVS recognizes veterinary surgeons with postgraduate clinical or research qualifications who make a significant contribution to their specialist field. Approved individuals can use the title RCVS Specialist in Veterinary Behavioural Medicine. The RCVS additionally accredits Advanced Practitioners in Companion Animal Behaviour; these are veterinarians who have met specific postgraduate study and experience requirements to be approved for this title, but who do not meet the requirements for specialist status.

uses to organize puppy groups on its behalf, or a person or organization staff recommend to owners seeking dog training or grooming. Similarly, veterinary surgeons should carefully assess those to whom they refer behaviour cases, particularly as duty of care is retained when referring to non-veterinary behaviourists (Box 7.1) (see Chapters 21 and 22). If clients raise concerns about a professional or paraprofessional, these should be thoroughly investigated.

Box 7.2. Behavioural organizations relevant to veterinary staff.

- **Animal Behaviour and Training Council**: www.abtcouncil.org.uk
- **British Veterinary Behaviour Association** (members have access to the Journal of Veterinary Behavior: Clinical Applications and Research): www.bvba.org
- **European Society of Veterinary Clinical Ethology**: www.esvce.org
- **European College of Animal Welfare and Behavioural Medicine** (ECAWBM): http://www.ecawbm.com/
- **RCVS Specialist Register of Behaviour Specialists and Advanced Practitioners**: www.rcvs.org.uk
- **Fellowship of Animal Behaviour Clinicians** (CPDIFAB Clinicians: Non-veterinary practitioner organizations accredited by the ABTC): https://fabclinicians.org/handouts/
- **Association of Pet Behaviour Counsellors** https://www.apbc.org.uk/vet-team/)

Encouraging staff enthusiasms and specific initiatives

In the behaviourally aware veterinary practice, a number of additional practical issues need to be addressed. While everyone within the clinic team should be proactively involved in identifying and preventing behavioural problems, it is desirable to identify one or two individuals with a special interest in this area and encourage further development. For example, a veterinary nurse may be provided with a specific CPD budget to pursue behaviour-related studies. This can lead to setting up and running individual projects, such as puppy groups, introductory sessions for kitten owners (see Chapter 12) or more detailed behaviour advice sessions to clients (see Chapters 17–20) (Box 7.1).

This will motivate staff members and offer a valuable service to owners. It is essential for staff to be patient, tactful and understanding and to work within their level of knowledge and competence. They should also be given sufficient time to address these cases adequately and be provided with the right environment in which to deal appropriately with pets that may be challenging within the practice environment and owners who are often upset or embarrassed. Such conditions may not be easily met in a busy clinic, but it is critical to enable them to be achieved if behaviour problems are to be addressed within the practice. In this area a 'little knowledge' or an incomplete approach can be dangerous. Lack of a thorough understanding of the situation and failure to adequately address a behaviour problem can lead to escalation, additionally compromise the animal's welfare and put people and their pets at risk, potentially with consequences of injury and liability. It is therefore essential to ensure staff members involved in more than dispensing basic 'first aid behavioural advice' are well qualified, adequately supervised, supported and resourced.

In addition, owners' needs for behavioural advice can arise at any time. Therefore, when individual interests in behaviour are encouraged and incorporated into the clinic's daily regimes, suitable cover for times when these staff members are not available is required.

Building a Practice Resource Base

Another essential resource that should be available to and used by all staff members is a series of well researched, regularly updated advice sheets dealing with commonly encountered problems, such as destruction, house training and soiling and basic control issues (available at: https://fabclinicians.org/handouts/, accessed 29 April 2022). These, where appropriate, can be given to owners as a 'first aid measure' together with a list of relevant books and websites (Box 7.3) to which they can refer for general, good-quality behavioural information; more serious cases should be 'referred' to an appropriate veterinary surgeon or clinician. Good pre- and post-operation or hospitalization written advice, which includes behavioural issues, especially if inpatients are returning to a multi-pet household, should be included in this essential practice resource (Lindley and Watson, 2010). Many practitioner organizations accredited by the ABTC have material for veterinary practices to utilize, as well as hand out to owners, available for download on their websites.

Box 7.3. Books and Websites.

Essential for the practice bookshelf

Ellis, S. and Sparkes, A. (2016) *ISFM Guide to Feline Stress and Health: Managing Negative Emotions to Improve Feline Health and Welfare*. International Society of Feline Medicine, Tisbury, Wiltshire, UK.

Hedges, S. (2021) *Practical Canine Behaviour for Veterinary Nurses and Technicians*, 2nd edn. CAB International, Wallingford, UK.

Horwitz, D.F. and Mills, D.S. (2009) *BSAVA Manual of Canine and Feline Behaviour*, 2nd edn. BSAVA Publishing, Gloucester, UK. Includes a CD with handouts for distribution to owners, e.g. 'How to find a good dog trainer'.

Mills, D., Braem, D.M. and Zulch, H. (2013) *Stress and Pheromonotherapy in Small Animal Clinical Behaviour*. Wiley-Blackwell, Oxford.

Rodan, I. and Heath, S. (2016) *Feline Behavioral Health and Welfare*. Elsevier, Oxford.

Yin, S. (2009) *Low Stress Handling, Restraint and Behaviour Modification of Cats and Dogs*. Cattledog Publishing, Davis, California.

Useful additions

Atkinson, T. (2018) *Practical Feline Behaviour: Understanding Cat Behaviour and Improving Welfare*. CAB International, Wallingford, UK.

Bradshaw, J., Casey, R. and Brown, S. (2012) *Behaviour of the Domestic Cat*. CAB International, Wallingford, UK.

Landsberg, G., Hunthausen, W. and Ackerman, L. (2011) *Behavior Problems of the Dog and Cat*. Elsevier, Oxford.

Overall, K. (2013) *Manual of Clinical Behavioural Medicine for Dogs and Cats*. Elsevier, Oxford.

Useful books for staff and owners

Bradshaw, J. (2012) *In Defense of Dogs*. Allen Lane, London.

Bradshaw, J. (2013) *Cat Sense: How the New Feline Science Can Make You a Better Friend to Your Pet*. Basic Books, New York.

Ryan, S. and Zulch, H. (2014) *No Walks? No Worries! Maintaining Wellbeing in Dogs on Restricted Exercise*. Hubble and Hattie, Poundbury, UK.

Zulch, H. and Mills, D. (2012) *Life Skills for Puppies: Laying the Foundation for a Loving, Lasting Relationship*. Hubble and Hattie, Poundbury, UK.

Zulch, H. and Mills, D. (2015) *Helping Minds Meet: Skills for a Better Life with Your Dog*. Hubble and Hattie, Poundbury, UK.

Useful websites for staff and owners:

- **International Cat Care:** www.icatcare.org
- **Cats Protection:** www.cats.org.uk
- **Cat Professional:** www.vetprofessionals.com/catprofessional
- **Dogs Trust:** www.dogstrust.org.uk and www.dogstrustdogschool.org.uk
- **Blue Cross:** www.bluecross.org.uk
- **The Blue Dog:** www.thebluedog.org – aimed at educating parents and children about the safest way to interact with their dog
- **House Rabbit Society:** www.rabbit.org
- **Rabbit Welfare Association & Fund:** www.rabbitwelfare.co.uk
- **Canine Arthritis Management:** https://caninearthritis.co.uk
- **Sound Therapy for Pets**: www.dogstrust.org.uk/help-advice/dog-behaviour-health/sound-therapy-for-pets

Publicly displayed posters from a reliable source are useful for informing clients and staff, thus improving patient care. For example:

- **Sophia Yin:** https://drsophiayin.com/blog/entry/free-downloads-posters-handouts-and-more/
- **Fellowship of Animal Behaviour Clinicians:** https://fabclinicians.org/info-for-owners
- **Association of Pet Behaviour Counsellors:** https://www.apbc.org.uk/posters/

Arranging Referrals

When a behavioural referral is arranged, a standard protocol, with which all staff members are familiar, must be in place. Referrals for problematic behaviour should be treated in a dedicated, diligent manner with good channels of communication established with the behaviourist. The protocol should be applied whether this person is a member of the veterinary profession or an associated paraprofessional body, such as the Fellowship of Animal Behaviour Clinicians (FAB Clinicians), with referral letters and medical details quickly and efficiently supplied. Any reasonable help an owner seeks in setting up a behaviour appointment should be readily forthcoming.

It is the veterinarian's responsibility, or that of a designated member of staff, to obtain suitable professional help for the client. No owner should be left to choose someone from a proffered list of those locally available, for example. In addition, it is critical that a recommended behaviour colleague has suitable qualifications and the experience necessary to treat the species, breed of pet and type of problem in question (Box 7.1). Those whose primary interest is dogs may be unsuited to dealing with cats or small prey animals, while the cultural background of the clients and sex of the professional must sometimes be taken into consideration to achieve owner compliance.

Important Policy Considerations

Recognizing signs of anxiety and fear

For many pets (and owners) visiting the veterinary clinic is a stressful experience. Reactions will vary depending upon the pet's species, breed, personality, life history and previous experiences in a clinical setting. When a variety of species are treated, it is therefore important that staff in all areas of the clinic are familiar with the signs that indicate anxiety and fear in all these different types of pet (Figs 7.1 and 7.2). Failure to do so, and to act promptly and appropriately to alleviate the situation, can mean that distressed patients remain in circumstances with which they are not coping well and which have a detrimental emotional impact upon them.

Once negative associations are established, continued or repeated exposure to stressful experiences increases the likelihood that responses will intensify and trigger more readily. Animals also become more anxious about the clinic generally after a negative experience and more likely to anticipate other similar unpleasant experiences. For example, many cats quickly learn to avoid their carriers when these appear at home and dogs become anxious when recognizing the clinic's location.

When anxiety and fear are not recognized and effectively dealt with in the clinic, these emotions can potentially undermine the welfare of the individual and can make handling and treating patients difficult, and sometimes risky for owners or staff. The danger also exists that the behaviour of fearfully aggressive animals will unnecessarily compromise the welfare of other patients.

A carefully considered practical management plan, to which adequate resources are committed, is required to ensure that staff are specifically educated to recognize the species-typical signs associated with these negative emotional states. It is particularly important that they are able to identify subtle, easily overlooked indicators that a pet is anxious or fearful, know what is required to improve matters and are willing and able to act quickly to adjust clinical programmes, treatment plans and handling procedures accordingly. For more information on specific signs see Chapters 2–5, 8–12 and 14.

Additional policy considerations

There are several other aspects of veterinary practice where evidence-based but accessible policies should inform newly arrived colleagues and the advice that every member of the clinic gives to clients. These include protocols that address the following.

- The specific needs of young animals with a view to ensuring that their early experiences of the clinic are positive.
- The need to integrate into vaccination protocols information about how to ensure appropriate early social and environmental experience to avoid the development of problematic behaviour.

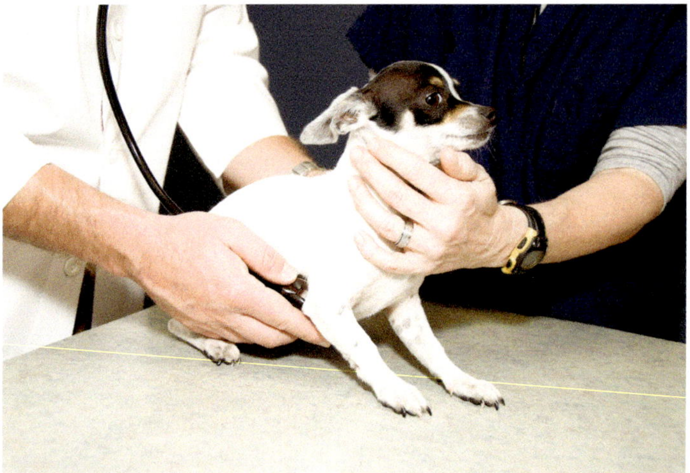

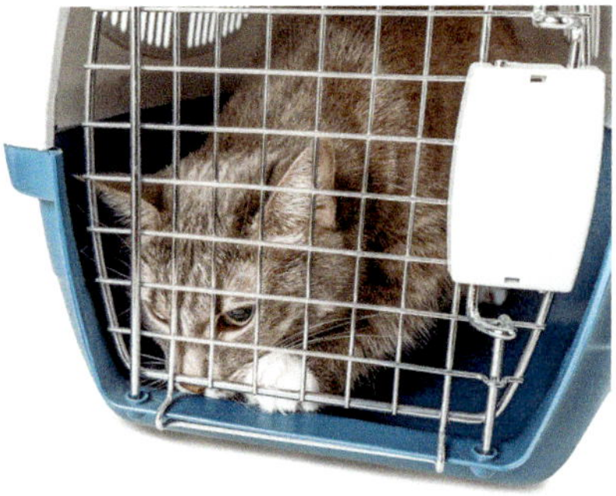

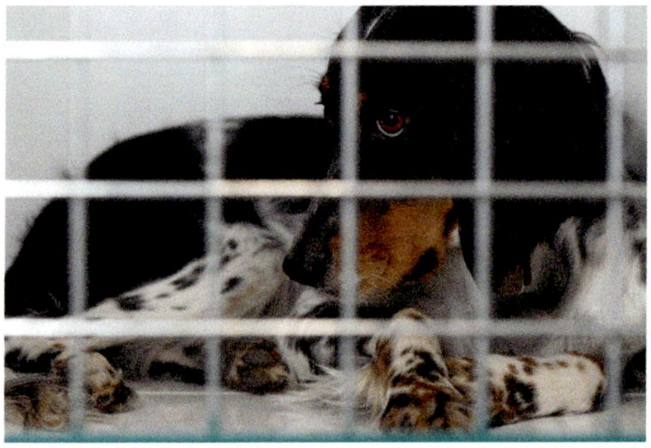

Figs 7.1, 7.2, 7.3. Throughout the clinic, staff should be able to identify even subtle signs of anxiety and understand how to act promptly to improve the situation from the patient's perspective. (Images: iStock.)

- Neutering – where science-based protocols are required for both routine de-sexing and neutering in relation to problematic behaviour.
- The specific requirements of 'special needs' patients, for example newly re-homed or elderly pets, those with disabilities or cognitive impairment and animals with a history of clinic-related distress.

Special Requirements and Handling Needs of Young Animals

The early period of any animal's life is crucially important in establishing its ability to live comfortably in the world it will inhabit lifelong. During the socialization period therefore, whatever the species of pet, thorough yet sensitive preparation for the future is essential. Limited exposure to people, other animals, the domestic environment, the outside world and everyday activities is likely to have a detrimental effect upon a pet later in life. Frightening experiences or poor handling, especially when associated with fear or pain, may well create negative associations that can potentially undermine the individual's welfare. These could also cause difficulties in managing the animal and providing it with good-quality husbandry and healthcare throughout its life. Negative experiences may additionally contribute to problematic behaviour.

Breeders and owners, with the assistance of veterinary clinic staff, have a critically important role to play in producing well-adjusted pets. Young animals need to be at ease with all aspects of daily life in domestic environments, so that they can adequately cope with new experiences, unfamiliar people and novel environments as adults. Everyone who interacts with a young animal in the practice needs to be aware of the potential for each new experience to have a profound impact on later behaviour. Exposure to any stressful or negative experience therefore needs to be dealt with promptly and appropriately to reduce the negative emotional impact upon the pet.

The ideal is not only to prepare an animal for future experiences, but also to avoid inadvertently exposing it to experiences that induce fear responses. If this is unavoidable, dedicated efforts should be made to minimize the negative impact. This can usually be achieved by ensuring that any first encounter with a novel situation or individual is made as positive as possible through a combination of forward planning, patience, good observational skills and sensitive handling. Nowhere can such an approach be more productive in avoiding unnecessary stress and future problems than when introducing young pets to the veterinary clinic environment.

However, no matter how well prepared an individual has been, or how dedicated everyone's efforts are during initial veterinary visits, it is important to remember that learning is not confined to the early months and socialization period but continues throughout life. Positive associations established by well-conducted initial introductions and experiences can be undermined by subsequent poor, frightening or painful events (Fig 7.3). A pet's ability to cope and its 'good behaviour' should never be taken for granted, whatever the circumstances.

Well-informed management techniques and sensitive handling should continue throughout life, with intensified and dedicated efforts being made to maintain previous beneficial learning whenever circumstances change and the animal is especially vulnerable. For example, social or environmental upheaval can undermine the well-being of any pet, with the likelihood that emotional effects will be intensified if it is very young or particularly well bonded to the people involved. Accident and illness, especially when they occur at a sensitive phase of the animal's life or require extensive medical care or hospitalization, can also have a negative impact and may even lead to subsequent behaviour problems. Some 'at risk' patients need targeted 'special measures' (see later in this chapter).

Building and Maintaining Positive Associations with the Veterinary Clinic

Individual requirements will vary and an in-depth discussion is beyond the scope of this chapter, but the suggestions in Table 7.1 are for building and maintaining positive associations with the practice.

Integrating Information on Appropriate Early Experience with Vaccination Protocols

When owners embark upon their pet's initial vaccination course, veterinary staff emphasize the importance of avoiding potential sources of infection. The duration of relative isolation required will depend upon the nature of the vaccine, the manufacturer's recommendations, the species, age and general health of the animal plus local conditions, such as the incidence of the infectious diseases against which the vaccine confers protection.

Each veterinary clinic will therefore have its own recommendations for puppy, kitten and rabbit owners, regarding timing of vaccination and when it is safe to allow a pet unrestricted access to the outside world. This is an area where a well-informed, carefully considered practice protocol and individual professional judgment are needed. The importance of socialization in all animals destined to become pets and the need to habituate them early to as many individuals, environmental stimuli and experiences as possible is well recognized (see Part I and further chapters of Part II).

Because of this and due to the requirements of the Animal Welfare Act (2018), it is not sufficient to adopt a policy of complete isolation within the home. The individual's need for appropriate psychological, behavioural and emotional development requires consideration. This issue has especial relevance in relation to successful puppy rearing. A failure to prepare dogs adequately for their future role as companions in potentially busy, noisy, highly populated and stressful places can result in an increased risk of fearful or aggressive behaviours, with some becoming generally 'unfit for purpose' as pets.

Table 7.1. Building and maintaining positive associations with the practice.

Requirement	Occasion	Suggestions
Positive first visit	In general	Young animals should visit when the practice is quiet and staff have time to devote to them
	Pre-vaccination, nurse-led appointment	Provide information on socialization, environmental introductions, vaccines and parasite control. Owners should be shown how to introduce gentle, hands-on contact in association with tasty treats If the young animal is still relaxed, introduce other areas of the clinic and other staff individually Observe carefully for signs of stress and ensure no one crowds the animal If necessary, act promptly to reduce the pressure upon the pet
	First vaccination	Discussions should *precede* vaccination and be accompanied by treats The aim is to create positive associations with the clinical setting and to reward calm, cooperative behaviour Restraint should be avoided or minimal (Fig. 7.4) Puppies should be lured into a 'sit' or 'down' by the owner or nurse and vaccinated while receiving treats On the way out, while other species leave as quickly as possible, the relaxed pup should enjoy another period of 'fuss', training and treats from staff in the reception area Unless demanded by legislation, avoid inserting a microchip during the first visit to the practice
	Second vaccination	Physical restraint should only be used if essential. It should never be regarded as the 'norm' Ensure treats are given throughout handling, preferably immediately following each cooperative activity
Handling	Every contact with a conscious patient is an opportunity for learning about humans and the veterinary environment	All staff should be competent with simple, reinforcement-based training techniques Lure canine patients into positions for examination wherever possible Reinforcing appropriate behaviour emphasizes and demonstrates to owners the pet's continuous need for guidance and feedback regarding the choices it makes in complex and challenging environments
Socialization and habituation	Puppy Class (Chapter 12)	Use an accredited puppy trainer Advise on continuing socialization, basic training and ensuring the pet is comfortable being left alone Reinforce the message that the practice is there to help with behaviour as well as health
	Kitten, rabbit and parrot introductory sessions	Provide species-appropriate behavioural advice (Chapters 8–12; Fig. 7.5)

Continued

Table 7.1. Continued.

Requirement	Occasion	Suggestions
Further social and environmental competence	Puppy Training Class: in-house or external provider	Use an approved, accredited puppy trainer
Continued positive associations	Subsequent visits	For dogs, encourage regular social visits to the practice Always observe carefully for signs of anxiety or fear, and deal appropriately with the situation if difficulties arise Keep restraint to a minimum Always reward a pet's calm, cooperative behaviour with food, toys or attention
	Adolescent and juvenile care for dogs	A Sixth-Month or Adolescent Check-up is advisable, as there are often significant behavioural changes between vaccination and first annual booster Ask pertinent questions about behavioural development at every appointment and act promptly to address any emerging issues

Fig. 7.4. It is in everyone's interests for time and patience to be expended on conditioning and maintaining positive associations with the clinic. Owners should be made aware of the value of gentle restraint techniques and shown how to lure their young animals into cooperative behaviours. (Images: iStock.)

Fig. 7.5. When animals need transporting in a carrier, practice policy should include advice to help reduce clinic-related stress, such as choosing a suitable, appropriately sized model, conditioning positive associations with it at home, and covering the container when in transit. (Images: iStock.)

When practice protocols are formulated, a balance must be struck between the potentially competing needs of avoiding exposure to infectious agents and instigating an ongoing process of introducing the pet in an appropriate, non-threatening manner to a wide range of individuals, environmental factors and experiences. For example, unless the local situation is particularly difficult, puppies could accompany owners out of the house to experience their new environs when carried and well protected throughout from contact with potential sources of infection. Obviously, care must be taken during these trips to ensure the safety and comfort of the pet. This may best be achieved when it is confined in a suitable carrier. Similarly, a programme of short car journeys should be instigated as soon as a puppy has recovered from the transition to its new home, with outings being short so that it will not need to be put down to relieve itself. Also, in dog-free gardens, puppies can be introduced early with gradually increasing levels of exposure to commonly encountered phenomena such as children's play, bicycles and skateboards.

Owners and breeders must receive informed, coherent advice

When receptionists, nurses, veterinary technicians and clinicians fully understand the importance of socialization and appreciate the complexity of balancing behavioural development with infection control, they can explain why this is not a black-and-white issue and offer practical solutions to potential difficulties.

For instance, where disease risks are high, rather than taking their puppy out it may be more appropriate to suggest that clients ask a number of visitors of both sexes, various ages and differing appearances to come to their house. Supervised, gentle handling by children is especially important. If this is not possible, owners could alter their own appearances by 'dressing up' in bulky clothing, various styles of head gear and so on. In addition, they may be able to expose their puppy to a range of new experiences within the home, for example by playing recordings of sounds such as traffic noises, barking dogs or children's voices. A diverse range of novel smells and substrates might also be introduced to simulate aspects of the outdoor environment that will later be frequently encountered. Similarly, visits by other dogs can be helpful, as long as they are sociable, appropriately behaved, fully vaccinated, wormed and treated for external parasites.

Similar issues will be raised with kittens and young rabbits, even if an individual animal appears destined for a lifelong indoor-only lifestyle, as there are still likely to be times when the pet has to cope with change and novelty.

Time should always be allowed at some point before or during vaccination visits to make owners fully aware of those aspects of behavioural development that need to be considered. Evidently some duplication may occur when puppy groups or kitten information sessions are organized in-house. However, not all clients will be willing or able to attend these sessions; some pets may not be fit to do so on all occasions; and some owners may be reluctant to voice concerns or lack of understanding in a public arena. Duplication of information is useful, since clients do not always pick up all aspects of advice straight away. Repetition ensures that key messages are understood. For example, people may grasp the importance of socializing their pet but overlook the need for sensitivity, monitoring the animal's response and gradually introducing any potentially novel experience. It is important to ensure that repetition is always consistent so that the same message is repeated and reinforced. If slightly different advice is given on different occasions, clients will find this confusing.

Behaviourally Aware Breeding 'Practices' Are Essential

Veterinary clinics that deal with breeders have an especially important role to play in helping these clients to produce behaviourally well-adjusted animal companions, whatever the species involved. Understandably perhaps, many breeders have concerns about prevention of infectious disease and put less emphasis on socialization and habituation to a domestic environment. This can result in pets with a narrower range of early experiences that have difficulty fitting into domestic environments and lifestyles. Given our current state of knowledge, this is unacceptable and veterinary practices need to be fully involved in ensuring that they can enlighten, advise and support their breeders in this respect.

It is also customary for owners to be given vaccination information and written instructions when they collect their new pets from a breeding establishment or rescue centre. Where veterinarians provide advice for these establishments and are highlighting the importance of experiential complexity on behavioural development, it is also important to ensure that this information is passed on to people when they purchase or re-home a young animal. This should include a brief behavioural history with, for example, a checklist of people, environmental stimuli and other animals thus far encountered, together with any individual sensitivities or problem areas. For dogs, recommending that breeders use the BVA:AWF/RSPCA puppy contract is also important, to make sure purchasers have information about the health and parentage of their puppy.

Identifying and Addressing Individual High-Risk Circumstances

When new animals of any age are first introduced to the practice, opportunities should be created for owners to discuss their pet's source, background and behavioural history and to highlight any individuals, situations or events that cause signs of anxiety or fear. Such discussions, conducted by either a veterinarian or a veterinary nurse with behavioural knowledge, can be included in the initial health check for rescued animals, for instance, and vaccination appointments for youngsters. It is essential that this becomes an acknowledged part of the practice routine, even when it means setting additional time aside, perhaps at minimal cost or no charge.

If social and environmental circumstances are identified as causing an individual pet anxiety or fear, those staff involved must also know how to deal appropriately with the situation. A suitably educated staff member may be able to investigate the behavioural issues in-house adequately and offer sufficient, appropriate advice to resolve the animal's difficulties. Alternatively, especially if significant concerns are raised, referral to a qualified behaviourist with relevant experience should be arranged. In such circumstances, the earlier that meaningful intervention is initiated, the better. It should never be assumed that any animal is too young or too old for a behaviour problem to be addressed. Successful outcomes can be prejudiced by unnecessary delays.

A particularly important aspect here is the tendency for puppies and kittens to be retained on their original premises until the end of the 'socialization period'. This is often for perfectly valid reasons. For example, prospective owners may have arranged to go away, move home, or engage in building work, or breeders could be unsure which individuals from a particular litter they want to keep until the animals are more mature. These individuals will again be at greater risk when finally entering their new home unless appropriate socialization and habituation have been undertaken at the breeding establishment.

Some 'At Risk' Patients Need Targeted 'Special Measures'

As previously discussed, everyone involved in the care of pets needs to be aware that a good start in life is essential and that learning about new events continues beyond the so-called 'socialization period'. Loss of previous learning engendered as a result of appropriate social and environmental encounters is a risk not uncommonly seen in animals that are ill or injured when young. Those in this category that are particularly vulnerable include patients afflicted with a series of minor ailments that interrupt their normal lives and deny them the opportunity to have continued exposure to a range of environments and experiences; pets hospitalized at especially sensitive behavioural stages, say immediately after they have been rehomed; and those that must be confined for long periods after surgery. Under these circumstances it is not good enough to leave things to chance or to assume that clients will know what they should do to ensure their pets do not 'lose ground' in terms of behavioural development. Detailing practical measures that clients can take to avoid future problems related to fear and anxiety is a responsibility of the veterinary personnel treating them.

Neutering: the Importance of Considering the Individual when Developing Practice Policies

Castration of male dogs, cats and rabbits and spaying of females is now routine in small animal practice. As well as controlling reproduction and unwanted litters, there are other legitimate medical reasons and some sound behavioural imperatives for neutering in all these species. For example, intact male dogs and cats are more inclined to roam, thus increasing their risk of accident and injury; the smell of mature tom cats' urine is unpleasant from the human perspective; and male-to-male feline aggression is significantly reduced by desexing. This also helps to decrease the incidence of fight-related wounds, infections and the spread of feline immunodeficiency virus (FIV).

Pseudopregnancy and mammary disease, including neoplasia, are a risk in entire bitches, does and queens, while uterine disease such as pyometra and cancer can affect older females. Aggression associated with resource acquisition and maternal defence of offspring or substitute individuals or objects in breeding females can also be avoided by neutering.

However, unless there is a carefully considered practice protocol in place, the danger arises that owners are given differing, sometimes conflicting, advice by different people at different times. This is a particular concern where poor timing of surgery undermines its value (delays that lead to unwanted pregnancies, for instance) or gives rise to medical complications, such as spaying bitches in pseudopregnancy; or where unrealistic impressions of the behavioural benefits of the procedure are created. These may include the expectation that castration of dogs will always reduce or eliminate urinary marking, mounting behaviour and aggression directed towards people or other dogs. Such behaviours can be influenced by learning as well as by other motivations, such as anxiety and frustration (Fig. 7.6).

Devising practice policy in relation to neutering, which should be based on current research and should consider behavioural issues as well as medical factors (for position statements, refer to the BSAVA website at www.bsava.com), is the responsibility of principal veterinary surgeons. It is also important that everyone understands that while it is desirable and good practice to have such documentation for staff to refer to, veterinarians should always be allowed to use their discretion when assessing individual patients, especially regarding their life histories and behavioural needs.

Fig. 7.6. The motivation(s) underlying frequent urine marking should be accurately identified and addressed, particularly if the problem is seen indoors before castration is recommended. (Images: iStock.)

Neutering as 'therapy' for undesired behaviour

Pre-operative checks for routine surgery done in advance of the surgery present the ideal opportunity for asking some basic questions about management and control and discussing in more detail any behavioural concerns clients may have. If these physical checks take place on the day of surgery, a policy should be instigated of having conversations around behaviour beforehand, perhaps at a nurse-led clinic or on the telephone with a nurse or receptionist when booking. If in the first instance time does not allow for the necessary comprehensive discussion, a return call should be arranged.

Whatever scheme is adopted, a definitive system with which all members of staff are well acquainted is essential. This is important, because some people still believe neutering is indicated as a solution to a range of undesired behaviours, including inability to control a dog, general boisterousness, urinary marking or mounting behaviour. While there may be a degree of improvement in behaviours that have an element of sexual motivation, behaviour problems are often much more complicated than a superficial review suggests. Learned behaviour and inadvertent owner reinforcement frequently play a part in maintaining such behaviours and more detailed discussion may well indicate that the underlying motivations owe more to anxiety or frustration than hypersexuality. Removing the primary source of sex hormones is therefore unlikely to wholly resolve the problem, address the contributing factors, or improve the owner's understanding or the animal's welfare. Resentment against the professionals and practice involved is an understandable owner response when subsequent anticipated positive results fail to materialize, despite the pet undergoing a surgical procedure and the client having financial outlay and inconvenience.

Poor advice can put people and animals at risk

Of particular concern here is the outdated but still prevalent view that neutering, especially castration of both dogs and rabbits, is the principal solution to aggression towards conspecifics and people. While it may sometimes form part of a behavioural modification programme aimed at resolving such problems, in many cases, especially those caused by underlying fear and anxiety, neutering can be associated with deterioration of behavioural signs. In addition, human-directed aggression may well put veterinary staff and owners at risk when handling affected animals during and after surgery and cause their pets further welfare compromise. The veterinary experience and hospitalization can be especially challenging for these animals.

Therefore, it is not advisable for owners who present with aggressive pets to be offered de-sexing as a first-stage intervention. Receptionists and support staff, who often simply book routine neutering appointments when asked to do so by owners, should be made fully aware of the need to enquire into the reasons for the operation and to ask some searching behaviour-related questions. Any concerns should be referred to a veterinary surgeon. This is particularly important when an animal is mature and such an owner decision seems 'out of context'. With any individual, when aggression is highlighted during pre-neutering checks, unless there is an obvious medical imperative that argues against delay, surgery should be put on hold while the behaviour is properly dealt with.

The situation is somewhat different with females, because maternal aggression or the defensive behaviour associated with resource acquisition during pseudopregnancy may well resolve completely after ovario-hysterectomy. However, even these patients' owners should be carefully questioned before anyone assumes that this is an entirely straightforward situation. If other potentially contributing motivations or factors are highlighted, such cases should be considered sufficiently significant to warrant specific behavioural intervention prior to, or in addition to, surgery. Timing of surgery is also significant and everyone who is involved in booking neutering appointments should be properly briefed so that females in season or pseudopregnancy, for example, are not inadvertently spayed. See Chapter 17 for more information on neutering and behaviour.

Considering the Specific Requirements of 'Special Needs' Patients

'Special needs' patients, a rapidly expanding category, often fare badly in a veterinary setting. This group includes pets with physical conditions that cause them to be especially sensitive, apprehensive or self-protective and those with behavioural traits that make them easily disconcerted and disorientated when confronted with the unfamiliar, busy environmental conditions encountered in the clinic. They are commonly time consuming to deal with, sometimes challenging to handle safely and, because no one perceives their difficulties, not uncommonly inadvertently disregarded in welfare terms.

It is easy to overlook just how stressful a veterinary practice can be, for example to newly re-homed animals that have passed through a number of hands or those poorly socialized when young. Pets with previous negative experiences during veterinary treatment, particularly related to intimidating handling or extensive treatment requiring hospitalization and frequent return visits, also fall into this group, as do those with disabilities, for instance canine amputees, younger animals with sensory impairment or breed-specific morphology that restricts sight or breathing. For such patients and for elderly dogs suffering from orthopaedic conditions, dodging people and other pets and negotiating steps and slippery floors can be painful, difficult and frightening.

Aged animals are also more prone to anxiety than youngsters. They generally cope less well with changes in circumstances and contact with unfamiliar individuals, particularly if their sight and hearing are deteriorating. Orientating themselves in unfamiliar situations may be troublesome and challenging, while veterinary practice visits can be especially problematic for those additionally afflicted with cognitive decline. Due to increased confusion and disorientation, such changes in location and upheavals to daily routines may also result in an inability to settle on returning home. This can put extra strain on owners already struggling with the increased management demands associated

Table 7.2. Strategies for supporting patients with special needs.

Special need	Potential reactions and problems	Suggested action and measures
Impaired vision; reduced mobility. Aged individual, especially if suffering cognitive impairment	Enhanced startle response and likelihood of defensive or aggressive reaction Distress Disorientation on return home	Support programme to establish familiarity with practice and handling techniques, for example non-clinical visits to condition positive associations; owners to regularly undertake simulated veterinary examinations at home (when and where safe to do so) Arrange appointments at quiet periods when staff have more time and the social environment is less challenging Minimize journey from car to consulting room; for example, allow access through staff entrances Avoid or minimize exposure to dimly lit areas of clinic Supply hygienic, non-slip mats and portable ramps Make sure animal has enough room to move comfortably, at a pace it can cope with Ensure as much continuity (e.g. of personnel and places) as possible, to avoid confusion Patience in approach essential: avoid precipitate, rapid actions; move slowly and calmly; clearly signal your intentions; allow pauses in examination for readjustment Change locations within the clinic as little as possible; for example, take blood tests in consulting room, not prep room (if this is normal regime)
Most 'special needs' patients	Easily disconcerted or intimidated; ready panic response Accelerated negative attitude to clinic/veterinary personnel Fear development to veterinary practice or medication administration and homecare	Consider longer appointments or out-of-normal consulting hours at no extra charge, or home visits Consider behavioural and cognitive status when deciding upon need for interventions and regular checks – compromise if necessary, on welfare grounds, in consultation with owner Aim for continuity of care – assign specific staff members to individual pets Make special arrangements if pets must be hospitalized; for example, allow owners to bring animal in later than usual to avoid lengthy stay in kennels before operation or blood test, or to be present on recovery from general anaesthetic or collect early If hospitalized – avoid washing all familiar-smelling bedding at once; ask owners to supply scent-impregnated items regularly to 'top up'; if possible and appropriate clinically, ensure continuity of food, cat litter, etc.

Continued

Table 7.2. Continued.

Special need	Potential reactions and problems	Suggested action and measures
Poorly socialized or poorly habituated animals; pets with previous negative veterinary experiences	Likelihood of fear or anxiety and defensive or aggressive reaction	Do not bring animal into clinic until staff are ready; for example, leave in car, use mobile phones to call owners Consider the use of medication to facilitate handling and reduce negative emotions and unwanted learning Assign staff members of the gender the animal is most accustomed to or least worried by Observe carefully and liaise with owners re increased negative responses. For example, if the pet's relationship with its female owner deteriorates because of ongoing handling or medication at home, have male staff deal with it in the clinic Change 'context' within clinic if increased negative reaction is associated with one consulting room If appropriate, establish programme of desensitization and counter-conditioning to veterinary clinic – if necessary, involve a qualified behaviourist

with these patients and disrupt existing relationships with other household pets.

There is an especial need, therefore, for practice staff to be aware of and sensitive to such patients and their owners. Management understanding and support are also required so that regimes remain flexible enough for 'tailored care packages' to be instigated for these 'special needs' pets. It is desirable to explain potential problems to owners, obtain in advance as much information from them as possible about their animals' difficulties, sensitivities and requirements and, where practicable and safe to do so, involve them in all aspects of handling.

What is required will vary with individual animals and their circumstances and an in-depth discussion is beyond the scope of this chapter. However, appropriate measures could include those outlined in Table 7.2.

When all members of staff are fully aware of the signs of stress exhibited by the pet species that their practice treats, they should easily be able to identify the circumstances that individual animals are likely to find challenging or distressing and recognize that on occasion any animal can fall into the 'special needs' category (Figs 7.1 and 7.2). Everyone should be constantly alert and prepared to adopt specific measures to deal with difficulties whenever they arise. For example, even confident, well-adjusted dogs are likely to be unnerved in a waiting area when another dog is out of its owner's control, barking and lunging at pets and people; while cats and small prey animals do not fare well when placed on the floor in busy areas or when subject to intrusive attention from unfamiliar people and/or dogs.

In this respect, as well as having an established policy in place to deal with distressed animals and people, individual initiative should be encouraged. Well-ordered waiting and kennel areas are especially important to protect all patients from the negative effects of behaviourally troubled, and troublesome, individuals. For example, reception staff could readily intervene to politely ask clients with dogs that are struggling to remain calm to wait outside, in their vehicles or an empty room. Owners of cats and small animals should be advised of an appropriate way to protect their pets from environmental stressors, such as placing them in quiet, elevated areas (see Chapter 6).

Conclusion

Veterinary staff undoubtedly deserve good conditions and a safe working environment. Pets with specific behavioural sensitivities, such as a nervous reaction towards the veterinary clinic, can be challenging and distressing for both staff and owners to handle. Aggressive animals may also pose actual physical risks to human safety. Therefore, it is a professional obligation for veterinary surgeons, nurses and managers to recognize the importance of companion animal behaviour and problems related to it as a welfare issue that deserves dedicated attention as a matter of practice policy.

Additionally, recognition by everyone in the veterinary team of indicators of anxiety and fear in patients of all species, and that this requires compassionate handling, is an essential skill that requires provision of adequate training and funding. It is also essential that management protocols and daily clinic regimes allow the flexibility necessary for reasonable measures to be taken when dealing with challenging pets in the surgery. Such protocols should always include a properly designed and well-resourced collective initiative to assist owners in preventing and dealing with problematic behaviour.

Reference

Lindley, S. and Watson, P. (2010) *BSAVA Manual of Canine and Feline Rehabilitation, Supportive and Palliative Care Case Studies in Patient Management.* BSAVA Publishing, Gloucester, UK.

(See also booklist in Box 7.3.)

8 What Every Puppy Owner Needs to Know

David Appleby

Soon after obtaining a puppy, the majority of new owners visit a veterinary practice for routine checks and vaccinations, offering the opportunity for the practice to provide guidelines for the puppy's socialization and habituation (Box 8.1) and helping it to develop appropriate behaviour. Due to the importance of the puppy's early weeks to its development and the fact that new owners tend to be enthusiastic about giving them the best start in life and are easily encouraged to do what needs to be done, these are important consultations.

Initial consultations are important. Many dogs are euthanized, rehomed, abandoned or surrendered to a shelter as young adults because of behaviour related to fear or anxiety, or other emotional states that owners find challenging to manage. Many more remain with their owners, but their welfare is compromised because everyday events cause them to lose emotional homeostasis. In a significant number of such cases defensive behaviour develops as a coping mechanism, which has implications for the safety of people (including practice personnel) and other animals and for the quality of life and the experience of pet ownership for the client. In the majority of cases the problems, and potential loss of income for the practice if the animal is disposed of, could have been prevented if the owners had received information about the importance of providing puppies with appropriate experience during early development to help them develop confidence (Murphree, 1973) and appropriate skills for domestic life.

With a growing population of knowledgeable dog owners, it is possible that the new puppy owner is aware of the importance of early experience and how to optimize it. Many will have been given, at the very least, verbal information about socialization by their puppy's breeder (Box 8.2), but encouragement from health care professionals in veterinary practice to prioritize it as an important project will be beneficial in motivating them to take it seriously. However, many owners will not be aware of the facts, or the information they have needs augmenting. In such cases identification and intervention are a priority.

If your practice runs puppy parties (see Chapter 12) or puppy socialization classes, or has links to a trainer who provides them, it is not enough to advise new owners to attend them to obtain the information there. Despite the benefits (Sterry *et al.*, 2005), many clients cannot attend or are unable to start as early as best practice, and opportunities during an important period of development may be lost if guidelines are not given. As it will only be possible to cover key points during routine consultation, it is important to provide reminders and more detailed information in the form of handouts, booklets, etc. Knowledge of some of the research on which the rationales are based will make it easier to provide convincing arguments as to why the owner should be proactive during these discussions.

DOI: 10.1079/9781800621312.0008

The Sensitive Period

There is a sensitive period during which experience forms the foundation for the development of confidence. The end of the neonatal period at 2.5–3 weeks to between 12 and 14 weeks is frequently cited in the literature. It is important to be aware that the research on which this is based indicates that near-total stimulus deprivation during the early weeks of life results in a declining willingness to approach novel stimuli from 5 weeks of age. If first exposure was not until 12–14 weeks, only avoidance behaviour occurred. Therefore, the sensitive period, in terms of when willingness to approach occurs, may be much shorter than the literature suggests (Pluijmakers *et al.*, 2010). Researchers Scott and Fuller (1965) concluded that, before 8 weeks, puppies are particularly ready to be accustomed to people and other experiences. Some of their research team (Freedman *et al.*, 1961) kept puppies and their dams in 0.3 ha fields. At 2, 3, 5, 7 and 9 weeks of age puppies from each litter were removed from the field for a week of testing before being returned. During this time indoors the puppies had three half-hour periods a day of being played with, cared for and tested. How avoidance behaviour develops was studied in daily tests lasting 10 minutes, during which the amount of time a puppy took to approach a passive reclining experimenter was recorded. Due to underdevelopment the 2-week-old puppies only displayed limited and random behaviours. However, the 3-week-olds approached immediately and spent most of the first 10-minute period pawing, mouthing and biting the researcher. The 5-week-old puppies did not exhibit approach behaviour at first but were similar to 3-week-

Box 8.1. Socialization and habituation.

- **Socialization**: the process whereby an animal learns how to recognize and interact with the species with which it cohabits or with which it has social contact.
- **Habituation**: the process whereby an animal becomes accustomed to stimuli that are not salient and learns to ignore them. For example, puppies are habituated to traffic, making it unsurprising that dogs are often said to have no road sense.

Box 8.2. Choosing a puppy from a breeder who provides the optimal environment for the development of confidence.

Most owners do not discuss with the practice the issues related to rearing a puppy before they obtain it. If they do, it may be beneficial to ask first-time owners to consider the level of commitment and suitability of the breed and gender for their needs and circumstances.

Breeders should select the dam and sire best suited to produce puppies of good temperament but they should also provide a stimulus-rich environment. As the puppy and its littermates grow, the breeder should gradually increase the amount of handling and general interaction the puppies have with the breeder and with other people, to accustom them to people of different appearance and behaviour.

Preferably the puppies should have been raised in a busy domestic environment with extra items occasionally introduced to provide variety. For example, recordings of environmental stimuli could be played, and scents from the outside world could be imported on cloths. Being raised in a domestic environment has the additional benefit of puppies becoming accustomed to a range of human behaviour beyond what might be experienced during special socialization sessions of puppies in separate housing.

Prospective owners can optimize the development of their puppy's confidence by obtaining it before 8 weeks of age, if it is raised in a non-domestic environment. If raised in a domestic situation, 8 weeks is the optimal homing age (Jokinen *et al.*, 2017). Up to a point, this is balanced by the level of experience and stimulation available in the breeder's home.

Prospective purchasers should seek a breeder who will let them see the puppies with their dam in their living quarters prior to homing The behaviour of any other dogs in contact with the puppies should also be observed and its influence on their learning considered. Searching questions should reveal the breeder's awareness of the need for a puppy's environmental enrichment. The behaviour of all the puppies should be observed and they should appear to be content and confident.

olds before the end of the first 10-minute test. However, the 7-week-old pups were described as initially frightened and wary and on average took two 10-minute test sessions to approach. The 9-week-old puppies required three sessions. Puppies left in the field until 14 weeks of age did not approach and were described as remaining fearful.

Michael Fox (1971) found that puppies with no experience other than the pen they were kept in, when introduced to a test area containing unfamiliar objects for only half an hour at 5, 8, 12 and 16 weeks, explored and showed increasing interest in more complex objects. Puppies kept in the pen until 12 or 16 weeks of age did not explore, withdrew or did not enter the test area. Some were described as being catatonic with fear.

In the real world of companion rather than laboratory dogs, the data of 820 adult dogs referred to the author for treatment of behaviour problems were studied for correlations between maternal environment, experience after homing, age of homing as puppies and behaviour related to fear and aggression in later life (Appleby *et al.*, 2002). Adult dogs showing symptoms of avoidance behaviour or aggression were compared with dogs referred for problems hypothesized and shown statistically to be unrelated to early environment or early socialization. The maternal environment and age of this second group of 82 dogs was compared with another control group of 124 randomly selected puppies up to 24 weeks old and they were found to be similar.

The study showed that dogs referred for avoidance behaviour were found to be significantly more likely (68.2%) to have been raised in a non-domestic maternal environment, such as a barn, kennel or special puppy room, than dogs in the comparison group (51.2%). Additionally, 39.9% had not been exposed to busy (urban) environments after homing, compared with 23.2% of the comparison group. Of the dogs referred for the problem of aggression towards unfamiliar people, 66.8% had been obtained from a non-domestic maternal environment. This compared with 51.2% of dogs in the comparison group. Dogs referred for this problem were also more likely to have lacked exposure to busy environments after homing (40.9%) compared with dogs in the comparison group (23.2%).

The influence that the maternal environment has on the development of behaviour is likely to be affected by the age at which the puppy leaves it. The data from Appleby *et al.* (2002) suggested that inadequate experience in the maternal environment had a significant effect on puppies obtained by their owners at 8 weeks or over but not on the puppies obtained before 8 weeks. A study of 5067 adult dogs, 93% of which were raised in a domestic environment, suggested that homing after 8 weeks is associated with higher instances of problematic behaviour in adult dogs towards familiar and unfamiliar people, and unfamiliar dogs even if raised in a domestic environment (Appleby *et al.*, 2013).

Conversely, it is preferable for a puppy to be retained in a stimulus-rich maternal environment rather than move at an earlier age to an environment that does not provide stimulation. Optimizing the puppy's confidence requires that the breeder expands its experience beyond their home at a rate that would have occurred had the puppy been homed where more varied experience is possible.

The importance of exposure to urban and other busy environments on the development of confidence occurs regardless of whether the puppy will actually live there. In busy environments they are likely to experience a wide range of stimuli and variations of them. The more non-threatening stimuli and variations that puppies learn are normal, the greater their confidence and ability to cope with things they meet in the future will become (Sherman and Mills, 2008).

An ongoing process that needs repeating

Although the early weeks of a puppy's life form the sensitive period, during which experiences have greater effect on the development of canine confidence than those that occur in later life, regression can occur if exposure is not repeated. This was suggested when well-socialized puppies between three and four months of age were placed in a kennel environment and left in virtual isolation until six to eight months old. It was found that they were shy of strangers, and even of their caretakers if they had not handled them much during that time. Additionally, they were fearful if removed from their familiar surroundings. In contrast, well-socialized puppies of six months or older left in the same surroundings for two months did not lose their confidence (Fox, 1978). Although extreme, this research affects puppy owners because it suggests they should plan to provide their puppy with adequate experience and to repeat those experiences periodically until at least 6 months old to optimize the puppy's potential for confidence in adult life. Other work suggests repetition should continue until maturity (Woolpy, 1968).

The Sensitive Period and Emotional States

Due to rate of brain development and myelinization of the CNS, the period around 3–5 weeks is associated with parasympathetic predominance (Fox, 1971, 1978) (Box 8.3). As a result, stimuli encountered during this period are likely to be associated with a positive emotional state and become part of a set that, when present, helps the puppy maintain a positive emotional state (Pluijmakers *et al.*, 2010). These stimuli can be social or non-social, including locations, but dependency on social stimuli is more likely to occur (Elliot and Scott, 1961; Cairns and Werboff, 1967). Due to the puppy's limited perception and mobility, its maintenance stimuli are initially limited to the dam, littermates and nest site. As the puppy develops and encounters new stimuli that may be included in the maintenance set, the value and the importance of existing stimuli can change. Changes occur in response to the stimuli's salience, availability and the degree of exposure the puppy has to them. The degree to which the puppy depends on its maintenance set and which stimuli in particular varies according to the puppy's need, which may change from one context to another.

The maturation and integration of various brain structures during the 5–7-week period increases the potential for a negative emotional reaction to novel stimuli (Fox, 1971). A broad and varied maintenance set enables the puppy to cope with this disruption and in doing so helps it learn how to organize its behaviour to return to a positive emotional state. The resultant sense of control helps the puppy develop the capacity to maintain a positive emotional state in the world as it has experienced it so far. For this to generalize it is necessary for the puppy to have a broad range of experiences. When this is done effectively there are benefits for the puppy's welfare and low levels of emotionality and reactivity that result in behaviour which owners find acceptable and manageable.

Box 8.3. Describing the parasympathetic versus sympathetic effect to clients.

When explaining to the client how the autonomic nervous system is associated with emotions, it is best to describe it as divided into two mutually exclusive parts and get them to relate to it through their own experience.

- **Activation** (A) (sympathetic) is associated with negative emotional states and increased respiration, heart rate and blood flow to the muscles, preparing the body for flight or fight. Owners are normally able to relate to co-occurring symptoms such as pacing, panting, increased toileting, loose motions and loss of appetite, in anxious dogs. Loss of sleep and restlessness may be examples they can relate to in their own experience.
- **Relaxation** (R) (parasympathetic) is associated with low heart and respiration rate and positive emotional states.

Disruption of emotional homeostasis

As we have seen, inadequate experience can result in novel stimuli causing disruption of emotional homeostasis. This is also true of familiar stimuli that do not perform to expectation, such as familiar people acting in an unusual way. There is an innate predisposition for some potentially threatening stimuli, such as sudden loud noises, to cause a fearful response (Gray, 1987). Contextual clues that allow the puppy to predict that any of these may happen can stimulate a negative emotional response. Sufficient maintenance stimuli can modulate these feelings but loss of stimuli from the maintenance set, even in the absence of potential threats, can result in a sense of loss of control and loss of emotional homeostasis. The extent to which this occurs depends on the importance of the stimuli to the individual. A common early example of this occurs when a puppy is homed. The maintenance set that developed at the breeder's is disrupted and it may be distressed during its first nights in the new environment. However, the level of distress may be less if the breeder kept the puppies in a domestic circumstance, because of similarities between existing maintenance stimuli and things encountered in the new home. The puppy may also settle in its new environment more easily if dog-appeasing pheromone (DAP) is used.

Dog-appeasing pheromone

This natural trigger for a relaxed emotional state requires no learning. Three to five days after parturition, an oily secretion is released from the sebaceous glands in the cleft between the bitch's mammary glands. This automatically stimulates a parasympathetic response (see Box 8.3), because the nerve pathways from the Organ of Jacobson, activated by the pheromone, are either directly connected to the amygdala through the limbic system or directly to the autonomic nervous system. In addition to aiding bonding to the dam, the presence of the pheromone helps the puppy be confident when it starts to explore its environment. Here it learns the properties of new stimuli and associates them with its emotional state (Box 8.4).

Box 8.4. Situations where dog-appeasing pheromone may be helpful to puppies.

The synthetically reproduced version of dog-appeasing pheromone, ADAPTIL™ (CEVA), available as a diffuser that is plugged into an electric socket, a collar or as a spray, can be used in any situation in which it is anticipated that the puppy may need support. Examples include the following.

- Helping the puppy settle into its new home.
- Accustoming the puppy to periods of time on its own.
- The presence of unfamiliar stimuli, such as new environments and unfamiliar visitors.
- The presence of frightening stimuli, such as sudden loud noises (fireworks, thunderstorms, gunshots).
- Changes in familiar environment or routine, such as the arrival of a baby or redecoration.
- Car travel, for example the puppy's first journey to its new owner's home.

General Principles Required for Introducing a Puppy to the World

Intensity of exposure and supporting maintenance stimuli

Whenever possible a puppy's introduction to new stimuli should be gradual, so that the experience does not exceed its capacity to cope. This will allow it to maintain emotional homeostasis and learn behavioural responses to the situation that the owner will find desirable (Box 8.5). In many cases the precaution of proceeding gradually will prove unnecessary but it is advisable to err on the side of caution.

So that unfamiliar stimuli are more easily associated with a positive emotional state, puppies should be supported by the presence of sufficient maintenance stimuli when they are exposed to them. In some situations it may be beneficial for the owner to stimulate a positive emotional state before increasing the gradient of challenge to the puppy's confidence. Examples include: placing titbits on the surgery table before first examination; asking visitors the puppy is not familiar with to feed it food treats; or playing with the puppy with a toy in the vicinity of potentially challenging stimuli.

Having looked at the theoretical aspects, some practical examples follow.

The new home

The process of developing the puppy's confidence in its owner's home can be started during visits to the breeder before homing. As long as the puppy is in its familiar surroundings it should be relaxed, making it easier for it to associate the sight, sound and scent of the prospective owners with a parasympathetic state. For the same reason, cloths impregnated with the scent

Box 8.5. Gradual introduction.

Gradual introduction to environmental stimuli can be achieved in a number of ways:

1. The primary principle is that puppies should be allowed to become familiar with new experiences at their own pace. This allows them to control their emotional state and choose whether to approach, withdraw or wait and observe for longer.
2. Gradually reduce the distance from the stimuli.
3. Gradually increase the number of stimuli. For example, start with one visitor to the home and progress to more on subsequent occasions.
4. Gradually increase the salience of the stimuli. For example, an unfamiliar person may be relatively passive, avoid eye contact etc. at first, gradually becoming more animated or gradually interacting with the puppy. Another example of increasing the salience is to increase the volume of sound.
5. Gradually progress from familiar environments to unfamiliar environments. For example, having accustomed a puppy to meeting people in the familiar home environment the owner can progress to meeting them away from home.

These criteria are interrelated. For example, if the puppy has been accustomed to a stimulus that has remained at a low level of salience as it has got closer and that salience is to be increased, it may be appropriate to return to a greater distance first.

If the puppy shows a fearful response, the owner should proceed as follows.

1. Try not to rush to reassure. Reassurance may reinforce whatever behaviour the puppy is displaying as a coping strategy. They need to consider if it is an appropriate and proportionate response.
2. Remain close by or allow the puppy to be with them, as knowledge of this will increase the puppy's confidence that it can cope.
3. Avoid pressurizing the puppy to approach what it is fearful of, as this could reduce its sense of control over its emotional state and compound its anxieties and fears.
4. If the stimulus is benign, such as an object in the street, rather than truly aversive, such as a sudden loud noise, the owner should start the process of exposing the puppy to it again. If it does not recover or if it shows fear of aversive stimuli, it may be appropriate for the puppy to be referred for specialist help. (See Chapter 21.)
5. Reward the puppy every time it does not react to stimuli similar to the original stimulus and stimuli associated with it. This will help to develop positive associations, which are incompatible with feelings of anxiety and fear.

of the prospective owners and their home can be taken to the breeder's environment in an airtight container to be removed by the breeder at intervals and left with the puppy. A few days before the puppy moves to its new home an ADAPTIL diffuser can be introduced to the area in which it will sleep and rest, so that when the puppy arrives the presence of the pheromone should enhance its sense of security.

When the puppy arrives, it will be beneficial for it to be accompanied by cloths carrying the scent of the breeder's environment to be introduced to the new home. The presence of objects that were familiar at the breeder's home, such as toys, should also help the puppy settle. During the initial settling-in period, noise levels, such as children playing, should be increased only gradually. Ideally the puppy will have already been accustomed to domestic sights and sounds, such as noisy mechanical equipment, but they should only be gradually introduced in the new home. This can be achieved by allowing the puppy to investigate them when they are not in use and, when they are in use, initially keeping them at a distance from the puppy or closing doors between them to reduce the volume.

Other dogs at home

If the owner already has a dog, they should introduce the puppy to it in the garden or other safe location away from resources the older dog may try to protect. Once the initial acceptance has been made by the older dog, the two should find their own level and settle down without too much intervention. Owners should intervene if the puppy's attentions are too much for the older dog to cope with, or if the older dog is hostile to the puppy or overly physical in play. There is a fine line as to when intervention is appropriate, because the older dog must be able to communicate to the puppy when it does not want to interact, otherwise the puppy may grow to display a lack of inhibition when interacting with the older dog, which could generalize to other dogs encountered during walks. If, as it matures, it finds its attentions are ignored, the fact that its expectations are not met can result in frustration and anger that can predispose frustration-related aggression.

Cats and other pets

If the owner has a cat, the cat should be provided with places to hide, surfaces it can climb onto out of the puppy's reach and an area of the home where it feels secure and the puppy cannot follow. The integrity and location of the housing for pets that do not normally have liberty within the home should be checked to ensure that the puppy cannot pose a threat or cause stress through showing interest. Pets that are potential prey species should also have the opportunity to withdraw behind a visual barrier. The puppy should be kept under control and rewarded each time it does not react to the cat moving about or the activity of other pets.

Accustoming the puppy to people

It is important that the puppy should experience as much of the diversity of human appearance and behaviour as possible so that it learns that the variations are normal and either habituate to them or to develop positive associations. One of the benefits of good puppy socialization classes is that the appearance of the people attending can be varied by dressing them in various guises. Socializing with people should start with introduction to individuals. They should be asked to remain relatively passive until the puppy shows it is not experiencing any anxiety. If the puppy is anxious, asking the individual to sit on a chair or crouch down rather than bending over it may help and food treat titbits can be given to the person for them to give to the puppy, or a toy can be used so they can initiate play. However, this approach should be used with caution if the puppy is still exhibiting fear, as it is important not to put the puppy into a situation where the desire for the food or toy conflicts with the desire to avoid the person. This situation can potentially lead to less positive experiences for the puppy.

Due to the dog's natural tendency to develop defensive behaviour towards people outside of its social group at the boundary of the owner's home, it is beneficial for owners to train their puppies against it. For example, they can regularly take their puppy to the threshold of the home to meet delivery people, such as those who deliver the post or milk. If these people are given food treats to give to the puppy, as long as the puppy is comfortable to take them, it will be less likely to display problem behaviour towards them as it matures. The puppy should also be accustomed to visitors of both sexes and all ages. These can also be associated with food treats to help develop a positive association with people entering the home in the presence of the owners. If the food treat is given only when the puppy is calm, it will help to reduce the risk of developing excitable behaviour when visitors arrive.

The puppy should be accustomed to the presence of children and being handled by them but the owner should supervise to ensure that the children interact with it appropriately. Ideally the owner should also arrange to meet someone with a baby regularly so that the puppy can get used to the sight and sound.

Accustoming the puppy to busy urban and other environments

Exposure to urban environments should start after the puppy has been accustomed to wearing and then walking on the leash, and to traffic noise, and has had some experience of meeting people or, if not possible, experience of ignoring them in quieter locations. The presence of the owner will act as a maintenance stimulus, as will the presence of other family members or a canine companion. The use of ADAPTIL in the form of a collar may also help. Although the presence of another dog can help initially, the puppy will need opportunities to meet experiences on its own so that it learns to cope as an individual rather than stand, literally or metaphorically, behind its companion when stressed.

It is best for the owner to start in quiet side streets or suburban areas or, if not possible, at quiet times, such as very early morning, in the evening or Sunday morning. Once the puppy has become accustomed to the quieter areas the owner can start to take it to locations that are busier but at quiet times of day. This is so that the level of challenge is low and the puppy has the chance to associate the location with remaining in a positive emotional state. At a rate the puppy can cope with, the timing can be gradually changed so it experiences busier conditions.

Although puppyhood exposure to busy environments seems to be particularly important to the development of canine confidence, the need to provide experience of other environments, such as the countryside, must not be overlooked.

Socializing with other dogs

One of the reasons owners often choose not to attend puppy socialization classes is that they believe their puppy will be socialized through interacting with dogs encountered during exercise. However, if those dogs have poor communication skills, they can have a detrimental effect on the puppy's development. Furthermore, a bad experience with such dogs can result in fear and the development of defensive behaviour towards other dogs.

Removing puppies from their mother and littermates before 18 weeks of age creates a problem because their canine social skills and bite inhibition in particular (Dunbar, 2004) are still developing. This means that good puppy socialization classes are invaluable, as they provide the ideal environment for puppies to continue to learn how to read and interact with other puppies of different appearance (see Chapter 12).

In addition to the other key points related to gradual introduction to stimuli (see Box 8.5) it is important that a puppy is rewarded for remaining relaxed in the presence of other dogs when play is not desired by the handler. When the puppy is allowed to play, the handler should practise calling it back to them for rewards, sometimes also doing a little training, sometimes putting it on the lead before allowing it to play again. This will have the effect of training the puppy not to always expect to play or otherwise interact with other dogs. Its low expectation will help to reduce the potential for frustration-related problems related to other dogs in later life. Training the puppy to return to its owner when interacting with other puppies will increase the potential for owner control where other dogs or other distractions are present.

Conditioning the puppy to cope with being left alone

The importance of making gradual variations to a puppy's environment also applies to accustoming it to being without a major part of its maintenance set: its owners. Before the new owners accustom the puppy to being on its own, for its own safety they should get it used to having a bed it likes to rest on in an indoor kennel, puppy pen or in a chew-proofed room that can be closed off using a child- or pet-gate. The purpose of using a gate rather than just closing a door is that it facilitates a more gradual transition to the puppy being on its own that is less likely to induce stress.

The puppy should be provided with stimuli to predispose a parasympathetic state, such as cloths freshly impregnated with the owner's scent, ADAPTIL, fresh chew items or toys it has to get food out of. The owners can initially accustom the puppy to separation from them by sitting in the same room or just inside the next room if they are using a gate. If the puppy remains relaxed, they can gradually increase the distance over a number of training sessions to the point where they can be out of sight. A similar approach can be adopted for accustoming the puppy to being on its own at night or, if an indoor kennel is being used, the puppy can be gradually moved further from the owner's bedroom until it is in the area where they would like it to sleep.

Checklist

To ensure the best chance of developing a puppy's sound temperament and capacity to cope in all circumstances, its social and environmental exposure should be systematic. To aid in this it is beneficial for the owner to have a checklist of places to take their puppy and experiences it should be accustomed to. The date these are done should be recorded and that they should be repeated made evident. For example:

Location	Date				
Bus travel					
Railway station					

Training and the Puppy's Relationship with its Owners

The puppy's good relationship with its owners is central to developing its confidence, as it may depend on them for emotional support. Therefore it is important that owners should be consistent in the way they interact and the behaviours they allow. Owner inconsistency, when the puppy jumps up for attention, for example, can result in poor training outcomes and the puppy experiencing disquiet because it cannot predict its environment. In turn this could be detrimental to its overall emotional stability.

The potential for the puppy to feel disquiet can be reduced if it is taught cues so it can discriminate between when a behaviour is allowed or not allowed. The puppy–owner relationship can also be enhanced if, rather than trying to stop the behaviours they do not want, the owners train the behaviours they do want, such as to sit during greeting. Training in general is beneficial to the owner–puppy bond because it results in higher levels of owner control and satisfaction. Activities that are done together as a unit, such as search games and retrieves, are good for the relationship and give the puppy outlets for normal behaviour.

The principles for training any behaviour are the same as those for training tricks and it is important that the owner does not consider training tricks as fun involving rewards and other training being synonymous with obedience and requiring overbearing or punitive techniques. Obedience is an unhelpful word to use with clients, because it implies that if a puppy or dog has not responded as desired it has disobeyed, whereas the truth is that either the training was not good enough, or some environmental stimulus is inhibiting the puppy's ability to respond as taught.

The technique probably most often used to train puppies is called lure and reward (see Box 8.6 for examples). When training starts the owners should ensure that there is minimal distraction, that the puppy has little difficulty performing the response and that it is required for a short duration. As training progresses, these criteria can be increased but not simultaneously.

Other general rules for training are as follows.

- Train on a little-and-often basis.
- Stop each session while the puppy is still enthusiastic and having fun.
- Develop the training in little steps.
- Use a range of rewards, the puppy's favourites being kept for the best responses and the occasional special reward to keep up its motivation.
- If training is not going well, do something the puppy can already do well and reward. Finish the session and reconsider the third point before the next.

There are aspects of puppy training that are likely to be of particular concern for owners and about which they may seek advice from practice staff. These include:

- Chewing
- Play-biting and mouthing
- House training.

Chewing

Owners can prevent the development of inappropriate chewing by using (once the puppy has learnt to accept being enclosed) an indoor kennel, puppy pen or an adapted chew-proof area of the home when it is not possible to supervise their puppy for short periods of time and overnight. However, an indoor kennel has the advantage that it can be moved from one location to another, even in the owner's car. Once the puppy has been accustomed to being relaxed in this when left, the owners can be confident that it is safe, as has not got access to electric cables and is not damaging their property or associating inappropriate materials, such as shoes or furniture, with chewing. All of this will be particularly important once the puppy starts teething.

Box 8.6. Using lure and reward to train a puppy to sit and lie down.

Sit

The owner should hold a small food reward between finger and thumb. After bringing it close to the puppy's nose they should slowly raise it to lure the puppy's head upward as it follows the hand. The owner should wait until the puppy sits, which it is likely to do because as it looks up at the food its hind-quarters will go down. When it sits it should be praised and given the titbit. After a few repetitions, the puppy will learn to sit to get the titbit (the reinforcer) without delay.

Once the puppy has learnt to sit quickly to get the titbit the owner can say 'sit!' as the puppy starts the action of sitting. Through repetition the puppy will learn to associate 'sit!' with the action it has to perform to get the reward. The action of the hand can then be gradually phased out or developed into a hand signal.

Down

To use the technique to train their puppy to lie down, the owner should gradually lower the food reward to floor level. The puppy may worry at their hand in an attempt get the titbit but the owner should not let go until the puppy is lying down, so that through repetition it learns that the way to get the titbit is to lie down. The owner can then add the word 'down!' as it starts to lie down. The hand lure can be gradually phased out once the puppy has learned the association.

The puppy will need to have some things it can chew, so chew-items designed for dogs to chew on safely should be left with it and made available at other times. This will help it exercise the chewing it needs to perform and learn to associate chewing with those items.

Play-biting and mouthing

As puppies grow up, adults and litter mates alike become increasingly intolerant of their sharp teeth when they go through the natural process of mouthing and play-biting. On finding this, puppies learn that other individuals react negatively and learn to control the strength they use. When a puppy is introduced into the family, this learning process is normally incomplete and the family must take over where the puppy's mother and littermates left off.

The simple way is for the person concerned to respond with a verbal signal and then freeze or walk away, ignoring the puppy for a few moments. In this way the puppy should learn to limit the strength of its bite in play and that to bite hard results in the withdrawal of interaction. Sometimes, however, preconditioning is required because the puppy continues to bite despite the family's best efforts, sometimes because it turns the sequence into a game, or does not cope well with the frustration of the person withdrawing.

Referring the owner to a suitably qualified trainer for advice tailored to their puppy should be considered if basic training approaches are not working.

The potential for the puppy to initiate play-biting can be reduced if it is given other activities to fill its time and use up its energies. Distraction can be used in situations where the behaviour needs directing away from an individual.

House training

A puppy's substrate preference for urination and defecation is normally established by around 7½–8½ weeks (Overall, 1997). The preference is developed by associating the substrate with the feeling of relief and is therefore self-reinforcing.

Clients who have obtained their puppies from environments where access to the substrate they would like their puppy to use, such as grass, has not been given may have to develop their puppy's preference for it. To do this they need to deny the opportunity to use the established substrate choice until that association fades and has been superseded by new and stronger association.

Of course, new-born puppies' defecation and urination are stimulated by their dam, which consumes the waste. As the puppies become more mobile, they start to move away from the nest to relieve themselves. The older they become, the further away they want to go and the puppy in the average-sized home will, with a little help and the right opportunities, easily develop a preference for relief behaviour outside.

It is important to ensure that owners understand that their puppy should not be punished when it has an 'accident'. Practices such as rubbing its nose in the offending mess or showing it the puddle and shouting will induce anxiety and the potential for reduced elimination control rather than learning and are detrimental to the puppy–owner relationship. Owners sometimes also need to be made aware of the misinterpretation of their puppy having what they think is a look of guilt, for example their puppy moving with a lowered body posture when the owner has noticed that it has urinated in the home. This is not a sign of guilt but anticipation of the owner's response, an association developed through previous learning.

Puppies can be trained to relieve themselves on newspaper placed near their sleeping area. This is a more appropriate approach if newspaper was also provided for the puppies by the breeder. Once the association between toileting on the paper in that location has been established, the paper can be gradually moved nearer the door and eventually placed outside. Subsequently the amount of newspaper can be gradually reduced.

A better way of developing house training is to prevent the puppy fouling in the locations where it is not desirable as much as possible. When supervision is not possible for short periods of time and overnight, the use of an indoor kennel or puppy pen will restrict the available space to an area close to its bed, which may inhibit toileting and, as a consequence, the puppy will learn to wait to go outside. The owners will have to make a point of retiring to bed late, getting up early and perhaps at intervals during the night to take the puppy out.

As soon as the puppy has woken, finished eating or is let out of its pen/indoor kennel, it is almost certain to want to relieve itself and should be taken to the area the owners want it to use, helping associate that place with toileting behaviour. At other times the owner should give opportunities every 1–2 hours but should also learn to identify when the puppy needs to relieve itself. They may see increased restlessness, sniffing, circling and finally movement towards a squatting or crouching position. They should immediately interrupt these behaviours by lifting the puppy and carrying it or, if it is old enough, by encouraging it to the door that gives access to the garden or other area they want it to use.

If the puppy does have an 'accident' inside, it is sensible to reduce the possibility of a repeat performance by removing the smell using a proprietary enzymatic cleaner, such as Urine Off™.

It is beneficial for some owners, such as those who do not have a garden or yard, to train their puppy to relive itself while it is on a leash at a given a signal. This training can also be useful when any puppy is taken away from home and is required to relieve itself in a location or on a substrate not previously associated with toileting behaviour. The signal for toileting is introduced by saying a specific word or phrase just before the puppy starts the action of relief. Through repetition the puppy will associate the sound with the action and the owner can use it to cue the behaviour when they anticipate their puppy has a need.

The Importance of Owners Understanding Canine Body Language and Communication

A puppy's body language, facial expression and vocal communication are clues to its confidence and emotional state. Observing their puppy and learning to interpret its behaviour provides owners with some insight into how it feels at any given moment and what it is likely to do next.

Dog language is multifactorial, complex and constantly varying in response to the puppy's perception, level of arousal and mood state. Despite the fact that humans have changed the physical appearance of many breeds, making some less easy for owners to read than others, there are general trends that are easy for owners to identify. For example, confidence tends to result in puppies having a taller body and head height, mid to raised tail carriage, confident stare and forward-facing ears (or at least the base of the ears). Lack of confidence typically results in a lowered posture and ear carriage, avoidance of eye contact (social situations), raised paw, licking of lips, turning away from social stressors, etc.

An aspect of puppy behaviour easily misunderstood by owners occurs when the puppy seems very excited, overactive or overly pleased to greet new people or dogs. Although these behaviours can have alternative explanations, they can be indicators that the puppy is stressed by the situation. For examples, over-activity and jumping up at unfamiliar people can be indicative of anxiety; arousal, excitement and pulling on the lead towards other animals or in response to movement of joggers, bicycles etc. can be associated with frustration.

Most often the well-socialized puppy is one that is confident, calm and collected. It is this behaviour that owners should aim for and reward, including those occasions when that decision is to do nothing. If this approach is applied to developing the puppy's confidence and behaviour appropriate for the circumstance, it will observe the contextual clues in the world around it and respond with the self-control its owners would like it to have.

References

Appleby, D.L., Bradshaw, J.W.S. and Casey, R.A. (2002) Relationship between aggressive and avoidance behaviour by dogs and their experience in the first six months of life. *Veterinary Record* 150, 434–438.

Appleby, D., Sandbacka-Saxén, S., Jokinen, O., Burgherr, T. and Valros, A. (2013) Homing age influences the prevalence of problematic adult canine behaviour. In: *Proceedings of the 9th International Veterinary Behavioural Meeting*, Lisbon, Portugal, pp. 69–70.

Cairns, R.B. and Werboff, J. (1967) Behavior development in the dog: an interspecific analysis. *Science* 158, 1070–1072.

Dunbar, I. (2004) *Before and After Getting Your Puppy*. New World Library, Novato, California.

Elliot, O. and Scott, J.P. (1961) The development of emotional distress reactions to separation in puppies. *Journal of Genetic Psychology* 99, 3–22.

Fox, M.W. (1971) *Integrative Development of Brain and Behaviour in the Dog*. University of Chicago Press, Chicago, Illinois.

Fox, M.W. (1978) *The Dog: Its Domestication and Behaviour*. Garland, S. & P.M. Press, New York.

Freedman, D.G., King, J.A. and Elliot, O. (1961) Critical period in the social development of dogs. *Science* 133, 1016–1017.

Gray, J.A. (1987) *The Psychology of Fear and Stress*. Cambridge University Press, Cambridge.

Jokinen, O., Appleby, D., Sandbacka-Saxén, S., Appleby, T. and Valros, A. (2017) Homing age influences the prevalence of aggressive and avoidance-related behaviour in adult dogs. *Applied Animal Behaviour Science* 195, 87–92.

Murphree, O.D. (1973) Inheritance of human aversion and inactivity in two strains of the Pointer dog. *Biological Psychiatry* 7, 23–29.

Overall, K.L. (1997) *Clinical Behavioral Medicine for Small Animals*. Mosby, St Louis, Missouri.

Pluijmakers, J.T.M., Appleby, D.L. and Bradshaw, J.W.S. (2010) Exposure to video images between 3 and 5 weeks of age decreases neophobia in domestic dogs. *Applied Animal Behaviour Science* 126(1), 51–58.

Scott, J.P. and Fuller, J.L. (1965) *Genetics and the Social Behaviour of the Dog*. University of Chicago Press, Chicago, Illinois.

Sherman, B.L. and Mills, D.S. (2008) Canine anxieties and phobias: an update on separation anxiety and noise aversions. *Veterinary Clinics Small Animal Practice* 38, 1081–1106.

Sterry, J., Appleby, D. and Bizo, L. (2005) The relationship between measures of problematical behaviour in adult dogs and age of first exposure outside the first owners' home and attendance at puppy classes. In: *Proceedings of Companion Animal Behaviour Therapy Study Group Study Day*, Birmingham, England, pp. 21–24.

Woolpy, J.H. (1968) Socialization of wolves. In: Masserman,, J.H. (ed.) *Animal and Human: Scientific Proceedings of the American Academy of Psychoanalysis; Science and Psychoanalysis,* 12. Grune and Stratton, New York, pp. 82–94.

9 What Every Kitten Owner Needs to Know

Jon Bowen

Problem Incidence

Whilst it is not desirable to alarm potential kitten owners with statistics about the difficulties they may face with their pets in the future, veterinary practices do have a responsibility to give owners some information so that they can make a decision about whether a kitten is the right choice of pet for their lifestyle and circumstances.

Most data on the prevalence of problem behaviour relate to presentation rates at the clinics of veterinary behaviourists. However, this does not give a good indication of the problems that pet owners will genuinely face, as the referral population tends to be made up of those types of cases that are most 'problematic' to owners, rather than those that are necessarily most common.

In one study, undertaken in Spain, rates of canine and feline behavioural problems were surveyed in a non-referral population (Fatjo *et al.*, 2006). Veterinary surgeons were asked to report on the incidence of a range of behavioural problems they saw in practice. In cats, indoor elimination, furniture scratching and excessive vocalization were, in order, the three top-ranked behavioural complaints, whereas in dogs the top-ranked were destructiveness, aggression and indoor elimination.

Interestingly, all three of the top-ranked behavioural problems identified in cats are normal behaviours expressed in a manner that the owner regards to be problematic. This makes it even more important that these problems be understood in advance by kitten owners, so that they can adapt the home environment to meet the needs of their pet and reduce the risk of problem behaviour developing.

This same Spanish group carried out a retrospective analysis of a referral population of feline behaviour problem cases, looking for potential risk factors. In this referral population, aggression was the commonest problem, most often targeted at other cats. Most cases of aggression towards the owners were identified as play-related aggression or petting-related aggression, with a strong overlap in incidence between these two problems. Compared with a control population, the referred population more frequently came from pet shops and had less outdoor access.

There is a general association between the number of cats resident in a household and the incidence of behavioural problems such as inter-cat aggression and spray marking. Owners run an increased risk of problems if they have a large collection of cats, especially if there is an insufficient level of resource provision and space for them.

Socialization and Rearing

The aim of socialization is for the cat to develop a multispecies identity that enables it to tolerate contact with people and other domestic species, such as dogs (Fig. 9.1). Socialization

DOI: 10.1079/9781800621312.0009

Fig. 9.1. Well-socialized cats are more likely to tolerate contact with people and other domestic pets, such as dogs. (Image: iStock.)

and habituation during the sensitive period of development have been shown to be important factors in emotional development and the prevention of behavioural problems in the dog and cat.

The sensitive period of cats was identified by Karsh and Turner (1998) as starting at 2 weeks of age and ending at 7 weeks of age. During this period kittens habituate to the stimuli in their environment and form social bonds with the species that they come into contact with. The daily, and total weekly, amount of human handling, as well as the number of human handlers, during this period has been shown to relate to altered approach behaviour to familiar and unfamiliar people, as well as attempts to escape from people (Collard, 1967; Karsh and Turner, 1998). Higher levels of handling were associated with better adult outcomes in terms of fearful behaviour around people. Well-socialized cats are more likely to tolerate, or enjoy, human contact, but potential owners should be made aware that not all cats will want close contact with people, such as sitting on a lap (Fig. 9.2).

Fig. 9.2. Owners may have an expectation that their cat will enjoy close human contact, such as sitting on a lap. It is important to emphasize to potential owners that even well-socialized cats may prefer not to have close contact of this kind. (Image: iStock.)

In a study by Casey and Bradshaw (2008), a group of kittens given enhanced socialization was compared with a normally socialized group (control) and the effects on behavioural problems, fear responses and the emotional bond with the owner were followed up at 1 year of age. Enhanced socialization included increased handling, manipulation, play and times being taken away from the litter, which extended to the 9th week of age.

This enhanced socialization was associated with lower rates of fear responses towards people at 1 year of age, compared with the control group. The owners of these kittens also reported higher scores for 'emotional support', which was a previously validated measure of the bond between pet and owner.

In a number of studies, regular gentle handling of kittens during the first 6 weeks of life has been associated with precocious development and novelty seeking, with kittens being more likely to approach novel objects and people (Meier, 1961; Wilson *et al.*, 1965).

In dogs, puppy socialization classes are now commonplace. Similar classes can be run for kittens. In Australia these were pioneered under the title 'Kitten Kindy'. The recommendation is that socialization classes be restricted to healthy kittens of 7–14 weeks of age (Seksel, 2001). The intention of these classes is to increase exposure of kittens to a wide variety of stimuli, including different people, and to enable them to engage in social interaction with each other. Although the sensitive period of kittens is at an end, during this period kittens can still learn by classical conditioning and positive reinforcement, as long as new events are presented very carefully and associated with a positive outcome. Social interactions between kittens need to be monitored very carefully in this type of class, as there is a risk of kittens having a negative experience and sensitizing them to new social contacts later in life. If an appropriate kitten class is not available, new kitten owners can still follow a similar programme themselves by following Kitten Kindy guidelines.

Despite the fact that kittens, like puppies, can live independently of their mother from as little as 8 weeks of age, it is general practice amongst the breeders of pedigree kittens to home them at 13 weeks of age or older. This is largely due to concerns about infectious disease and vaccination status. However, it places an additional responsibility on the breeder of such kittens to engage in a programme of enhanced socialization and habituation equivalent to Kitten Kindy. Any mismatch between rearing and domestic environment for these kittens is likely to lead to difficulty adapting to the home environment, increased anxiousness and fearfulness.

Key points

- Kittens are independent of their mothers from 8 weeks of age, and should be homed at around this age rather than later.
- There is no demonstrable health benefit to homing at 12–13 weeks or later, as opposed to at 8 weeks.
- The sensitive period ends at approximately 7 weeks of age.
- After this age, kittens are still highly receptive to learning and can benefit from appropriate kitten socialization classes or an equivalent enhanced socialization programme, as long as this is carried out with great care and new experiences are positive for the kitten.
- Increased socialization and habituation during the sensitive period lead to better adult behavioural outcomes and a stronger pet–owner bond.

Hybrid Cat Species

The domestic cat is a subspecies of *Felis silvestris*, which is a small felid that inhabits a wide range, including Northern Europe, Africa and the Middle East. Within this species there is also a range of body types, with individuals from hotter climates being thinner and those from cold or temperate climates being stockier and more heavily built. This variation also appears to extend to the ability to be domesticated; the North European wildcats, such as the Scottish wildcat, show no propensity for domestication. The reason for this variation in character may relate to the ecological niches available to these species, with colder climates favouring solitary behaviour and hence a lower requirement for sociability. So, it is likely that the origin of the domestic cat is within the more slightly built, easier-to-domesticate cats found in hotter climates (i.e. *Felis silvestris lybica*).

There is a growing movement amongst the cat fancy to hybridize the domestic cat with wild felids such as the Asian leopard cat, margay and serval. The resultant hybrid cats may be larger, more strikingly marked and have different behaviour. In some cases the species hybridized with the domestic cat have little overlap in distribution and habitat with *Felis silvestris*. For example, the margay is a native of the Americas, a region of which *Felis silvestris* was not a native. In others, such as serval, the distribution of the two species does have some overlap.

In all cases, however, these are cat species that have inhabited regions where humans have been present for thousands of years without any indication of a process of successfully domesticating them. It is irresponsible to deliberately introduce genetic material into the domestic cat population from other cat species that have never been properly domesticated.

Anecdotally, hybrids such as the Bengal (domestic cat crossed with wild Asian leopard cat) appear to be more aggressive towards neighbourhood cats. This represents a welfare problem not only for neighbourhood cats, but also for hybrids that may find domestic living highly stressful.

The risk of human injury is also an important consideration. In a study of animal bite injuries at University of California Los Angeles (UCLA) Hospital Emergency Department, Kizer (1979) found that 20% of cat-related injuries involved the head and neck, with 60% of those injuries involving the eye and surrounding tissues. Twenty-nine per cent of cat-bite–scratch victims returned with complications, compared with only 5% of dog-bite victims. *Pasteurella multocida* was the most common pathogen cultured, being found in 50% of dog and 80% of cat injuries. These risks have been confirmed in more recent studies (Esposito *et al.*, 2013).

Hybrids such as the Savannah (domestic cat crossed with wild serval) can reach weights in excess of 15 kg, the size of a small to medium-sized dog. Cat owners frequently misinterpret feline social behaviour and get scratched or bitten when handling their own domestic cats, and cats may be very challenging to handle in veterinary practice.

Key points

- Hybrid cats are genetically distinct from the domestic cat, which is a subspecies of *Felis silvestris*.
- The wild species hybridized with domestic cats do not have any history of domestication.
- The precise nutritional, environmental and social needs of hybrid cats are not fully understood, so husbandry can be a challenge.
- They may pose a threat to other pets (especially domestic cats), people and wildlife.
- The status of hybrid cats as wild animals varies between countries.
- In the author's oppinion, prospective kitten owners should be discouraged from purchasing hybrid cat species to keep as pets, whether they allow them outdoor access or not.

The intentional breeding and pet keeping of larger, and therefore potentially more dangerous, hybrid cat species is a potential public health risk.

Elimination and Toilet Training

For approximately the first 3 weeks of life, kittens do not have voluntary control over elimination; urination and defecation are stimulated by the kitten's mother grooming its perineal area. From 3 weeks of age kittens begin to develop voluntary elimination and this takes over from maternal stimulation by about 7 weeks of age. If kittens are hand reared, elimination will need to be stimulated manually. For the first few weeks of life the kitten's mother will consume any waste products passed by the offspring in order to maintain hygiene around the nest. This change in elimination behaviour is in parallel with developments in the sensory and locomotor systems; by 7 weeks of age the kitten has full adult locomotion and sensation.

Kittens learn adult feeding, predatory and elimination behaviour by observation of an adult 'demonstrator' cat. Speed of learning is determined by the level of attention the kitten pays to the demonstrator, with kittens showing a greater level of attention to their mother than other female cats. Kittens learn substrate and location preferences for elimination sites. The choice of location relates to features of security and privacy, and distance from resting, feeding and hunting sites.

In order to model this behaviour during kitten-hood, the mother must be able to demonstrate a suitable range of preferential behaviour, for example moving well away from the nest to use a private location to eliminate. If the location and substrate choices available to her are limited, such as only having a litter tray close to the nest site and filled with newspaper rather than litter, then the kittens will develop potentially problematic preferences. This may be more serious if kittens have been homed at 12 or more weeks of age, when inappropriate preferences will have been more strongly ingrained.

Kittens should be provided with at least one shallow-sided litter tray, deep filled with a fine, mineral-based cat litter that is easy to dig. Silica gel-based litters should be avoided for the first few months, as they can be harmful if swallowed. Since the kitten's vision and mobility are limited to begin with, enough litter trays should be installed to give the kitten quick access to a latrine wherever it is. The number of litter trays can be reduced as the kitten becomes more mobile and able to navigate its environment.

If the kitten has already had experience of a litter tray, it is sensible to use a similar tray in the new home, with similar litter. If, however, the litter the kitten has previously used is not of a preferred mineral-based type, then a small amount of the preferred litter can be scattered on the existing litter to allow the kitten to adapt to the new texture.

Once kittens are able to go outside, the owner should create one or more outdoor toilets for it. This is a hole approximately 30 cm square and 30 cm deep, filled with soft playground sand. These toilets should be located in a border of the garden where they are sheltered from view to allow the cat some privacy. Scattering a small amount of used litter over the sand will help to attract the kitten to use it as a toilet. Outdoor toilets like this are particularly important in gardens that have a clay soil that retains water and is difficult to dig, or where there are few suitable flowerbeds.

Eating and Drinking

Prior to weaning, kittens are completely dependent on their mothers for food. This makes them vulnerable to problems if the mother is malnourished. In one study, kittens born to mothers that were 50% food deprived went on to have underdeveloped cerebral, cerebellar and brainstem regions despite recovering their general growth rate on *ad lib* solid food from week 6 (Smith and Jansen, 1977). They show impaired learning, antisocial behaviour to other cats and increased fearfulness, as well as delayed development of a range of normal behaviours, including locomotion, play and predation (Simonson, 1979).

Kittens only start to eat solid food at around 4 weeks of age (Martin, 1986), which signals the start of weaning. Kittens are generally attracted to eat novel foods, but they also learn food preferences by observing their mother. Kittens are more likely to eat food that they have seen their mother eating, even if the food is not conventionally part of a cat's diet (Wyrwicka, 1978). This modelling of food preference is most active toward the end of weaning, at around 7–8 weeks of age.

The age of weaning varies under the influence of a number of factors, including availability of food from the mother. Early weaning occurs if the mother has longer and more frequent periods of separation from the kittens starting at 4 weeks. Weaning is delayed if kittens are left with their mothers and solid food is not available (Tan and Counsilman, 1985).

Weaning also affects behavioural development, with object play beginning at an earlier age if weaning is advanced. This has implications for a kitten's ability to live independently, and therefore be homed at a particular age.

Adult cats will normally eat 10–20 small meals each day, and the most suitable presentation of food to adult pet cats is *ad lib* without the anthropocentric ritual of 'mealtimes'. Kittens should only be homed after weaning is complete, at which point they should have free access to food. This should include some moist food (e.g. canned), as kittens may be unable to chew dried food easily. There is no need to introduce meal-feeding with a gradual decrease in meal frequency, as would be the case in a puppy, since the objective is not to reach the point of offering one or two fixed mealtimes per day. Activity feeding, using toys that dispense food during play, can be introduced as a supplementary source of food after weaning is complete; in order to avoid malnutrition, kittens should continue to have easy access to food from a bowl until their physical development is complete. Activity feeding is beneficial to young cats, as it provides a source of reinforcement for the maintenance of active object play into adulthood, which can help to prevent obesity.

Hand rearing kittens is a specialized task, carrying a high risk of malnutrition, aspiration pneumonia and the development of behavioural problems.

As with latrine sites, since kittens cannot fully navigate their environment or move long distances to begin with, it is necessary to provide several feeding and drinking sites so that the kitten is never more than a few metres from somewhere to eat and drink.

> **Key points**
>
> - A kitten only develops normally if its mother is properly fed.
> - Weaning will be delayed if kittens are not offered *ad lib* access to solid food from the age of 3–4 weeks, and the mother is able to spend periods away from the nest.
> - Delays in weaning will delay behavioural development, especially play.
> - A pattern of *ad lib* feeding should be established early in life. Meal feeding should be avoided.
> - Activity feeding can be introduced at an early age, but not as the only source of food.

Play Behaviour

Ask any kitten owner what they find most attractive about their new pet and they will most likely focus on its playfulness, but play can be a source of problems, most often destructiveness and play-related biting and scratching. This is a period during the cat's life when client education can create an appreciation of the true nature of cats.

In general, any behaviour that consumes time and energy must be of benefit to the animal. Play behaviour is no different, and yet the function of play behaviour is still not fully understood. The potential benefits of play include motor training, cognitive training and socialization, and in cats the metabolic costs of play are actually very low (Bradshaw, 1992).

The potential benefits of motor training are easily understood; the animal is able safely to rehearse sequences of behaviour that may be essential to locomotion, predation and self-defence. Cognitive training and increased physical contact between kittens are of equal importance. Play can be divided into three types: (i) social play; (ii) object play; and (iii) locomotor play.

Object and social play are composed of a set of specific manoevres, and there is significant overlap (Fig. 9.3). Many of these behaviours (such as scoop, toss, grasp, mouth and bite) are self-explanatory. 'Poke' and 'bat' are forepaw prods that make contact with an object from a vertical or horizontal direction (respectively). In a 'vertical stance' the kitten rears up on its hind legs with its forelegs extended. A 'side-step' is when a kitten arches its back, curls its tail upwards and walks sideways. In a 'face-off' the kitten hunches forward and raises a paw toward another kitten, often with its tail thrashing. The 'stand-up' behaviour involves the kitten standing over an object or another kitten, lunging with its mouth open or with a raised paw.

Social play

Social play begins at around one month of age and continues until weeks 12–14, when it

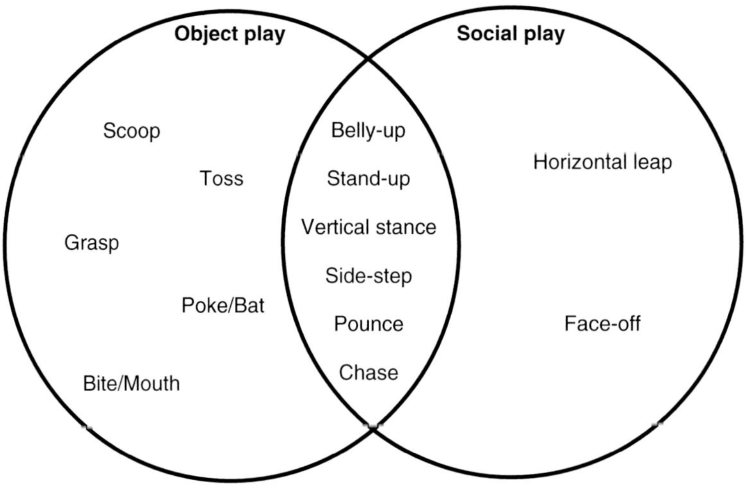

Fig. 9.3. Venn diagram of play behaviour. (Author's own figure.)

gradually reduces. A bout of social play most often starts with a pounce, belly-up or stand-up behaviour. This will usually provoke a response from the other kitten. These behaviours are a consistent feature of play throughout the first 6–12 weeks of life. As the kitten develops, other play behaviours develop. For example, vertical-stance behaviour increases, and side-step behaviour decreases, between weeks 6 and 12. This may be because it is less successful (75%) in getting a play response. A side-step usually provokes an identical side-step from the opponent, but this progresses to a play interaction on only 75% of occasions, compared with 90% for pounce, belly-up or stand-up.

During play, role reversal is an important part of preventing progression into aggression. Behaviours such as belly-up provoke a complementary behaviour, such as stand-up. Kittens will switch roles throughout play, each taking the role of protagonist until play terminates with a chase or vertical leap behaviour.

Kittens also use a 'play face' to communicate the lack of aggressive intent during play; this is seen as a half-open mouth. The play face is usually signalled at the beginning of play and during a belly-up. During play, kittens are careful not to escalate the intensity of play, and the play behaviours themselves are quite ritualized.

Play leads to conflict increasingly often as kittens get older (Voith, 1980) and as males incorporate sexual behaviour into play. From the 19th week, sexual behaviour of male cats begins to develop, including mounting, neck grasps and ano-genital sniffing. At this time, male social play changes to include this behaviour, which elicits aggression from females.

Single-born kittens do not engage in any more object play than litter-born kittens, indicating that these types of play have different motivations and cannot substitute for each other. However, they will direct social play at their mother, as she is the only suitable social partner.

In a domestic setting, kittens will seek out opportunities for social play with adult cats and dogs, if other kittens are not available. This can be very stressful for elderly or debilitated adult cats, which should be a consideration when getting a kitten. One option is to home two kittens as playmates, in order to reduce pressure on adult animals. Due to the differences in play behaviour between male and female kittens, same-sex pairs of kittens are often the best choice.

Object play

Object play begins slightly later than social play and reaches a peak at about 8 weeks of age (Barrett and Bateson, 1978). Kittens will investigate any novel object using all senses and from a number of angles before play starts. This is followed by a pounce or leap and then the kitten will poke, bat, grasp and toss the object. About a third of play begins with a poke or a bat, as this causes the object to move. These behaviours are then commonplace during play, making up about a third of all behaviour during object play. In adult cats, a very similar pattern of play is seen during the post-hunting manipulation of prey.

Object play is often focused on lightweight objects that are easily moved by a poke or bat. This will elicit a sequence of further pokes and bats, tossing and scooping to maintain the movement of the toy.

There are visible differences in object play between male and female kittens as early as 7 weeks of age: males spend twice as much time in object play as females, even though females are more generally active. In mixed litters, the play of females is 'masculinized' compared with single-sex female litters: they engage in more object play.

Locomotor play

Locomotor play begins at around 7–8 weeks and involves rehearsal of general locomotor activity, such as climbing and balancing (Martin and Bateson, 1985). It is an essential part of learning to use the 3-dimensional environment, and is strongly influenced by modelling of the mother's behaviour. Locomotor play is self-rewarding and kittens that have other opportunities for social and object play will preferentially engage in locomotor play when suitable opportunities arise. Kittens climb higher and become increasingly adventurous after 7 weeks of age, as they develop greater skill.

The implications for kitten owners are two-fold: (i) that opportunities for locomotor play are essential and owners should provide cat furniture that allows kittens to rehearse a complete range of climbing and balancing; and (ii) that owners should expect kittens to attempt to climb anything they encounter in a domestic environment. It is pointless to attempt to discourage this, so the home needs to be made safe.

Predatory behaviour and play

Some cat owners are keen to prevent their pets from developing predatory behaviour. This may extend to limiting certain kinds of object play and to attempts to socialize kittens with rodents and other small mammals in order that they will live alongside and play with these species without harming them.

To become an effective predator, a cat must learn not only the skills needed to catch and handle prey, but also how to identify suitable prey and apply a suitable strategy for catching it. Some of these skills are learned during object play and through the presentation of immobilized prey by the mother. More complex skills, such as prey identification, are developed through observation. Owners may assume that unless a young cat has had all of the necessary experience during kitten-hood, then it will not become an effective hunter. This is not the case; young cats can learn through a very small number of observations of adult cats. For example, kittens that do not have opportunities for object play go on to meet the same level of predatory ability as conventionally reared kittens at 6 months of age (Caro, 1980). Cats appear driven to meet certain milestones, particularly predatory behaviour, regardless of precise pattern of early experience. This makes sense, as predatory behaviour is so essential for survival.

In experiments in which kittens were reared with rats in the same cage (as if they were littermates), many kittens became tolerant of rats, but some still went on to catch and kill them (Kuo, 1938).

Problems with play

While humans can provide social interaction for kittens, our size and body shape have such little correspondence with those of the cat that it is impossible for us to engage in social play with a kitten.

However, we want to become involved in play and the temptation is to use hands and feet to lure kittens into interacting with us. Rather than providing social play, this makes us targets for locomotor and object play.

Whereas play with inanimate objects is limited in intensity by the lack of response that they provide, the movement of human feet and hands provide considerably more stimulation to a kitten, leading to more intense biting and scratching.

Since biting behaviour is under local reflex control, once a cat is holding something in its mouth any struggling or movement will trigger a more powerful bite. It is easy for the cat to cause unintentional injuries during play. Cat bite injuries more often lead to complications, including abscesses and septicaemia because cats harbour pathogenic bacteria in their mouths. For these reasons, owners should never play games that involve directing object or locomotor play to themselves.

Key points

- Locomotor, object and social play are all valuable to kittens.
- Locomotor play is encouraged by the kitten's mother, so opportunities for locomotor play are essential in the rearing environment.
- Locomotor play is stimulated by the presence of opportunities to climb and balance.
- Object play is stimulated by novelty, so kittens should be provided with an ever-changing array of small, lightweight objects that encourage object play.
- In single-kitten households, social play is likely to be directed towards other pets, including adult cats and dogs. This can be stressful for adult animals, especially if they are elderly or unhealthy.
- If kittens are homed in pairs, it is better to take two kittens of the same sex rather than a male and a female. This avoids conflict due to sex differences in play behaviour.
- Owners should never play games that encourage kittens to target play to hands or feet.

Stages in Development

For a summary of key stages in development, see Fig. 9.4. Between days 9 and 14 the kitten's eyes and ears are open and beginning to become

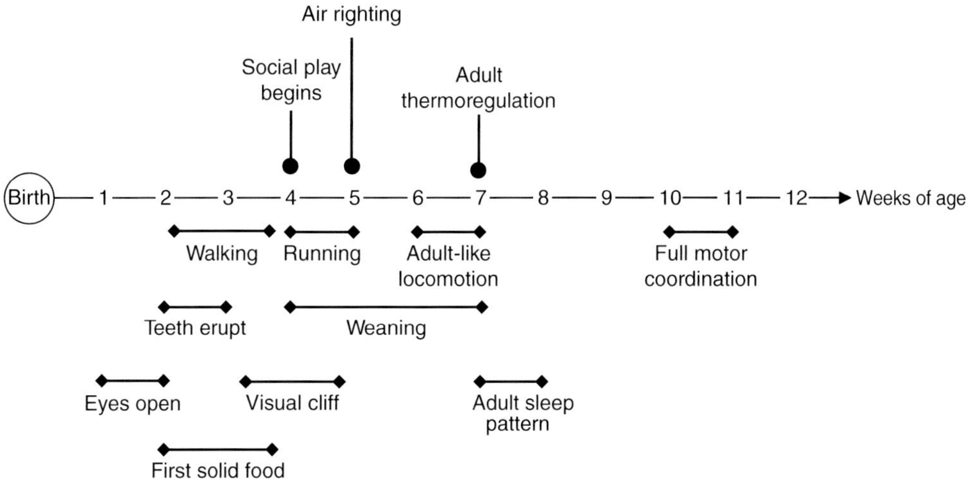

Fig. 9.4. Developmental timeline. (Author's own figure.)

functional, and the kitten becomes able to walk. Even at this point the kitten's vision is not fully functional, as it is not until weeks 12–14 that vision is properly developed (Ikeda, 1979). For example, kittens are not able to reliably detect the edge of a table until around the 5th week of age (so-called visual cliff). The famous ability of cats to right themselves as they fall (air-righting) does not begin to develop until day 30 and is only complete after a further 2 weeks. Kittens cannot balance well enough to walk along and turn around on a narrow plank until week 11 (Villablanca and Olmstead, 1979).

Key points

- Kittens should be allowed to develop at their own rate; it is dangerous for them to be put in situations that are beyond their developmental stage.
- For example, they should not be put on elevated surfaces until they have the ability to see the edge of a table and air-right.

Kitten Selection

There are currently no validated tests for assessing kitten temperament that have any predictive value for adult temperament or behaviour. Prospective kitten owners are therefore left to make a choice based on: (i) breed; (ii) personality and condition of the kitten's parents; (iii) the rearing conditions of the kitten; and (iv) the kitten's stage of development at the time of homing.

Unfortunately, evidence supports the fact that appearance is a key characteristic of cat and kitten selection. Colour, size and weight are three of the most important features influencing the choice of cat, with appearance accounting for three-quarters of the variance in liking the cat (Turner and Bateson, 1986).

Breed

There is some evidence of variation in behaviour and temperament between breeds of cat, both from studies of the relative incidence of problem behaviour and studies of development. Anecdotally, indoor spray-marking and compulsive behaviours (such as wool-sucking) appear to be more common in oriental breeds, such as the Siamese and Burmese. Feline hyperaesthesia also appears commonest in these breeds, but it is also found in lines of other breeds.

In one study involving 736 referral cases (Bamberger and Houpt, 2006), Siamese cats were more commonly evaluated for aggression and ingestive behaviour problems and Persian cats for elimination problems (correcting for the relative population sizes of these breeds). Domestic shorthair cats were seen less often than expected overall, and specifically for aggression and ingestive problems. Male cats were generally over-represented.

Siamese cats also have a developmental defect in their visual system, which prevents the development of normal binocular vision. As a result, cats of this breed often have pronounced esotropia (convergent squint) and a defect of depth perception. Although this cannot be corrected, most cats learn to compensate.

Compared with oriental breeds, Norwegian forest cat kittens showed differences in patterns of activity and exploratory behaviour (Marchei *et al.*, 2009). Takeuchi and Mori (2009) asked veterinary surgeons to rank 12 cat breeds for each of 11 characteristics, including friendliness, activity, novelty seeking, playfulness, aggression to cats, aggression to humans, nervousness and inappropriate elimination. Sex differences were also modelled, with males being identified as ranking significantly higher for aggression to other cats, general activity, novelty seeking and excitability compared with females. Females ranked significantly higher for nervousness and inappropriate elimination. Two personality factors were identified: (i) aggressiveness sensitivity; and (ii) vivaciousness. Aggressiveness sensitivity was associated with aggression to people and cats, nervousness, timidity and excitability. Vivaciousness involved playfulness, novelty seeking, general activity and affection demand. Four breed groups were identified:

- High aggressiveness sensitivity, low vivaciousness: Abyssinian, Russian Blue, Somali, Siamese, Chinchilla.
- Low aggressiveness sensitivity, high vivaciousness: American short-haired cat, Japanese domestic cat.
- Mild aggressiveness sensitivity, very low vivaciousness: Himalayan, Persian.
- Low aggressiveness sensitivity, low vivaciousness: Maine Coon, Ragdoll, Scottish fold.

The most desirable combination of traits would appear to be low aggressiveness sensitivity and high vivaciousness. It is interesting that this combination was associated with the American short-haired cat and Japanese domestic cat (non-pedigree cat). Although the American short-haired cat is classed as a pedigree breed by the American cat fancy, its breed description includes a very wide range of coat colours and appearances, which suggests a lack of genetic bottlenecks due to selective breeding. These cats are therefore the closest equivalents of the non-pedigree domestic short-haired cat that is most popular in Europe.

Another important issue relating to breed is the impact that it can have on the level of outdoor access offered to cats. Pedigree cats can be valuable and breeders may even coerce purchasers of kittens to agree not to allow kittens outside.

Personality and condition of the parents

The tom-cat is not involved in the rearing of kittens, but his genetic contribution is important. With pedigree cats, it is possible to get information about the tom and previous litters he has sired. In many cases, particularly with domestic short-haired cats, the sire of a litter is likely to be an immature male that has not yet been neutered, or a local stray tom-cat. In the former, adult personality has not yet been established so it cannot be used as a guide. Unless they live in a rural area where they do not pose a nuisance and are unlikely to be trapped and neutered, stray adult tom-cats are likely to have remained intact as a result of avoiding human contact. This can be an indication of personality traits that are not altogether desirable in kittens sired by such a male.

Kittens model much of their behaviour on that of their mother, so her influence is particularly important. Prospective kitten owners should look for personality traits in the mother that most closely resemble those which are desirable in their new kitten. If the mother is nervous, avoidant, or defensive, then the kittens may develop the same traits either as a result of genetic factors or by the mother demonstrating behaviour that is then modelled by her kittens.

Since the health status of the mother, and in particular her nutritional status, has been shown to have a profound effect on kitten development, prospective owners should reject kittens of mothers that appear thin or unhealthy.

Rearing environment

The rearing environment is important in three main ways.

- It should provide the right conditions for development of the kitten so that it becomes independent from its mother by the time of homing.

- It must provide a suitable range of experiences that enables the kitten to become habituated to a wide range of stimuli that it will encounter in adult life, including contact with humans and with other domestic species (such as dogs).
- It should be as similar as possible to a typical domestic environment, so that transfer to the new home is not challenging for the kitten.

It is sensible for prospective owners to check the rearing environment before making a commitment to purchase a kitten. The ideal rearing environment is a domestic home in which kittens are exposed to a normal range of stimuli, activities and people.

Requirements for normal development include:

- Regular, gentle handling by people.
- Regular access to solid food from 3–4 weeks of age (preferably *ad lib*).
- Opportunities for the mother to come and go freely from the nest.
- Plenty of small toys to encourage object play.
- Apparatus for kittens to climb and balance on (i.e. cat furniture).
- Access to a suitable litter tray.

Kittens should encounter a wide range of different kinds of people, including children. However, children should not be allowed to pick up kittens; this will prevent them from handling kittens roughly or placing them in situations that risk injury (such as on top of furniture that a kitten could not climb itself) (see Fig. 9.5).

Stage of development at the time of homing

Kittens should reach independence from their mothers at around 8 weeks of age and this should be considered the minimum age for homing. Allowance should also be made for individual variation. Homing a kitten before it is ready risks problems with adapting to the new home.

In general, kittens should not be homed unless they have reached the following developmental milestones.

- Object, social and locomotor play: kittens should be old enough to play games with each other and with objects and be beginning to climb.
- Eating of solid food: kittens should not be reliant on their mother for food.

Fig. 9.5. It is important for children to interact with kittens, but not to pick them up. Being picked up can alarm a kitten, and if the kitten is dropped it may be injured. (Image: iStock.)

- Voluntary urination and defecation: if kittens still rely on their mother to stimulate elimination, then they are too young to be homed.
- Confidence in exploring the environment away from the mother.
- Lack of fear when interacting with unfamiliar people (particularly the new owner's family).

Unless these conditions are met, it is possible that the kitten is not ready to be homed, or it is not of the age being claimed for it. This should be seen as an indicator that the whole litter needs to be kept together for longer.

Key points

- There is no obvious benefit to choosing an exotic pedigree kitten, as domestic short-haired cats have a good balance of temperament and are also likely to be reared in the kind of non-specialized conditions that most closely resemble those of a domestic home.
- The minimum age for homing is 8 weeks, but the decision to home kittens should be guided by their level of development.
- The rearing environment should be seen as a reflection of the effort the breeder has made to producing suitable kittens.
- The breeder has a great responsibility to handle, socialize and support the development of kittens, especially if they are homed late.

Neutering

Neutering has a number of benefits. In males it reduces their territorial size to around that of a female's and prevents roaming. Intact males are highly aggressive to outsiders and are therefore a threat to other cats in an area. Neutering appears to reduce this. The pungency and frequency of male urine marking is also reduced by neutering.

The effect varies with the timing of neutering. If performed during the first year of life, spray marking and inter-cat aggression are significantly reduced in male cats.

Apart from the prevention of sexual behaviour, including calling and sexual marking, the effect of neutering on female cats is limited. There is no evidence to support the idea that allowing a female cat to have a litter of kittens is beneficial to her either physically or psychologically.

In the UK, kittens are usually neutered from 4 months of age onwards, because at this time the reproductive organs are well developed enough to be identified and removed easily.

Preparing the Home for a New Kitten

Kittens alternate between short periods of activity and sleep throughout the day. At 8 weeks they will be at the peak of their need for object and locomotor play. This means that the owner will need to make preparations for the kitten to play and climb. This can include providing specialized cat furniture that is easy for the kitten to climb, as well as limiting access to other less desirable climbing opportunities (for example, knotting long curtains so that they do not reach the floor for the kitten to climb, and clearing shelves of easily damaged mementos).

Expensive toys are generally unnecessary, as kittens are highly attracted to small, lightweight toys that move easily when poked or batted, such as table-tennis balls, and feathers. Kittens, like adult cats, become rapidly habituated to toys and so it is important to provide a constantly changing selection of play items. Larger objects may be unsuitable for play in young kittens but provide a setting for interactive play. Owners should also use fishing toys, and similar items, to play games with the kitten without becoming a target themselves.

Kittens do not engage in as much oral exploratory behaviour as puppies but they will chew novel items, including the leaves and flowers of plants. Some common plants have been found to be toxic to cats, including azaleas, daffodils, rubber plants, weeping and variegated figs, poinsettia and some lilies. These are best removed from the environment before the kitten arrives.

It is best to prepare the home's living room as the main space accessible to the kitten, so that it is exposed to the maximum amount of human contact when people are around. That room should have at least one latrine location in the form of a litter tray, with a litter material that is familiar to the kitten.

Letting Cats Outside

Giving a cat outdoor access enables it to perform a wider range of normal behaviours, including exploring and establishing a territory, social interaction with other cats and hunting. Allowing the cat to have outdoor access enables it to follow a normal circadian pattern of activity.

The main risks of outdoor access are disease, road traffic accidents (RTAs) and injury by other cats. With modern vaccination and antibiotics, cats can be protected against all of the common infectious diseases except feline immunodeficiency virus (FIV).

Road traffic accidents are a common source of injury in cats. The risk of an RTA is 1.9 times greater in male than female cats and varies with age. Cats of 7 months to 2 years of age are three times more likely to be involved in an RTA than cats younger than 6 months of age (Rochlitz, 2003a). The risk of an RTA generally decreases by 16% per year of age. RTAs commonly occur close to a cat's home, are associated with living in an area with a high density of traffic and occur more often at night (Rochlitz, 2003b).

Other factors, such as the density of the cat population in an area and the availability of resources in the cat's own garden, would be expected to have an effect on rates of RTAs, but these have not been studied.

Injuries by other cats are also commonplace and are the main route by which infectious diseases such as feline leukaemia virus (FeLV) and FIV are transmitted. In a retrospective study by the author (see Chapter 3), injury rates correlated strongly with how frequently the owner reported seeing non-resident cats in the garden, as well as the number of different individual cats seen visiting. There was no association between injury rate and the features of the garden, or the type of home location, even though certain garden features were identified as attracting cats to the gardens of non-cat owners. The indication is that the rate of injuries is most likely related to overpopulation.

In that study, a subpopulation of about 16% of households experienced particularly serious problems with neighbourhood cats terrorizing their cats. Householders from this group regularly saw three or more individual non-resident cats in their garden and these cats were seen more than twice each week, compared with two different cats seen once a week or less for the less severely affected cats.

This raises the issue that prospective kitten owners should consider whether the cat population in their area is already too high to cope with the addition of another cat. This is especially true if the owner already experiences problems with cat-related injuries to their existing cats; there is no evidence that households with multiple cats are any less susceptible to home invasion and nuisance from non-resident cats.

The age at which it is safe for cats to go outside depends on the disease risk in an area and the developmental stage of the kitten. Most vaccination courses are complete by 14 weeks of age and kittens can then be chaperoned for daily outdoor access until it is clear that they are safe to be left outside alone (Fig. 9.6).

Another aspect of outdoor access is the potential for neighbourhood cats to enter the home. In a study by the author (see Chapter 3), nearly 19% of 734 cat owners whose cats had outdoor access had experienced a problem of neighbourhood cats entering the home to fight with resident cats and nearly 26% reported a problem of neighbourhood cats entering the home to steal food. For homes where cats had outdoor access using an insecure cat door, the percentages were 25% and 39%, respectively. Given that only 45% of cat owners in the study provided *ad lib* food for their cats, it seems likely that the primary motivation for home invasion is to gain access to food.

Once a cat has found an opportunity to obtain food from a home, it is likely to be persistent. The best way to avoid this is to start with a secure cat door that blocks any non-resident cat from entering the home and to keep food far enough away from the cat door that it is undetectable from outside. For cat owners who are concerned about a cat wearing a collar, secure identity chip-activated doors are now available.

Key points

- Before getting a kitten, consider whether the local area is already over-populated with cats.
- Provide an enriched garden environment for cats in order to reduce the reason for them to travel farther afield.
- Consider restricting outdoor access at night for young male cats that live in very traffic-dense areas.
- Kittens should not be allowed outside until they are vaccinated and then they should be supervised outdoors until they have a full pattern of adult locomotion.
- Use a secure cat door to prevent other cats from gaining access to the home.

Fig. 9.6. Kittens can begin to enjoy supervised outdoor access once vaccination is completed at around 14 weeks of age. Free access should only be permitted once a kitten shows complete confidence in the outdoor environment. (Image: iStock.)

References

Bamberger, M. and Houpt, K.A., (2006) Signalment factors, comorbidity, and trends in behaviour diagnoses in cats: 736 cases (1991–2001). *Journal of the American Veterinary Medical Association* 229, 1602–1606.

Barrett, P. and Bateson, P. (1978) The development of play in cats. *Behaviour* 66, 106–120.

Bradshaw, J.W.S. (1992) *The Behaviour of the Domestic Cat*. CAB International, Wallingford, UK.

Caro, T.M. (1980) Effects of the mother, object play and adult experience on predation in cats. *Behavioural and Neural Biology* 29, 29–51.

Casey, R. and Bradshaw, J.W.S. (2008) The effects of additional socialisation for kittens in a rescue centre on their behaviour and suitability as a pet. *Applied Animal Behaviour Science* 114, 196–205.

Collard, R.R. (1967) Fear of strangers and play behavior in kittens with varied social experience. *Child Development* 38, 877–891.

Esposito, S., Picciolli, I., Semino, M. and Principi, N. (2013) Dog and cat bite-associated infections in children. *European Journal of Clinical Microbiology and Infectious Diseases* 32(8), 971–976.

Fatjo, J., Ruiz-de-la-Torre, J.L. and Manteca, X. (2006) The epidemiology of behavioural problems in dogs and cats: a survey of veterinary practitioners. *Animal Welfare* 15, 179–185.

Ikeda, H. (1979) Physiological basis of visual acuity and its development in kittens. *Child Care and Health Development* 5, 375–383.

Karsh, E.B. and Turner, D.C. (1998) The human–cat relationship. In: Turner, D.C. and Bateson, P. (eds) *The Domestic Cat: The Biology of its Behaviour*. Cambridge University Press, Cambridge, pp. 159–177.

Kizer, K.W. (1979) Epidemiologic and clinical aspects of animal bite injuries. *Journal of the American College of Emergency Physicians (JACEP)* 8(4), 134–141.

Kuo, Z.Y. (1938) Further study of the behavior of the cat toward the rat. *Journal of Comparative Psychology* 25, 1–8.

Marchei, P., Diverio, S., Falocci, N., Fatjó, J., Ruiz-de-la-Torre, J.L. and Manteca, X. (2009) Breed differences in behavioural development in kittens. *Physiology & Behavior* 96, 522–531.

Martin, P. (1986) An experimental study of weaning in the domestic cat. *Behaviour* 99, 221–244.

Martin, P. and Bateson, P. (1985) The ontogeny of locomotor play in the domestic cat. *Animal Behaviour* 33, 502–510.

Meier, G.W. (1961) Infantile handling and development in Siamese kittens. *Journal of Comparative Physiology and Psychology* 54, 284–286.

Rochlitz, I. (2003a) Study of factors that may predispose domestic cats to road traffic accidents: part 1. *Veterinary Record* 153(18), 549–553.
Rochlitz, I. (2003b) Study of factors that may predispose domestic cats to road traffic accidents: part 2. *Veterinary Record* 153(19), 585–588.
Seksel, K. (2001) *Training Your Cat*. Hyland House, Melbourne, Australia.
Simonson, M. (1979) Effects of maternal malnourishment, development and behavior in successive generations in the rat and cat. In: Levitsky, D.A. (ed.) *Malnutrition, Environment and Behavior*. Cornell University Press, Ithaca, New York, pp. 133–160.
Smith, B.A. and Jansen, G.R. (1977) Maternal undernutrition in the feline: brain composition of offspring. *Nutrition Reports International* 16, 497–512.
Takeuchi, Y. and Mori, Y. (2009) Behavioral profiles of feline breeds in Japan. *Journal of Veterinary Medical Science* 71(8), 1053–1057.
Tan, P.L. and Counsilman, J.J. (1985) The influence of weaning on prey-catching behaviour in kittens. *Zeitschrift für Tierpsychologie* 70, 148–164.
Turner, D.C. and Bateson, P. (1986) *The Domestic Cat: the Biology of its Behaviour*. Cambridge University Press, Cambridge, UK.
Villablanca, J.R. and Olmstead, C.E. (1979) Neurological development in kittens. *Developmental Psychobiology* 12, 101–127.
Voith, V.L. (1980) Play behavior interpreted as aggression or hyperactivity: case histories. *Modern Veterinary Practice* 61, 707–709.
Wilson, M., Warren, J.M. and Abbott, L. (1965) Infantile stimulation, activity and learning by cats. *Child Development* 36, 843–853.
Wyrwicka, W. (1978) Imitation of mother's inappropriate food preference in weanling kittens. *Pavlovian Journal of Biological Science* 13, 55–72.

10 What Every Rabbit and Rodent Owner Needs to Know

Emma Magnus and Anne McBride

As discussed earlier in this book, UK keepers are legally responsible for the welfare of animals in their care throughout all life stages. Legislation aside, such provision is ethical and underpins the term 'responsible ownership'. Veterinary surgeons, breeders, retailers, rehomers, suppliers and other professionals have a duty to provide current, appropriate species-specific advice and must meet these standards themselves.

Physical and psychological health are inextricably linked and fundamental to good welfare and well-being (Yeates, 2018). Potentially undermining factors include: inherited and acquired disease or injury; inherited nervousness; and acquired anxieties and fears. Meeting an animal's psychological needs encompasses all Five Freedoms (FAWC, 2009): provision of a suitable diet, appropriate physical and social environment in which to perform normal behaviour, and the taking of all possible steps to ensure freedom from fear and mental distress.

The Five Freedoms are a useful start, but good welfare requires that overall positive emotional states occur substantially more than negative ones (Mellor, 2016). Indicators of positive emotions include behaviours shown when relaxed and comfortable, be that when resting or when the animal chooses to explore, play or be otherwise physically and mentally active. How we and other animals feel normally fluctuates between negative and positive in response to changes in the immediate physical and social environment. How an individual animal responds is determined by its species, breed, individual genetics, mental and physical health, lifelong experiences and learning, as well as the circumstances at the time (McBride and Hinde, 2022).

The normal healthy rabbit and rodent have been introduced in Chapter 4. This group comprises many species, most considered as 'captive wild' (Defra, 2012; Yeates, 2018), and readers were advised to further research species-specific ethology. Such knowledge is essential in preventing behaviour problems. Many problems arise because inadequate husbandry and management results in animals being fearful or frustrated. The current chapter provides general pointers on applying species knowledge to promote behaviour welfare. The first is suitability. Just as for dogs, cats and horses, the choice of small animal species and breed must be appropriate for the owner's lifestyle and circumstances.

Since 2011, the UK charity People's Dispensary for Sick Animals (PDSA) has conducted an annual survey of a nationally representative sample of UK pet owners, including rabbit owners. The data (PDSA, 2015, 2018) support those of Rooney *et al.* (2014). In 2015 the PDSA surveyed 1200 rabbit owners and 1677 owners of other small species and, in 2018, 483 rabbit owners. Comparing those two years, it is clear that there had been some improvements, though overall the Five Freedoms were still not being

 Companion Animal Behaviour Problems: Prevention and Management of Behaviour Problems in Veterinary Practice (eds. R. Casey, S. Heath and H. Zulch)

DOI: 10.1079/9781800621312.0010

met. In both years over 30% of rabbits were fed inappropriate (muesli) concentrates or insufficient hay, over 50% had inadequate exercise opportunities and 30% were housed in indoor cages or outdoor hutches that were too small. Though a social species, over 50% lacked suitable companionship, as did 25% of guinea-pigs and 53% of chinchillas (PDSA, 2015).

The PDSA (2020) surveyed 629 rabbit owners and the findings indicated improvements in rabbit care, with only 18% given muesli foods, yet 42% lived alone and 26% in inadequate housing. The Pet Food Manufacturers' Association (PFMA, 2020) estimated the UK pet rabbit population as one million. Extrapolating the PDSA 2018 and 2020 data means the welfare of hundreds of thousands of rabbits and other animals is needlessly compromised. This is particularly disappointing given the availability of good information about rabbits, including a well-publicized annual UK Rabbit Awareness Week (available at: https://www.rabbitawarenessweek.co.uk/, accessed 1 May 2022), and the first UK Guinea-pig Awareness Week (Westgate, 2021).

The situation for other species is worse, even for those that are suitable as pets (Ashley *et al.*, 2014; McBride, 2017a; Toland *et al.*, 2020). Many species are not suitable, as they have very specialized needs that are rarely met by owners or by those in the supply chain. Apart from human health considerations, zoonoses and injury, serious compromises to animal welfare are common, as evidenced by the number and variety of species abandoned or advertised online as unwanted (Ashley *et al.*, 2014: Anonymous, 2018).

Choosing Your Pet

Veterinary professionals, breeders, retailers and rehomers need to be proactive in advising and explaining why some species should not be considered as pets (Schuppli and Fraser, 2000). Appropriate information about how to meet the needs of individual species, directly or by referral to true experts such as exotic veterinary specialists and zoological collections, should be given. Ideally, this should be pre-purchase (PDSA, 2018). Pre-purchase advice must address the three common misconceptions about this animal group, namely: (i) they only live a short time; (ii) they are cheap to keep; and (iii) they are easy to keep, and thus make good pets for children.

Lifespans

The belief that small animals have short lifespans is one reason for their popularity as a children's pet, as children are anticipated to 'grow out of' or get bored with them. However, species lifespans differ considerably, ranging from a couple of years to over a decade (see Table 4.1 in Chapter 4). Most, if not the majority, probably die young due to preventable health and behaviour factors. UK veterinary professionals estimate 30% of rabbits seen in practice are overweight or obese. This is probably a significant underestimate, as 34% of rabbits were not registered with a vet. Of those registered, 58% had not had booster vaccinations, so had missed health checks (PDSA, 2018).

Cost

The monthly cost of keeping rabbits was underestimated by 77% of owners and 92% underestimated the average lifetime cost (PDSA, 2018). The RSPCA (2012) estimated the annual costs of providing for two healthy rabbits to be £1550, with a total of £15,500 over a 10-year lifespan. This can rise considerably where breed conformation predisposes to medical issues, including tooth malocclusion in dwarf rabbits and ear problems in lop rabbits (UFAW n.d.; CAWC, 2006) (Fig. 10.1).

- Dental: £60–200. For a rabbit with chronic dental problems the frequency of dentals could be every 4–8 weeks.
- Ear problems: common in lop breeds. Likely to cost £400–500 initially, but may require ongoing treatment and this cost would not include advanced imaging, such as CT scans, which are recommended. Costs could increase to £1000 if surgery and diagnostics such as a CT scan are required (RSPCA, 2012, p. 6).

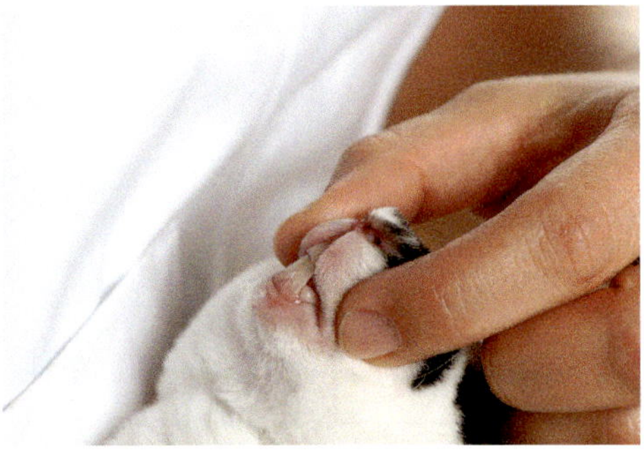

Fig. 10.1. Regular veterinary care is essential for all pets, including dental check-ups. (Photo: iStock.)

There is little doubt that costs of keeping other species are also underestimated.

Easy to keep and good pets for children

The notion of being 'easy' to keep implies that little needs to be done except feeding daily and regular cleaning. This is unlikely to be sufficient for good welfare.

Where pets are intended for children, the parents or guardians are legally responsible for ensuring that the animals' needs are met. As children develop, their interests and priorities change. Whilst pet ownership is to be applauded for the opportunities it can provide to teach children about empathy and responsibility, this will only happen if the adults involved are prepared to show by example. Regrettably, many children given small pet species simply learn that it is acceptable to ignore, neglect or discard (rehome) living beings once the initial attraction has worn off.

Copping (2009) estimated that 35,000 rabbits were taken to UK rehoming charities annually, most within three months of purchase. In 2012 this was reported to be 87,000, and reasons for relinquishment included housing problems, and that children were no longer interested (Ellis *et al.*, 2017).

One drawback of the internet is that it facilitates both impulse purchasing and easy advertising and disposal of animals no longer wanted. Rarely are any checks required to see if the new home will be suitable (Toland *et al.*, 2020; Blue Cross and Born Free Foundation, 2016) and, for this group, an internet sale may simply be a death sentence as they become food for the purchaser's other pets (Warwick, 2014).

Neville *et al.* (2018) reported their findings from a single UK rehoming online site, Preloved. In 2016, 7315 individual adult (over 16 weeks) rabbits were advertised. Most were under 2 years old, the average being 14 months. Given the expected lifespan of 10 years, these are young animals. The main reasons given were not enough time to care for it (21%, over 1500 rabbits) and the child had lost interest in it (10%, over 700 rabbits).

Clearly, realistic information about the animals' needs and associated time and money costs must be given to prospective owners (Warwick *et al.*, 2018). Examples of useful books and online resources are the Pet Detective series for children (Milne, 2015) and for adults and professionals the 'Easy, Moderate, Difficult or Extreme – EMODE' suitability as a pet assessment system (Warwick *et al.*, 2013) and available from the Emergent Disease website (Emergent Disease, 2013).

Breeders, retailers and rehomers should deter impulse buying and ensure that people can make informed decisions. Reserving animals for a minimum of 48 hours gives potential buyers time to consider the information provided and research the prospect more thoroughly.

Choosing the Right Animal(s): Health, Socialization and Company

Health

Animals should be in good physical health when obtained and have a veterinary health check-up within a day or two. If health problems are found, then veterinarians should inform the place from where the animal was acquired. Health issues may be an unfortunate occurrence for one individual animal or be indicative of a wider problem in the population held by the breeder, retailer or rehomer or further back in the supply chain.

Sadly, 'puppy farm' concerns are common in the trade of all small mammals and other exotic species. Issues include breeding from inappropriate or unhealthy parent stock (UFAW, n.d.), barren cages, poor handling and inappropriate transport/supply chains (Ashley *et al.*, 2014; Blue Cross and Born Free Foundation, 2016). These can lead to chronic, usually lifelong, damage to the psychological and physical health of the individual (Franklin *et al.*, 2010). Additionally, obtaining 'pets' from the wild is detrimental to wild populations. Potential owners should carefully investigate the source of their proposed pet to ensure that they are sourcing it from an appropriate breeding establishment.

Socialization and experiences during development

To maximize well-being, animals should be generally confident and not fearful of the presence of people. It is the responsibility of breeders to breed from stock that is physically and behaviourally sound. It is also their responsibility to ensure that both parents and offspring are well socialized to people and have had a range of positive experiences within their environment. Responsibility for continuing socialization and positive experiences extends to everyone in the supply chain and the eventual owner.

Experiences during sensitive periods of development, prenatally to adulthood, have profound and lasting effects on an animal's behaviour (Chapillon *et al.*, 2002). They define the animal's 'known/normal' world. In the wild this will be the family unit, the surrounding habitat and environmental stimuli. Experiences encountered for the first time later in life will be classed as 'unknown/not normal' and the animal will be anxious and likely fearful of them. This process is necessary for wild animals to learn what to like and what to fear; for rabbits and rodents this includes learning to fear predators but accept members of their own group. Animals lacking sufficient, appropriate experiences when young are likely to be anxious and fearful when adult, with associated behaviour problems and knock-on effects for physical health and lifespan (Cavigelli and McClintock, 2003).

Conversely, appropriate experiences during development leads to adults with better health (Laviola *et al.*, 2008), confidence and fewer behaviour problems (e.g. Cowley and Widdowson, 1965; Núñez *et al.*, 1997; Cloutier *et al.*, 2012) (Figs 10.2 and 10.3). They have greater resilience to cope with life's stressful events, including the presence of threats (Roy *et al.*, 2003) and recovery from surgery (Roughan *et al.*, 2015). Because these causal interactions are known, mice and rats are used extensively in human medical research investigating the effects of early experience on later psychological disorders (e.g. Franklin *et al.*, 2010; Vetulani, 2013).

Breeders, owners and all others in the supply chain should ensure that early experiences reflect the types of stimuli animals will encounter in later life, such as handling, other animals, noises, scents and objects. The presence of the (relaxed) mother and littermates will enable youngsters to feel more secure when new stimuli are introduced. Gentle exposure to noise can include low-volume recordings of household noises, dogs barking and a radio playing (both

Fig. 10.2. A compatible pair of dwarf hamsters. (Photo: iStock.)

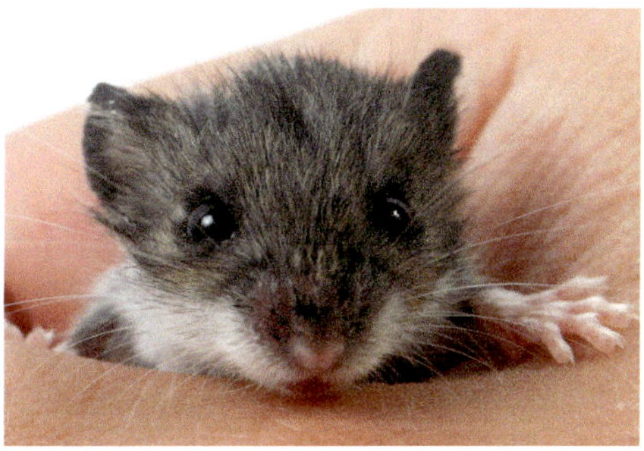

Fig. 10.3. Gentle handling when young increases tameness, confidence and health when adult. (Photo: iStock.)

speech and music stations). Placing novel objects in the home environment will introduce scent and texture, and trails of tasty bits of food leading to the object will help develop pleasant associations.

It may seem obvious, but for all species new stimuli should be introduced gradually, one or two at a time, rather than making very large changes to the environment. The process is intended to increase confidence, not overwhelm the animal.

Scent is the primary way in which rabbits and rodents identify individuals, including their own young. Objects that have been handled by different people help to increase socialization (Carter, 1972). Early gentle handling is also important for both altricial and precocial species. It is essential that before touching the young, hands are rubbed with some of the nest bedding and the mother is stroked so that the scent profile on all individuals is similar.

Rabbits handled between 10 and 20 days of age are more willing to approach unknown people at 49 days (7 weeks) old than unhandled rabbits. Seven weeks is a typical age to home rabbits (Der Weduwen and McBride, 1999). A similar effect was found by Pongrácz and Altbäcker (1999) with rabbits handled before their eyes opened. However, early handling of any altricial species needs to be done carefully so that the mother does not abandon or even cannibalize them. Thus, it is best done after the eyes open.

Company

Most small mammals are social. In the wild they have company of their own kind throughout their lives. Compatible company is just as important when kept by people. It means animals can perform natural social behaviours, owners can enjoy watching them and, if their human is busy, the animals will not be lonely (Fig. 10.2).

It is preferable that animals are acquired at the same time, ideally when young, so that they grow up together. Introductions during adulthood can be problematic and need to be done carefully (see Chapter 19).

Fig. 10.4. Holding an adult mouse. (Photo: iStock.)

One good way to avoid problems is to choose the individual animal wanted and then let it choose its compatible companion. Having chosen an individual, take time and watch it undisturbed in the group in which it is currently living. The animal(s) it chooses to be near when feeding, resting and which it grooms should be the one(s) taken as its lifetime companions. Good breeders and retailers should know their animals well enough to be able to help.

Alternatively, one may choose to get bonded adult animals in need of a home, perhaps through a rescue centre. Sometimes an individual that has not been socialized to its own species, or that has had bad experiences, may prefer human company. In these cases, the owner must be committed to spend substantial time with it every day throughout its whole life (Fig. 10.4).

As indicated in Chapter 4, less territorial species can live happily in same-sex groups, or harems. Where both sexes are territorial, as in rabbit and chinchilla, a neutered male and female is the best combination. Housing just males or just females together can cause aggression when they are adults (McBride, 2000). Many individuals suffer severe injuries due to incorrect pairings and many more are separated permanently.

Keeping neutered animals or same-sex groups, as appropriate, means that there are no unwanted offspring.

When animals are neutered, they will be separated for a period. To avoid permanent breakdown of a previously amicable relationship, advice on scent swapping and reintroduction should be followed (see Chapter 19) (McBride, 2009; Campbell, 2010).

Husbandry and Avoiding Behavioural Problems

Whilst a good start is integral, good psychological welfare is only maintained with suitable lifelong husbandry. Important aspects are diet, an environment that facilitates physical and mental activity, and appropriate interactions with humans and other animals.

Diet

All species have particular dietary requirements. For example, guinea pigs are unable to store vitamin C and must have this supplemented in the diet; and degu are intolerant of sugars and easily develop diabetes. The natural diet is the basis for understanding how to provide good nutrition in captivity. Owners should be given current scientific advice on appropriate feeding.

People become attached to their animals and wish to give them nice things. Often these 'treats' are food, though other treats include new toys or objects to explore. One problem with food is that we tend to rate it by how we humans would like to eat it, how it looks and how it tastes. Many would not consider a dried mealworm to be a tasty snack, although their pet hamster, gerbil, rat or mouse certainly would. Humans, as diurnal primates, are drawn to brightly coloured foods – fruit, fresh vegetables and salad. We and our pets share an attraction to sweet and rich foods, as found in young (salad) leaves, grasses and autumnal fruits, seeds and nuts. This is advantageous in the wild where sweetness is seasonal (spring and autumn). These are the times when wild animals need extra energy to rear young or lay down fat for the winter. However, in captivity, like us, our pets are more sedentary and many live indoors. Eating too many sweet and rich foods leads to potentially fatal or painful digestive, dental and other health problems. A further complication is portion size. People rarely account for their pets being so much smaller than themselves. Inappropriate diet and over-feeding are major factors causing poor health and reduced lifespan (Meredith and Delany, 2010; Prebble, 2014).

This group of species is broadly divided into two groups: herbivores (fibrevores) and omnivores. Herbivore diets look pretty boring to humans: dry brown hay, green grasses, vegetables and herbs, maybe some flowers and the occasional bit of fruit. An omnivore diet is more varied and similar to our own: herbs, fruit, seeds and animal protein sources, such as insects, eggs and meat. Whilst all eat plants, not all plants are safe to eat. Many house and outdoor plants can be deadly (Richardson, 2010; Way, 2016).

The natural diet of the herbivorous rabbit, guinea pig, degu and chinchilla is high in fibre and low in sugars. The animals are caecotrophic, catching soft, caecal pellets from their anus (see Chapter 4). Concentrated commercial food should only comprise 10% of a normal maintenance diet. They eat over the course of several hours and the rest of the diet should be *ad lib* provision of dried (hay) or fresh grasses and herbs (70–80%) and fresh green vegetables (20%) (McBride and Meredith, 2018).

Obesity, arthritis and misaligned/overgrown teeth all prevent the individual performing caecotrophy. This causes malnutrition and bottoms soiled by the uneaten caecal pellets. This can lead to fly strike (myiasis) and, within a few hours, a gruesome and painful death. The accumulated faeces attract flies who then lay their eggs. Within 24 hours, these hatch as maggots which literally eat the animal's bottom, causing open wounds and infection (Prebble, 2014; McBride and Meredith, 2018).

Inappropriate feeding also leads to behaviour problems. Aggression, chewing of non-food items (pica), fur-chewing (McBride *et al*, 2016) and repetitive (stereotypic) behaviours such as bar biting (Fig. 10.5) (Hansen and

Fig. 10.5. A bar-biting hamster. (Photo: Adobe stock.)

Berthelsen, 2000) can all be due to diet-related pain, malnutrition (e.g. of trace elements) or frustration. Prevention of and reducing such problems can be attained quite simply by giving the animals the opportunity to spend several hours a day in the normal behaviour of foraging and eating appropriate food.

Basically, food does not (should not) come in a bowl. For all species, foraging/hunting behaviours can be stimulated by using a variety of puzzle feeders. These can be homemade or commercial and can range from simply providing a toilet roll tube stuffed with hay, or folded up with a tasty morsel inside, to more complicated feeders such as the 'teach 'n' treat' feeder. Owners can be encouraged to be creative; for example, Hopping Mad (2011) offers a series of three *Boredom Buster* videos with ideas for feeding activities for rabbits, many of which can be adapted for other small mammal species.

All species need free access to fresh clean water. For human convenience this is often provided in bottle drinkers. The natural way for animals to drink is by lapping and it has been shown that this is the preferred method for rabbits (Drescher and Hanisch, 1995; Tschudin *et al.*, 2011) and probably also for other species. Where possible it is advised that both bottles and non-spill bowls are available.

Physical environment

The vast majority of animals that humans keep, including small mammal pets, spend most or even all of their life in a cage. Thus, cages must be sufficiently large and complex to enable the animal to have a life that is interesting and worth living.

Animals are active and curious. They move around their environment, foraging, feeding, exploring and socializing. They memorize important details such as pathways, obstacles, sources of food, water and shelter. Rabbits and rodents are able to exhibit a great turn of speed when they feel under threat. These animals have large home ranges (Table 4.1, chapter 4; McBride, 2017a). For example, wild rabbits and chinchillas traverse an area similar to that of two football fields per night whilst feeding, moving from one patch of vegetation to another. Rats can range 150 ft (50 m) or more from their nest in search of food, and mice up to 50 ft (16.6 m) (ISDH, n.d.). Yet most of our pets will spend their entire lives in cages that are orders of magnitude smaller than this (McBride, 2017a) (Fig. 10.6).

Space is a very important consideration. The basic principle is to provide as much as possible. In the UK, there are some legally prescribed minimum space requirements. The Animal Welfare Act Guidelines for rabbits (Welsh Government, 2009) state that hutches need to be higher than the rabbit at full stretch and long enough

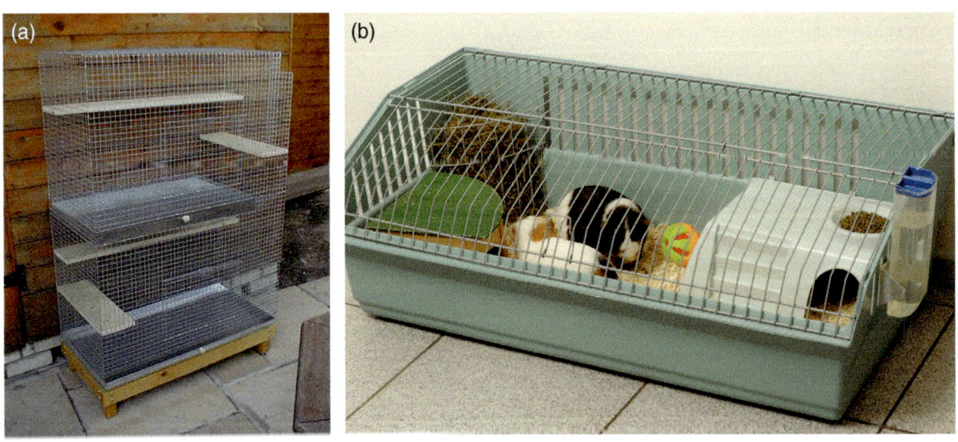

Fig. 10.6. Cages based on traditional farming and laboratory designs are too small and do not cater for the species' needs, as in these examples. **(a)** Traditional chinchilla cage, the 'Chillavilla': standard accommodation for two chinchillas. **(b)** Traditional indoor guinea pig cage.

for them to take three good hops in one direction. Many commercial cages are even smaller than this and 97% of veterinary professionals considered these should be banned from sale (PDSA, 2015). Problems of unsuitable size or design of cage are true for other species (Makowska and Weary, 2016; McBride, 2017a; McBride and Meredith, 2018).

For good physical and mental health, animals need sufficient space to be able to exhibit natural behaviour. They need to be able to stretch up to explore, to move slowly and fast, have platforms of different heights, and species-appropriate opportunities to burrow, climb, sand bathe and swim. Some good examples of creatively designed indoor guinea pig and chinchilla cages, giving space and stimulation, can be found at the following web addresses (accessed 1 May 2022): https://www.itsoverflowing.com/diy-guinea-pig-cage-ideas/; and https://chinchillachronicles.com/chinchilla-cage.

Surfaces must not be slippery (to avoid injury) and some completely dark (light-proof) sleeping areas must be provided. A basic design can be adapted for other species-specific requirements, such as bathing, digging and climbing structures. Cages should be checked regularly to ensure that they are escape-proof. Such designs would also be suitable for other floor-living species, such as rabbits and pigmy hedgehogs. Fig. 10.7 shows a good example of an outdoor guinea pig cage.

Ideally, the animals should have free access to this space for at least several hours every day. It is also important to remember that animals grow and size differences can be considerable, particularly in giant rabbit breeds. Where more than one animal is kept, there needs to be sufficient space for them all to run or stretch out simultaneously (Saunders, 2014; Makowska and Weary, 2016).

In addition to space, the location of the accommodation is important. Each species will have its own ambient temperature range. Overheating can be a particular problem, especially if cages are small, lack good ventilation, or are located in direct sunlight or by sunny windows or radiators. In the wild, tunnels, tall vegetation and rock crevices provide safe places with a stable, cool temperature to which an animal can retreat when it is too hot or cold.

These small animals are predominately crepuscular or nocturnal, but are often kept in rooms used by people day and night, and brightly lit at night. Most cages do not contain suitably dark, solid-walled resting places or tunnels that would lessen disturbance from light and noise, particularly ultrasound from computers, TVs, refrigerators, etc. Many animals suffer from disturbed sleep, which leads to health and behaviour problems, including aggression (McBride, 2017a).

Fig. 10.7. A good example of a creative outdoor guinea pig cage. (Photo: D. Emmerson).

Rabbits and rodents are destructive creatures. They will dig and chew anything – fabric, wood, wires, wallpaper. All accommodation therefore must be 'critter-proofed', devoid of sharp edges that can injure and objects that can be chewed and bits swallowed and potentially cause intestinal blocking (Buseth and Saunders, 2015).

The opportunity to dig is a requirement for all these species except the guinea pig. Digging boxes can be filled with a mixture of 'children's sandpit' sand and sterilized microwaved soil.

Chinchillas also need weekly access to a 'sand bath', filled with special, fine 'chinchilla' sand, which is made of ground pumice or silver sand. The fine particles remove grease and dirt from the coat and keep it in good condition. Other species, especially semi-desert species such as hamsters, degus and gerbils, also appreciate the opportunity of refreshing weekly sand baths (Meredith and Delany, 2010).

Rats, guinea pigs and mice can enjoy access to a 'swimming pool' of tepid water. This must have ramps so they can get out easily when they want, back onto some bedding to dry off without getting cold.

As discussed in Chapter 4, scent is significant in rabbit and rodent communication. Scent-marking objects, people and other animals denotes territory and identifies group members. Such markers are important to the animal's well-being and in reducing anxiety. It is strongly recommended that some used bedding and soiled litter is put back after cleaning the cage, so that it still smells like home (Hubrecht and Kirkwood, 2010).

Some species, like the rabbit and chinchilla, deposit scent with their faeces in a few prominent sites known as latrines. This habit means that they can be trained to use litter trays. Whilst most droppings are deposited in the tray, some are excreted as the animal moves around and owners can misinterpret this as a sign that their pet is losing toilet training.

Last but by no means least, these are prey species that are naturally fearful of rapid, sudden movements, bright lights and loud noise. Such stimuli cause them initially to freeze and then to flee rapidly to a safe and *dark* small space if the danger comes too close. Providing lookout places, tunnels and nest boxes is essential (Magnus, 2002; McBride, 2017a). Nest boxes containing suitable bedding and accessed via a dark, curved tunnel have been shown to reduce stress-related stereotypical behaviours (Waiblinger and König, 2004; Gross *et al.*, 2011; Hawkins *et al.*, 2018). There should be a sufficient number of nest boxes of appropriate sizes so that individuals can choose to be alone or to be together with their cage mates.

Mental stimulation

Providing a place to live that is spacious and interesting, along with appropriate feeding, will help to ensure that behavioural needs are met.

There is no hard-and-fast rule for encouraging physical and mental activities. Simple changes that increase the usable three-dimensional space can have a profound impact. Raised or hidden areas, partial barriers or hanging items can change an environment from a flat piece of flooring into a maze of interest (Buseth and Saunders, 2015). Provision of toys and new objects to investigate, climb on and over, changing tunnel routes, scatter-feeding and using puzzle feeders are simple means of keeping these animals mentally alert, active and healthy. These all provide interest to animal and owner alike and can be cheap and easy to make. To keep animals engaged, and therefore maximize attempts to keep their mind active, toys should be rotated so that each one retains novelty (Magnus, 2007).

Animals that are active are more interesting to watch, and the increased observation by owners means that they are more likely to notice when something is wrong and will seek timely veterinary help. Further, increasing owner engagement means the animal is less likely to be rehomed, or 'effectively abandoned' – that is, simply fed, watered and the cage cleaned until it dies.

Training the pet also enhances owner–pet relationships and helps prevent aggressive behaviour. Suitable ways of training are luring and clicker training (e.g. Orr and Lewin, 2006; Mackie and Patel, 2022). All the species can be taught a variety of useful cues, such as coming when called or going into a carrying basket, as well as various 'tricks', including fetching objects and doing 'agility' courses.

Small Mammals and Humans

Most species under discussion are social and require amicable contact with another member of their species or a human every day. Even the golden hamster, which is solitary, will happily interact with its owner.

Interacting with humans involves an enormous amount of trust by the animal and can very easily descend into distrust and fear. Prey animals are easily frightened and it is important to be quiet and move slowly around them. Speaking softly or quietly whistling will alert them that someone is coming, and be less frightening than a loud approach or a face or hand suddenly appearing.

The greatest issue affecting a small prey animal's relationship with a human, and particularly a child, is handling. Initially, the scared animal will try to move away and avoid being touched or picked up. If this is not recognized and the animal is chased, cornered and caught, it may well bite. The owner is then likely to become wary and interactions will reduce. The end result may be the animal being 'effectively abandoned', dumped, rehomed or euthanized.

Handling requires calmness, even if the pet curiously nibbles a finger, or struggles to get free, bites or kicks. Only adults or older children should pick up pets, as younger children are often too small to handle them confidently. A squirming or struggling pet may be squeezed or dropped. At best, this can cause acute pain and fear of being handled; at worst it can cause fatal internal injuries or breaking of the animal's fragile limbs or spine and chronic pain.

Regardless of any learnt fear or pain, being picked up is a frightening experience for prey species – in the wild this only happens when they have been caught by a predator. For all small animals, including dogs and cats, being lifted is likely to be uncomfortable and disconcerting, even frightening, as they are raised to a comparatively great height at some speed often with little or no warning. Good handling may be like travelling in a smooth elevator, whereas poor handling like being on a very scary roller-coaster.

With positive reinforcement all animals can be trained to willingly approach people and be lifted (Figs 10.8 and 10.9). No animal should be grabbed or picked up by the tail, ears or the loose skin on the neck (scruff), as this is painful and can cause injury.

Smaller species can be trained to get on the hand (Hurst and West, 2010), larger ones to stand still to be picked up appropriately (McBride, 2000, 2014). In all cases, the whole body is supported and held gently, but firmly.

Fig. 10.8. Whenever possible interact with rabbits, and guinea pigs, at ground level. (Photos: (a) iStock; (b) Shutterstock; (c) Phoebe with Ron the guinea pig. Photo: Hanna Armitage.)

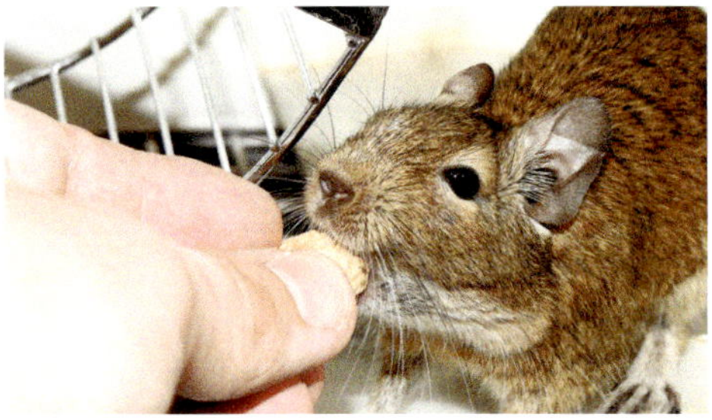

Fig. 10.9. Using food to help a degu make pleasant associations with human hands. (Photo: iStock.)

Larger animals like rabbits should be lifted by placing one hand gently over the scruff or between the front legs and chest area so that the weight is supported but it cannot move forward. The other hand supports under the rump (bottom) and is used to lift, whilst the hand on the scruff/chest adds support and balance.

Once lifted, the animal should quickly be brought close into the person's body for support (Fig. 10.10). If it begins to struggle it should be put back into the hutch or on the ground in a safe area as quickly as possible. Not doing so will simply increase the fear and probability of a bite.

Fig. 10.10. Carrying a rabbit with proper support of its body weight. (Photo: iStock.)

Small Mammals and Other Pets

The predators: dogs, cats and others

Many households keep both small mammals and predators, including dogs, cats, ferrets, snakes and raptors. Small mammals find even the scent of a predator frightening (Berdoy, 2002; McBride, 2017a) and predators naturally view the small mammal as food. Indeed, some keep rabbits and rodents to feed to their predator pets. Live feeding is unethical, and is illegal in the UK. The rabbit or rodent must be killed humanely first. Regardless of this, all owners need to take due precautions so that the predators are not frustrated and the small animals not frightened, badly injured, or worse.

These small species must always be supervised when out of the safety of their cage. Cages need to be secure from predators. Dogs and cats will stalk, stare at, chase and generally terrify small mammals in cages and may knock the cage down, releasing the occupant, who may then be injured or killed. Dogs can be trained to ignore animals in cages, hutches or runs (McBride, 2000), although some may remain very interested, especially if they have had experience of chasing or killing small animals.

Sometimes rabbits can coexist quite happily with dogs and cats. A dog that has lived peacefully with a cat or another rabbit previously may accept the new arrival. However, training of the dog and precautions should still be taken. Cats usually soon find the presence of a rabbit rather uninteresting, as they are more attuned to hunting smaller prey (Hall and Bradshaw, 1998).

Small mammals will be frightened of predator species and there is always a risk of injury or death. Some small animals will not cope well with the presence of a predator, regardless of the work put in by the owner. In such cases, there should be a total separation of the two at all times, and this may mean rehoming to a predator-free household.

Box 10.1. Twenty steps to healthy and happy small prey mammal (SPM) pets (from McBride, 2017a).

1 Ask your veterinarian for advice and sources of further information about meeting your pet's welfare needs: its diet, a suitable environment, how to keep it physically healthy and mentally active and happy and how to detect if it is unwell or in pain.

2 Think about where is best to keep your small prey mammals: they need low light and quiet, away from lots of human activity, and from machines that emit ultrasound including computers, TVs and refrigerators. Keep, and interact with, SPMs in quiet, low-light conditions with stable light–dark cycles. (Please note: pets kept outdoors, such as rabbits, will have natural daylight cycles.)

3 Set up a suitable home for them; this should be as large as possible and give them free access to enough space to enable them to move slowly and run quickly, jump, stretch to full height, and give climbing and digging opportunities. Ensure surfaces are not slippery.

4 Suitable cage furniture enables your pet to escape from light and ultrasound and frightening stimuli, include shelters, lookout places and **light-proof** nest boxes with access via a dark tunnel.

5 Provide lookout places and novel objects to provide exploration opportunities.

6 Provide variety in diet (suitable for the species) and mental stimulation in its presentation, such as homemade puzzle feeders.

7 Take time to learn how your pet communicates: you need to be able to recognize when it is happy, scared or in pain.

8 Keep small prey mammals in species-appropriate groups and try to acquire all the animals at the same time, in this way they are more likely to be compatible. It is worth spending a bit of time watching them in the pet shop to choose an animal you particularly like, and the one with whom it already bonded, that it chooses to sit next to or groom.

9 When bringing new animals home, allow a day or two for them to settle before trying to interact with then

10 When approaching your pet, talk softly or whistle so you do not startle them. Gently tap cages before moving, and carry them steadily. Simply being moved is stressful.

11 Using small pieces of food, train small prey mammals to approach to sit on your hand or be picked up and to enter carry cages on cue.

12 When holding small prey mammals, ensure the body weight is supported and the head remains upright to prevent it feeling it may fall. Science has shown this is less stressful for them.

13 Consider training your pet to do some tricks – this can be done with clicker training and can provide a lot of enjoyment for you and your pet.

14 Small prey mammals recognize cage mates and their owners by smell. They also detect danger by smell and recognize the scent of predators and intruders. Thus, it is very important to consider scent when we handle our pets in order to help them feel safe.

15 Try not to wear perfumes or use hand creams before handling your pet; simply have your natural hand smell.

16 Avoid transmission of scent from predators (e.g. dogs, cats, ferrets, snakes, birds of prey), and non-group members by washing your hands (even changing your top) before handling your small prey mammals.

17 Leave some used bedding in cleaned cages, so it still smells of home when you put your pet back.

18 If your pet needs to go to the veterinary surgery, take all its cage mates along too. This reduces stress of being alone and helps to keep the group scent and prevent any fighting.

19 Ensure group scent profiles remain similar should any animal need to be temporarily removed from the group. This is done by scent swopping and your veterinary surgeon can advise how this is done. This is important to reduce the chance of them fighting when you reintroduce them.

20 As prey species, these animals do not like to attract attention and are subtle in showing when they are unwell. Be observant: if your animal shows any change from its normal behaviour it may be ill or in pain or otherwise stressed and need to see the vet. This includes aggression, chewing the cage or cage bars, apathy or restlessness, over- or under-eating. Do not delay – by the time signs of illness are obvious it may be too late to help your pet.

Rabbits and guinea pigs

Keeping rabbits and guinea pigs together is risky and the victim is usually the guinea pig (Saunders, 2014). Guinea pigs are very susceptible to respiratory disease caused by *Bordetella bronchiseptica*, which they can catch from rabbits, dogs and cats who carry the bacterium with no signs of illness (Goodman, 2009).

When people pair a guinea pig and a rabbit, they often do not bother to neuter as there is no problem of unwanted babies. But an entire male or female rabbit can mount a guinea pig and cause injuries. Similarly, a rabbit that becomes territorial around the time of puberty may direct aggression towards the guinea pig.

In addition, because each species uses different vocal, visual and scent signals, they cannot communicate important messages such as submission, stress and wishing to be left alone. This can result in rabbits being frustrated and bullying the guinea pig, with the welfare of both being compromised. This is exacerbated if there is little space and few options for retreat and hiding, which would be a natural diffusion of a potential conflict situation.

Sometimes these two species have been kept together with few problems (McBride, 2000). As with keeping two or more members of the same species, the success or failure of mixed-species groups will be dictated by the housing, space, hormonal status of individuals, the behaviour and health of all the individual animals. Mixed groups that contain at least two compatible companions of each species, with, at least, the rabbits being neutered can work. The accommodation must be large with lots of tunnels, feeding places, etc. As each animal has company of its own kind and lots of things to do, it is likely to be less stressed. Having space and several options for avoidance reduces escalation and chances of injury should there be any disputes.

Summary

Small mammals are inaccurately perceived as cheap and easy to keep. Potential owners must be able to make a considered and informed decision before acquisition (Jessop and Warwick, 2014). Promoting emotions of pleasure and reducing psychological distress throughout an animal's life is core to welfare and preventing behaviour problems. This requires good standards of appropriate husbandry and management. This is the responsibility of breeders, retailers, rehomers, suppliers, owners or others, regardless of why the animal is kept. For small prey species this means considering how their size and being prey will influence their experiences of their world (McBride, 2017a,b).

This chapter has provided general principles regarding socialization, companionship, handling and physical and mental stimulation. Readers must research species-specific details about other factors, including diet, temperature and humidity. Welfare can be greatly enhanced by providing space, truly dark nest boxes and tunnels, digging and climbing facilities, and opportunities for foraging, exploratory behaviour and problem-solving. There is creative scope here both for owners and for manufacturers of cages and cage furniture.

Box 10.1 gives a summary of key points to assist owners to meet the welfare needs of small prey mammal (SPM) pets. It is written in owner-friendly language to aid veterinarians who may wish to use the text for their clients (McBride, 2017a).

References

Anonymous (2018) Public urged not to buy exotics online. Available at: http://mrcvs.co.uk/en/news-story.php?id=17562 (accessed 3 August 2018).

Ashley, A., Brown, S., Ledford, J., Martin, J., Nash, A.E. *et al.* (2014) Morbidity and mortality of invertebrates, amphibians, reptiles, and mammals at a major exotic companion animal wholesalor. *Journal Applied Animal Welfare Science* 17, 308–321.

Berdoy, M. (2002) *The Laboratory Rat – a Natural History*. Video. Available at: http://www.ratlife.org and via youtube.com (accessed 1 May 2022).

Blue Cross and Born Free Foundation (2016) One Click Away: an investigation into the online sale of exotic animals as pets. Available at: https://www.bornfree.org.uk/storage/media/content/files/Publications/ONE_CLICK_AWAY_REPORT_SPREADS.pdf (accessed 1 May 2022).

Buseth, M.E. and Saunders, R.A. (2015) *Rabbit Behaviour, Health and Care*. CAB International, Wallingford, UK.

Campbell, F. (2010) *Bonding Rabbits*. Coney Publications, Ipswich, UK.

Carter, C. (1972) Effects of olfactory experience on the behaviour of the guinea pig (*Cavia porcellus*). *Animal Behaviour* 20, 54–60.

Cavigelli, S.A. and McClintock, M.K. (2003) Fear of novelty in infant rats predicts adult corticosterone dynamics and an early death. *Proceedings of the National Academy of Sciences USA* 100, 16131–16136.

CAWC (2006) *Biotechnological Methods, to the Form, Function, or Behaviour of Companion Animals*. Available at: www.cawc.org.uk (accessed 10 March 2013).

Chapillon, P., Patin, V., Roy, A.V. and Caston, J. (2002) Effects of pre- and postnatal stimulation on developmental, emotional and cognitive aspects in rodents: a review. *Developmental Psychobiology* 41, 373–387.

Cloutier, S., Panksepp, J. and Newberry, R.C. (2012) Playful handling by caretakers reduces fear of humans in the laboratory rat. *Applied Animal Behaviour Science*,140, 161–171.

Copping, J. (2009) Welfare crisis among Britain's pet rabbits. Article in *Daily Telegraph* 4 April 2009.

Cowley, J.J. and Widdowson, E.M. (1965) The effect of handling rats on their growth and behaviour. *British Journal of Nutrition* 19, 397–406.

Defra (2012) *Zoo Licensing Act 1981 Guide to the Act's Provisions*. Department for the Environment, Food and Rural Affairs, London.

Der Weduwen, S. and McBride, E.A. (1999) Rabbit behaviour and the effects of early handling. In: *Proceedings of the 2nd World Meeting on Ethology and Mondial Vet*, Lyon, France, pp. 221–228.

Drescher, B. and Hanisch, A. (1995) Testing of various drinking troughs in consideration of the physiological drinking behavior of rabbits. *Deutsche Tierärztliche Wochenschrift* 102, 365–369.

Ellis, C.F., McCormick, W. and Tinarwo, A. (2017) Analysis of factors relating to companion rabbits relinquished to two United Kingdom rehoming. Journal of applied animal welfare science, 20(3), 230–239.

Emergent Disease (2013) Pets: easy or difficult to keep? Available at: http://emergentdisease.org/assets/documents/emode-brochure-screen.pdf (accessed April 2021).

FAWC (2009) *Farm Animal Welfare in Great Britain: Past, Present and Future*. Available at: https://www.gov.uk/government/publications/fawc-report-on-farm-animal-welfare-in-great-britain-past-present-and-future (accessed 22 August 2016).

Franklin, T.B., Russig, H., Weiss, I.C., Graff, J., Linder, N. *et al.* (2010) Epigenetic transmission of the impact of early stress across generations. *Biological Psychiatry* 68, 408–415

Goodman, G. (2009) Rodents: respiratory and cardiovascular system disorders. In: Keeble, E. and Meredith, A. (eds) *Manual of Rodents and Ferrets*. BSAVA, Gloucester, UK, pp. 142–149.

Gross, A.N., Engel, A.K.J. and Würbel, H. (2011) Simply a nest? Effects of different enrichments on stereotypic and anxiety-related behaviour in mice. *Applied Animal Behaviour Science* 134, 239–245.

Hall, S.L. and Bradshaw, J.W.S. (1998) The influence of hunger on object play by adult domestic cats. *Applied Animal Behaviour Science* 58, 143–150.

Hansen, L.T. and Berthelsen, H. (2000) The effect of environmental enrichment on the behaviour of caged rabbits (*Oryctolagus cuniculus*). *Applied Animal Behaviour Science* 68, 163–178.

Hawkins, P., McBride, A., Paiba, G., Rose, R., Prescott, M.J. *et al.* (2018) Report of the 2017 RSPCA/UFAW Rodent and Rabbit Welfare Group. *Animal Technology and Welfare* 17(2), 75–83.

Hopping Mad (2011) *Boredom Busters* series of videos. Available at: http://www.youtube.com/watch?v=IrDKTEcligM; http://www.youtube.com/watch?v=Za5kl5dnnFc; http://www.youtube.com/watch?v=5sk3XCzKJ1k (accessed 18 June 2022).

Hubrecht, R.C. and Kirkwood, J. (eds) (2010) *The UFAW Handbook on the Care and Management of Laboratory and Other Research Animals*, 8th edn. Wiley, Oxford.

Hurst, J.L. and West, R.S. (2010) Taming anxiety in laboratory mice. *Nature Methods* 7(10), 825–826.

ISDH (n.d.) *Rats and Mice*. Indiana State Department of Health. Available at: http://www.in.gov/isdh/23256.htm (accessed 22 August 2016).

Jessop, M. and Warwick, C. (2014) Pets: to keep or not to keep – perhaps EMODE has the answer. *Veterinary Times* 29 September 2014, 23–26.

Laviola, G., Hannan, A.J., Macri, S., Solinas, M. and Jaber, M. (2008) Effects of enriched environment on animal models of neurodegenerative diseases and psychiatric disorders. *Neurobiology of Disease* 31, 159–168.

Mackie, J. and Patel, C. (2022) Welfare-focused animal training. In: Rendle, M. and Hinde, J. (eds) *BSAVA Manual of Practical Veterinary Welfare*. BSAVA Publications, Gloucester, UK.

Magnus, E. (2002) *How to Have a Relaxed Rabbit*. Pet Behaviour Centre, Defford, UK.

Magnus, E. (2007) *Teach Yourself: Keeping a Rabbit*. Hodder Publications, London.

Makowska, I.J. and Weary, D.M. (2016) The importance of burrowing, climbing and standing upright for laboratory rats. *Royal Society Open Science* 3(6), 160136. doi: 10.1098/rsos.160136

McBride, A. (2000) *Why Does My Rabbit…?* Souvenir Press, London.

McBride, A. (2009) Rampaging rabbits: Minimizing stress in the veterinary surgery and rescue centre. *Rabbit Welfare Fund Conference: Rabbit Health Matters. Proceedings of the 2009 RWF Conference*, 28 February 2009, Nottingham, UK.

McBride, E.A. (2014) Normal behaviour and behaviour problems. In: Meredith, A. and Lord, B. (eds) *BSAVA Manual of Rabbit Medicine*. BSAVA Publications, Gloucester, UK.

McBride, E.A. (2017a) Small prey species' behaviour and welfare: implications for veterinary professionals. *Journal of Small Animal Practice* 58, 423–436. doi: 10.1111/jsap.12681

McBride, E.A. (2017b) Stress in the veterinary surgery – small mammals. *The Veterinary Nurse* 8, 376–381.

McBride, E.A. and Hinde, J. (2022) Behaviour. In: Rendle, M. and Hinde, J. (eds) *BSAVA Manual of Practical Veterinary Welfare*. BSAVA Publications, Gloucester, UK.

McBride, E.A., Magnus E. and Hearne, G. (2016) Behaviour problems in the domestic rabbit. In: Appleby, D. (ed.) *The APBC Book of Companion Animal Behaviour*, 3rd edn. Souvenir Press, London, pp. 174–192.

McBride, E.A. and Meredith, A. (2018) Guinea pigs, chinchillas and degus (caviomorphs). In: Yeates, J. (ed.) *Companion Animal Care and Welfare: The UFAW Companion Animal Handbook*. Wiley, Oxford.

Mellor, D.J. (2016) Updating animal welfare thinking: moving beyond the 'Five Freedoms' towards 'A Life Worth Living'. *Animals* 6(3), 21. doi: 10.3390/ani6030021

Meredith, A. and Delany, C.J. (2010) *BSAVA Manual of Exotic Pets*, 5th edn. Wiley, Oxford.

Milne, E. (2015) *Are Rabbits the Right Pet for You? Can You Find the Facts? The Pet Detectives Series*. 5M Publishing, Great Easton, UK.

Neville, V., Hinde, K., Line, E., Todd, R. and Saunders, R. (2018) Rabbit relinquishment through online classified advertisement in the United Kingdom: when, why and how many? *Journal of Applied Animal Welfare Science* 22, 105–115. doi: 10.1080/10888705.2018.1438287

Núñez, J.F., Ferré, P., Escorihuela, R.M., Tobeña, A. and Fernández-Teruel, A. (1997) Effects of postnatal handling of rats on emotional, HPA-axis, and prolactin reactivity to novelty and conflict. *Physiology and Behaviour* 60, 1355–1359.

Orr, J. and Lewin, T.A. (2006) *Getting Started: Clicker with your Rabbit*. Sunshine Books Inc., Waltham, Massachusetts.

PDSA (2015) *PDSA Animal Wellbeing (PAW) Report 2015*. Available at: https://www.pdsa.org.uk/media/2629/pdsa-paw-report-2015.pdf (accessed 3 August 2016).

PDSA (2018) *PDSA Animal Wellbeing (PAW) Report 2018*. Available at: https://www.pdsa.org.uk/media/4371/paw-2018-full-web-ready.pdf.

PDSA (2020) *PDSA Animal Wellbeing (PAW) Report 2020*. Available at: https://www.pdsa.org.uk/media/10540/pdsa-paw-report-2020.pdf (accessed 26 June 2022).

PFMA (2020) *Pet Food Manufacturers' Association Pet Population Report*. Available at: https://pfma-reports.co.uk/annual-report-2020/ (accessed March 2022).

Pongrácz, P. and Altbäcker, V. (1999) The effect of early handling is dependent upon the state of the rabbit (*Oryctolagus cuniculus*) pups around nursing. *Developmental Psychobiology* 35, 241–251.

Prebble, J. (2014) Nutrition and feeding. In: Meredith, A. and Lord, B. (eds) *BSAVA Manual of Rabbit Medicine*. BSAVA Publications, Gloucester, UK.

Richardson, V.C.G. (2010) *Rabbit Nutrition*, 2nd edn. Coney Publications, Ipswich, UK.

Rooney, N.J., Blackwell, E.J., Mullan, S.M., Saunders, R., Baker, P.E. *et al.* (2014) The current state of welfare, housing and husbandry of the English pet rabbit population. *BMC Research Notes* 7(1), 1.

Roughan, J.V., Bertrand, H.G.M.J. and Isles, H.M. (2015) Meloxicam prevents COX-2-mediated post-surgical inflammation but not pain following laparotomy in mice. *European Journal of Pain* 20(2), 231–240. doi: 10.1002/ejp.712

Roy, V., Lointier, L. and Chapillon, P. (2003) The emotional reactivity increase at mid-pregnancy is attenuated in female rats handled during their infancy. *Behavioural Brain Research* 145, 23–30.

RSPCA (2012) *The time and costs involved in keeping rabbits*. RSPCA companion animal pet care factsheet. RSPCA, West Sussex. LocateAsset?asset=document&assetId=1232729413756&mode=prd (accessed 22 August 2016)
Saunders, R. (2014) Husbandry. In: Meredith, A. and Lord, B. (eds) *BSAVA Manual of Rabbit Medicine*. BSAVA Publications, Gloucester, UK.
Schuppli, C.A. and Fraser, D. (2000) A framework for assessing the suitability of different species as companion animals. *Animal Welfare* 9, 359–372.
Shadow The Rat (2017) 10 tricks to train your rat/mouse first – Mini tutorials. Available at: https://www.youtube.com/watch?v=MRh4weM19qc (accessed 24 July 2018).
Smartmice (2009) Amazing mouse tricks. Available at: https://www.youtube.com/watch?v=z_pOvtOYulw published by http://www.facebook.com/smartmice (accessed 24 July 2018).
Toland, E., Bando, M., Hamers, M., Cadenas, V., Laidlaw, R. *et al.* (2020) Turning negatives into positives for pet trading and keeping: a review of positive lists. *Animals* 10(12), 2371.
Tschudin, A., Clauss, M., Codron, D. and Hatt, J.-M. (2011) Preference of rabbits for drinking from open dishes versus nipple drinkers. *Veterinary Record* 168(7), 190. doi:10.1136/vr.c6150
UFAW (n.d.) *Genetic Welfare Problems of Companion Animals*. Available at: http://www.ufaw.org.uk/genetic-welfare-problems-intro/genetic-welfare-problems-of-companion-animals-intro (accessed 22 August 2016).
Vetulani, J. (2013) Early maternal separation: a rodent model of depression and a prevailing human condition. *Pharmacological Reports* 6, 1451–1461.
Waiblinger, E. and König, B. (2004) Refinement of gerbil housing and husbandry in the laboratory. *Animal Welfare* 13, 229–235.
Warwick, C. (2014) The morality of the reptile 'pet' trade. *Journal of Animal Ethics* 4, 74–94.
Warwick, C., Steedman, C., Jessop, M., Toland, E. and Lindley, S. (2013) Assigning degrees of ease or difficulty for pet animal maintenance: the EMODE System Concept. *Journal of Agricultural and Environmental Ethics* 27, 87–101. doi: 10.1007/s10806-013-9455-x
Warwick, C., Steedman, C., Jessop, M., Aprena, P., Pilny, A. and Nicholas, E. (2018) Exotic pet suitability: understanding some problems and using a labelling system to aid animal welfare, environment and consumer protection. *Journal of Veterinary Behavior: Clinical Applications and Research* 26, 17–26. Available at: https://www.journalvetbehavior.com/article/S1558-7878(18)30036-4/fulltext (accessed 26 June 2022).
Way, T. (2016) *Gardening for Rabbits*. RWAF, Taunton, UK.
Welsh Government (2009) *Code of Practice for the Welfare of Rabbits*. Available at: http://gov.wales/legislation/subordinate/nonsi/animalwelfarewales/2009/3633485/?lang=en (accessed 22 August 2016).
Westgate, J. (2021) First Guinea-pig Awareness Week announced. *Vet Times*. Available at: https://www.vettimes.co.uk/news/first-guinea-pig-awareness-week-announced/ (accessed 26 June 2022).
Yeates, J. (2018) *Companion Animal Care and Welfare: The UFAW Companion Animal Handbook*. Wiley, Oxford.

11 What Every Parrot Owner Should Know

Clare Wilson

Information the Veterinary Surgeon Can Provide to Prevent Problem Behaviour

Although parrots do not require primary vaccinations as do puppies and kittens, owners will hopefully bring newly acquired parrots in for a routine check-over. This can be an ideal opportunity to discuss the parrot's needs to ensure good welfare and prevent the development of behavioural problems later in life. Recently weaned and hand-reared young birds are commonly easy to handle and generally do not present with behavioural problems at this stage. If there are already signs of problems, such as fearful or anxious behaviour, at this young age it is vital that these are addressed immediately. Problems are more likely to arise as birds reach adolescence, so giving preventive advice before maturity is more rewarding and effective than resolving established problems later. The information in this chapter is essential for understanding why behaviour problems have developed and working out ways to resolve them.

Understanding the parrot

The key to preventing the development of problems and treating those that are already established is to educate the owner and help them better understand their parrot. Once an owner can understand their bird's emotions and needs, they can form a trusting two-way relationship that is mutually beneficial and highly rewarding. The majority of parrots that are relinquished or euthanized due to behaviour problems have largely been misunderstood. The three most common problem behaviours are excessive vocalization, aggression and feather damage. These are all related to instinctive behaviours that need not develop into problems if managed appropriately. Understanding wild parrot behaviour and how that relates to the domestic setting is essential for prevention and successful behaviour modification.

Vocalization

As discussed in Chapter 5, although the companion parrot is living in a domestic situation it is a tame individual rather than a domesticated species. All types of vocal communication that occur in wild parrots (see Chapter 5) will therefore also occur within the home. Contact and alarm calls are by their very nature loud, to carry long distances, and owners must expect and accept a certain level of noise. The volume varies considerably between species, with Moluccan cockatoos, macaws and conures being renowned for being particularly loud. Careful reinforcement with attention for lower-volume communication with

DOI: 10.1079/9781800621312.0011

owner interaction can significantly affect the need for the parrot to use high-volume calls (Box 11.1). If owners are given appropriate advice when first acquiring a parrot, vocalizations can often be managed to an acceptable level. Prior to homing a parrot, owners should consider their close environment, such as their proximity to neighbours, and whether the noise of the louder species is compatible with their particular circumstances. The level of contact calling voiced by an individual parrot will depend on many factors but is particularly associated with the level of bond between the bird and its owner. A bird that is highly bonded is likely to use frequent contact calls when the owner is out of sight because, in the wild situation, bonded pairs would not usually separate. This behaviour will also be affected by the bird's ability to cope on their own. Unless a bird has been taught to enjoy being alone, it may become anxious and stressed when in isolation and this can result in an increased frequency and volume of contact calls. In an ideal situation parrots would never be left in social isolation and owners who are expecting their bird to cope alone should be encouraged to find a suitable avian companion for their pet.

Body language

With so much variation between species in terms of communication, it is vital that owners spend time learning about the individual bird. However, some generalizations can be made across species about the manifestation of fear and anxiety in birds which veterinary staff and parrot owners should be aware of. As with all social species, parrots have highly developed communication strategies for avoiding conflict. Research has shown that parrots with larger beaks tend to have more complex communication repertoires (Serpell, 1982). This is thought to be due to the fact that it is even more imperative that they avoid combat, because of the higher risk of damage.

Fear, anxiety, arousal and aggressive behaviour

Much of the popular literature about parrot behaviour suggests that owners should expect to be bitten and that biting is a normal part of owning a parrot. This is not the case: if owners have a thorough understanding of their bird's behaviour, they will not put their pet in a position where it feels the need to bite. Parrots will only bite if they are stressed (frightened or frustrated) and their welfare has been compromised. Trust and respect are vital between owner and pet, especially in a prey species. Biting is part of the normal behavioural repertoire, but, as with

Box 11.1. Advice for owners to prevent vocalization issues developing.

Responding to contact calls

- Ensure you talk to your parrot or make other sounds, such as whistling or singing, as you move about the house when the parrot is in another room. This way, the parrot will feel comfortable knowing that you are still present and will not feel the need to scream to regain contact with you.
- When your parrot calls to you, make sure you answer. Many people make the mistake of thinking they need to ignore these contact calls in case they reward the noise. They must understand that contact calls are a normal parrot behaviour and that if these are ignored the parrot is likely just to try harder and harder to make contact. If ignored, these calls can become the basis of a screaming problem.
- Reinforce quiet sounds. Although contact calls cannot be stopped altogether, parrots can learn to call quietly provided that quiet calls are reinforced. The owner should be advised to listen carefully to their bird and if a quiet contact call is heard, the owner should make sure that they reinforce this behaviour either by calling back or by returning to the bird as a reward.

Responding to alarm calls

- Identify the source of concern for the bird. If the bird does not settle, the intensity of the source should be reduced (for example: decrease volume of a sound, increase distance of an unfamiliar person, increase distance of a novel object, increase distance of a pet dog) and/or the bird should be allowed to show avoidance behaviour such as flying to a higher perch.

dogs, parrots have complex communication signals, with numerous lower-level signals to indicate perceived threat. If these signs are respected, the need for the parrot actually to bite is avoided. Therefore, one of the most important issues in aggression cases is to help owners understand their parrot's communication strategies. By recognizing when the bird is asking for space, owners can avoid escalation of aggressive responses. In conjunction with this, owners need to respect their parrot's choices. Many people think their bird should be consistently obedient to commands, with much of the popular literature encouraging them to 'dominate' and 'be the flock leader'. However, parrots are highly intelligent and having individual choice is an essential aspect of maintaining their emotional welfare. Owners should be trying to work in partnership with their companion rather than controlling and nagging. It is also crucial to remember that parrots are tame rather than domesticated (Chapter 5), hence socially mature birds are fully independent and not predisposed to accept a carer. There are clearly some situations where a bird must be trained to be obedient from a safety point of view, for example having a reliable recall or learning not to chew dangerous household items. However, in situations where the bird is not in danger, owners must respect that their bird needs to be able to make its own choices, for example whether it is ready to interact with a person, or sit on a particular perch.

Severe aggression is very rarely observed in the wild and seems to be a consequence of captivity. Given the choice, most parrots show a preference for avoidance behaviour rather than risking the danger of confrontation. In the domestic setting avoidance may not be possible for the bird, due to environmental confinement (e.g. cage or walls) or due to physical interventions (e.g. wing clipping preventing a flight response). It may also be that a new situation is too intense or too sudden for the bird and it may bite without the owner observing any apparent warning signals. Escalation due to continued threat is particularly likely in the case of sexual rivals. In the wild situation an unsuccessful rival would take flight and try his luck elsewhere, which is not possible for captive birds. Although rare, reproductive activity is the one situation in the wild where serious injury can be seen and parrots can get into combat over competition for mates, or protection of or competition over nest sites.

Parrots show aggression for a reason and their emotions must be respected. Despite differences in breed conformation, dogs have a fairly consistent repertoire of signals when they feel threatened (Chapter 2). Unfortunately, this is not as clear with parrots, due to interspecies and individual differences, but there are some signs that are often significant in all species. The best way for an owner to interpret the signals from their parrot is to observe the behaviours that precede and follow particular signals so that they can work out whether each signal is associated with relaxation or tension.

Tables 11.1, 11.2 and 11.3 describe a range of behaviours that can indicate relaxation, anxiety, fear and general arousal level as well as the potential for a bite. None of these behaviours is likely to occur in isolation and owners may need to assess a group of different signals to truly interpret what the parrot is communicating. Parrots may show some similar signs when highly aroused due to various motivations, so both the context and the full spectrum of signals must be examined closely. However, any highly aroused parrot is at risk of biting, whatever the underlying motivation, so it would be best to give such a parrot time and space to calm down before attempting interaction.

Parrots have striated muscle controlling the iris, so have voluntary control of pupil size rather than responding automatically to light intensity as in humans. When parrots are highly aroused they will rapidly change the pupil size, a behaviour known as eye pinning. This is purely a sign of high arousal and does not on its own indicate whether the parrot is in a positive or a negative emotional state. It is therefore important to interpret eye pinning in conjunction with the rest of the bird's body language. This is easier to observe in species with light-coloured irises such as the African Grey, macaws and mature Amazons. In other species, such as cockatoos, that have very dark irises it is much more difficult to observe eye pinning. Box 11.2 provides some advice for appropriate responses to a fearful parrot.

Table 11.1. Recognizing relaxation. (Data: author's own.)

Signs of relaxed parrot	Comments
Loosely ruffled feathers	Take care not to confuse this with a parrot who is feeling unwell
Body position central or leaning to investigate	Take care not to misinterpret leaning to investigate with leaning to bite. Feather and beak position must also be observed
Half-closed eyes	Alert and watchful parrots are keeping an eye out for potential threats. A parrot that feels safe and relaxed will half-close the eyes (Fig. 11.1)
Playing	Only a relaxed bird is likely to be interested in play although this needs to be interpreted in conjunction with the style of play and the rest of the body language, because if the 'play' is vigorous it may indicate frustration or re-directed aggression
Resting one foot tucked into the feathers	A bird that is relaxed and therefore not expecting a sudden escape will happily rest on one foot
Head bobbing	A happy or playful parrot will often bob the head up and down and many parrots greet people they are pleased to see with this gesture or show this when offered a favourite treat
Exploratory behaviour on the floor	Take care interpreting this in parrots with clipped wings, as the parrot may in fact be looking for an escape route to climb up higher to a less vulnerable location. This also varies with species, with some being more inclined to spend time on the ground than others
Turning the head on one side, upside down or head curled under with the neck arched upwards	Only a relaxed bird will move into these vulnerable positions. These are common ways a parrot invites a head scratch (Fig. 11.2)
Tail wag or shake	Some parrots will show this tail wag as part of a greeting behaviour
Wing flick	Gentle movement of the wing(s), which is an affiliative behaviour often seen during greeting
Bathing	Although an essential feather maintenance activity, bathing puts the bird in a potentially vulnerable position, particularly if it involves coming down low to the ground
Relaxed and low crest in cockatoos	Many species of cockatoos have a crest and the position of this indicates their arousal level (Fig 11.3)

Socialization

As for the other species discussed in this book, appropriate early experience is crucial in order for parrots to cope with the demands of living in a domestic setting. Influencing behavioural development is perhaps even more crucial in these non-domesticated creatures, which have no natural predisposition to reside with humans. A complex physical and social environment in early life ensures increased behavioural flexibility later (Box 11.3). Due to the high rate of neurogenesis early in life, a wider range of experiences has long-term effects on the complexity of neural networks in the brain. There has been limited study of critical learning periods in parrots but studies that have been done certainly indicate that early learning is just as crucial as it is in other companion species. Species are likely to be variable in terms of socialization, as the larger parrots with more complex behavioural repertoires take longer to mature and become independent: ongoing socialization and positive experience are likely to be especially significant in these birds. However, even ideal experiences during these crucial early months cannot affect the hard-wired innate behaviours of pet parrots, which are essentially genetically the same as their wild relatives. It is therefore crucial during early development to guide parrots to direct these behaviours to be expressed in contexts and on to targets appropriate to the domestic setting. This can aid the prevention of problems later on.

Encouraging owners to observe the body language of their parrot is an essential aspect of

Fig. 11.1. A parrot that feels safe and relaxed will half-close the eyes. (Photo: author's own.)

Fig. 11.2. Many parrots enjoy a head scratch and will turn their head as an invitation. (Photo: author's own.)

the socialization process. The aim should be for birds to be exposed to a variety of novel social experiences, objects, sounds and so on at an intensity that promotes interaction. If a bird is showing any signs of anxiety or fear, the situation must be altered to allow relaxation. This might, for example, involve increasing the distance between the bird and an unfamiliar visitor or placing a new toy on the other side of the room for a few days before placing it closer to the cage (see Box 11.2).

Chick rearing

Fortunately, the majority of pet birds entering new homes in the UK are now captive-bred. Although import of wild birds has been banned under EU law since 2007 (an extension of a temporary ban in 2005), unfortunately some wild-caught birds are still managing to enter the UK. There are of course still many birds already in homes and rescue centres who were wild-caught and it is vital that their background is known and understood when attempting to advise owners. The method of chick rearing is crucial to parrots being appropriately adapted to living in the domestic environment.

Parrots are generally cavity nesters. As soon as they are mobile, they tend to peep out of the nest and communicate with surrounding birds through vocalizations. Once wild chicks fledge, they experience a wealth of social experiences and live in a highly complex physical environment where they learn dexterity, food manipulation skills, flight and all the other activities essential to survival. If chicks are incubator-hatched or taken from their parents very early for hand-rearing, they may have very limited experience of their human caretakers and minimal environmental stimulation. This

Fig. 11.3. Greater Sulphur Crested cockatoo with relaxed crest and loosely ruffled feathers. (Photo: author's own.)

Table 11.2. Recognizing signs of increased arousal. (Data: author's own.)

Signs of parrot in negative emotional state or state of high arousal	Comments
Tightly sleeked feathers (with upright stance)	When in flight, the feathers are all smoothed down against the body. A parrot who is ready to take flight as an avoidance response will sleek the feathers in preparation for escape
Body or head leaning away, showing avoidance response	Given the choice, parrots will choose avoidance rather than confrontation (unless they have learned to use aggression straight away from previous experiences)
Holding the wings away from the body	As the level of arousal and aggression increases, the wings will be held further away from the body (Fig. 11.4)
Lowering the head	If the threat does not respond to the bird moving its head away it may lower its head to get into a position to bite (Fig 11.4)
Beak gaping	Parrots will usually give a warning of an imminent bite by first gaping their beak as if to bite and, if ignored, this may progress to a peck before a full-strength bite is inflicted
Wing flap	Wing flapping is often a threatening behaviour and this must be distinguished from the much gentler wing flick and from wing flapping related to preening and re-arranging misplaced feathers
Growling or hissing	Vocalization is not that common in aggression but growling may occur in frightened macaws, Amazons and African Greys and hissing may occur in frightened cockatoos
Strutting	This is often seen in territorial aggression and indicates quite a high level of arousal
Weaving from side to side	This can involve just the head or it may involve the entire body
Quick head movements	The parrot is surveying the environment, looking for an escape route
Cockatoo crest rising	An anxious bird will slightly raise the feathers at the back of the crest whilst keeping those at the front flat. As the bird becomes more aroused, either excited or aggressive, the crest becomes more raised.
Redirected aggression	As mentioned above, a parrot may aggressively 'play' with toys or show aggression towards other objects in close vicinity if they cannot easily reach the true target of their fear or emotional arousal

Fig. 11.4. Blue and Gold macaw with a lowered head, holding his wings away from the body and slight facial flushing indicating he may bite if approached. (Photo: author's own.)

Table 11.3. Recognizing escalation of arousal and possibly aggressive responses. (Data: author's own.)

Signs of increasing aggression	Comments
Feathers on nape of neck raised (early sign of aggression) As above plus feathers raised on back of head (higher level of aggression) As above plus feathers raised on shoulders and upper back (bite imminent)	All of these feather changes are unique to each bird so owners must observe and learn about the individual bird rather than treating this information as a gold standard. Generally, the more feathers that are raised, the more likely the bird is to be aggressive (Fig. 11.5)
In cockatoos the crest is raised to a greater degree as arousal increases (aggression or excitement)	A cockatoo that is highly aroused may well bite, whether it is aggressive or excited, so best to let them calm down and the crest fall to a relaxed position before interacting
Rapid eye pinning indicating a high arousal level (aggression or excitement)	See text for further explanation. A parrot showing this behaviour may well bite and interaction should be avoided
Tail fanned	Some birds, particularly Amazons but also some macaws and cockatoos, will fan out the tail feathers as a sign of aggression. As with eye pinning and raised crests, this can also be a sign of excitement so must be interpreted in conjunction with the rest of the bird's body language and the context
Wings held out wide with head lowered	A higher level of aggression if a threat does not move away (Fig. 11.6)
Facial blushing in macaws	Macaws are unique in having areas of bare skin around the beak and side of their faces. When highly aroused (excited or aggressive) this area can blush red (Fig 11.6). This will frequently be seen in combination with eye pinning. A bird that is this aroused may well bite and would be best left alone

can potentially affect their ability to cope with novelty and the domestic setting later in life.

Learning to cope with novelty is a key factor in early learning. In the domestic situation parrots are expected to deal with situations such as meeting unfamiliar people, being given novel toys or foods, having their home redecorated, interacting with other family pets, being taken to a carer when the owners are on holiday, visiting the veterinary surgery and so on. Although little research has been carried out on parrots, the socialization period for most altricial species begins as the senses become functional and the young animal can start to learn about their social and physical environment (see Chapters 2, 3, 8 and 9). It is likely that this process is similar in parrots. Parrot chicks generally open their eyes at 1.5–2 weeks of age but this is variable between species. When they first hatch, they are aware of the warmth and touch of their siblings and brooding parent and they have a feeding reflex.

Fig. 11.5. This African Grey is highly aroused and a bite is imminent if the person continues to approach. (Photo credit: A.G. Turner.)

Captive-bred birds are reared using a number of different methods. There is growing concern as to the most appropriate techniques for raising birds best able to cope as pets. Many birds are now hatched in incubators and then hand-reared. In some cases, chicks are reared as a group, but other breeders will raise chicks in isolation from other parrots. Breeders that allow parent rearing vary with respect to the age at which chicks are removed. Some remove them pre-weaning and hand-feed, whereas others allow natural weaning. Rearing methods can have serious implications for how the parrot can cope with domestic life and so it is crucial that owners are aware of the rearing background of a new pet. Parental separation is stressful and can have long-term effects on behavioural and physiological development (Fox, 2006). In 2014, the Netherlands banned hand-rearing of parrots due to concerns over welfare, health and

Fig. 11.6. Escalating threat signals of the wings held further away from the body, facial blushing and the head further lowered. (Photo: author's own.)

Box 11.2. Advice for owners to avoid the development of fearful and/or aggressive behaviour.

- The most common motivation for birds showing avoidance or aggression is an underlying fear or anxiety and if owners understand this they can be sensitive to their bird's emotions and reduce the chance of aggression developing or escalating. Fearful avoidance behaviour can also be successfully managed such that it does not escalate.
- If a bird feels threatened the owner should assess the situation and make appropriate changes to allow the bird to relax again. So, for example, if a bird is nervous about a new toy being hung in the cage, the owner can remove it and place it further from the cage to allow the bird time to get used to it. If the presence of a visitor causes the bird to show signs of fear or anxiety, the visitor can be taken to a different room or asked to sit further from the bird's cage, or if the bird is out of its cage it can be given a high perch where it feels safer to view its surroundings and adapt to the new situation.
- Never punish fearful responses. If an owner is bitten by their bird, it is crucial that they do not lay blame on the bird and become cross or tell the bird off. Aversive training will be discussed in more detail below but owners must be aware that punishment when the underlying motivation is fear will be counter-productive and is likely to exacerbate the fear response.

Box 11.3. Advice for owners: socialization and habituation.

- The young parrot needs to encounter plenty of novelty and variation as part of its normal daily routine. This provides vital learning for it to be well adapted to the demands of domestic life.
- When introducing the parrot to new situations (objects, visitors, sounds and so on) it is vital to observe body language and ensure that it stays relaxed. Signs of avoidance or anxiety must be respected and the parrot given a bit more time and space to assess the situation. Give the young parrot opportunities to explore novel objects when it feels comfortable.
- Cater for all aspects of physical and mental development through provision of opportunities for flight, climbing, swinging, perching, foraging, communicating and so on. The more opportunities to learn and grow in confidence, the better adapted the parrot will be for living with humans.

behaviour (Williams *et al.*, 2017). Research suggests that babies that are tube fed are more likely to be aggressive as adults, so owners choosing a hand-reared parrot should favour those that have been spoon- or syringe-fed (Fig. 11.7a). Hand-reared African Greys have been found to be more selective when forming human friendships than parent-reared birds (Schmid *et al.*,2006), and it is thought that minimal human contact during the hand-rearing process may make this more likely. Handling parent-reared chicks after eye opening has been shown to have great benefits in developing tame birds with lower stress responses to handling (Aengus and Millam, 1999; Collette *et al.*, 2000). Williams *et al.* (2017) reported higher incidences of stereotypies and less interaction with environmental enrichment in hand-reared birds, whether hatched in incubators or removed from the nest early, as compared with those reared by parents for longer periods. Hand-reared birds are at higher risk of developing aggression problems (Schmid *et al.*, 2006). This could be related to lack of learning about bite inhibition, which would usually be taught by the parents and siblings, and perhaps also frustration due to the bird finding it difficult to communicate with its human family and changes in dynamics when the bird reaches sexual and social maturity. These studies suggest that the ideal situation for rearing birds that are well adapted to domestic life may be to have parents who rear their chicks to weaning whilst human handlers regularly interact with and handle the chicks.

Heritability of some temperament and behavioural characteristics is recognized in many species and, although research in parrots is lacking, this is an important

Fig. 11.7. **(a)** Spoon or syringe feeding of hand-reared parrots should be encouraged rather than tube feeding. **(b)** Natural perches are ideal to allow chewing, foot exercise, exploration and novelty. **(c)** Encouraging the use of foraging feeders provides essential mental and physical stimulation. (Photos: author's own).

consideration. Feather-picking behaviour in captive Orange-winged Amazons has been found to have a significant genetic component (Garner *et al.*, 2006), suggesting that selective breeding may be beneficial in reducing the risk of such problems in captivity. Unfortunately, many breeding birds are rejected pets that have been relinquished to an aviary life for one reason or another. These birds are therefore not necessarily the ideal genetic stock for breeding the next generation of pets. Responsible breeders are starting to realize the importance of temperament of breeding birds and this is hopefully an area that will change dramatically in coming years.

Where breeders create a more natural environment for their nestlings, there is considerable benefit for behavioural development. Simple measures such as providing a towel for nestlings to snuggle into can create a similar feeling of reassurance to being brooded by a parent. Such comfort and security are essential for emotional well-being. Some breeders are starting to investigate the use of dark nest boxes for brooding young, rather than the commonly used clear plastic boxes, and it will be interesting to see whether this added security during early rearing has benefits for the later behaviour of these birds.

General Behavioural Requirements of the Parrot in the Home Environment

Husbandry requirements as related to behavioural needs

There are numerous texts that discuss housing, dietary and management requirements of parrots, so this section will concentrate purely on behavioural aspects. Space is of vital importance and the cage or indoor aviary should be as large as possible. In the wild, parrots fly long distances on foraging expeditions, so being confined to a cage is a stark contrast from their natural life. Ideally the confinement area must be large enough and shaped appropriately to allow flight, which in the larger species will mean an indoor or outdoor aviary rather than a cage. All parrots are likely to need confinement for a proportion of the day and so the area must be sufficient to provide for ample mental and physical stimulation. Provision of a complex physical environment that the parrot can enjoy exploring must be carefully balanced with leaving clear areas to allow flight (Box 11.4). Owners should provide a variety of types of environmental enrichment: this both allows the parrot freedom to make its own decisions and helps the owner to see which types of enrichment are most enjoyed, so that they know what to provide in the future. Natural perches are ideal (Fig. 11.7b) and fruit trees or others known to be non-toxic can safely be used. These provide not only perching and climbing opportunities but also opportunities for chewing. These can be replaced regularly, which also adds novelty.

Box 11.4. Advice for owners regarding appropriate stimulation.

- Daily access to a physically complex environment to allow climbing, swinging, flying, chewing and foraging.
- Cage/play stand locations must allow for social interaction and activity during the day, with time for quiet restful sleep in the evening and overnight.
- Social stimulation is crucial and will be discussed in more detail below.

Owners need to think carefully about the location of the enclosure. The three main considerations are: the social requirements of the parrot, their feeling of security and the need for 9–12 hours of sleep per night. The individual parrot's temperament will inform specific choices but ideally the cage should be in a busy area of the house where the parrot can be fully involved in family life but not close to throughways where it might feel vulnerable to potential predation. Research on Orange-winged Amazons showed that being placed in a doorway put the birds more at risk of developing feather plucking (Garner *et al.*, 2006). There must be provision of cover (preferably one side of the cage against a wall, some large toys to allow hiding and peeping, or perhaps a cover draped over part of the cage) and high perches to allow the parrot to feel secure. However, it must also be given opportunities to rest and should be in a quiet area, away from late-evening chatting, televisions, etc., in the evening and overnight. Depending on the layout of the home and the activity patterns of its occupants, it might be ideal to have the parrot's sleeping quarters in a separate room from the daytime active quarters, which may involve two cages, or a cage in the quiet area and a stand or play gym in the active area. This two-cage system also fits in with the natural behaviour of parrots roosting in one location and travelling to foraging sites during the day and helps the owner provide a predictable routine for their bird. The regular handling involved also improves the bird–owner relationship and reduces the chances of territorial behaviour developing around the cage.

Feeding should, as best as possible, replicate the natural behaviour of parrots as well as the relevant natural diet of the individual species. The diet is of particular importance, given research over recent years regarding the gut–brain axis and the connection between gut health and the general nervous system (Suchodolski, 2018). Providing easily accessible *ad lib* food, such as in a bowl, provides very little stimulation and also puts birds at risk of overeating and obesity. Parrots are naturally opportunistic foragers and introducing them to feeders that encourage this behaviour at an early age (or carefully as older birds) can give them hours of stimulation (Fig. 11.7c).

Wild parrots show a cyclical pattern to their daily activity: foraging and feeding early in the day, resting during the middle of the day and feeding again before roosting. Ideally owners should try to mirror this pattern by providing access to foraging activities in the early part of the day and again in the late afternoon. Parrots do continue to forage at lower levels throughout the day and should therefore ideally have continual access to foraging activities. Feather picking has been shown to be prevented and successfully treated by providing appropriate foraging activities and a physically complex environment for parrots. Severe feather picking can be very difficult to resolve fully and therefore it is highly beneficial to advise owners to use these enrichments on a daily basis to prevent the development of feather-damaging problems (Fig. 11.8). Owners also need to be prepared for their parrot to be very destructive when chewing. Provision of plenty of readily available chewable objects is essential. If the parrot attempts to chew a forbidden object, such as some wooden furniture, when out of its cage, the owner should calmly say 'no' and provide an alternative. Provided with consistent rewards for chewing acceptable items and consistent gentle interruption of forbidden items, parrots can easily learn which objects are accessible to them. Owners need to bear in mind that commercial toys are sometimes designed more for humans than their birds, and brightly coloured plastic 'indestructible' toys do not allow natural chewing behaviour.

Fig. 11.8. Providing appropriate chewing and destruction opportunities is vital in cases of feather plucking. (Photo credit: A.G. Turner.)

Bathing is another crucial behavioural requirement for parrots. In the wild, birds would shower in the rain or rub themselves on wet leaves rather than risk coming down to ground level. Domestic parrots often enjoy being sprayed with a fine-mist pump spray, such as those designed for spraying plants (Fig. 11.9a), and many also enjoy having a perch in the human shower. Those who are fortunate enough to enjoy access to an outdoor flight will be able to choose to shower in rainfall (Fig. 11.9b).

Continuation of socialization and habituation

Larger parrot species are dependent on their parents for a long time after leaving the nest. This can make an ideal period for taming a new pet bird whilst it is dependent on humans, without the need for potentially damaging early hand-rearing practices. Education of the newly weaned bird must be carried out with care so that the bird still learns to be independent and a socially competent and polite member of the human 'flock'. Whilst birds are learning to become independent, they seem to have a particularly high level of exploratory behaviour, which is thought to be related to learning to forage for themselves. This provides an ideal opportunity for the new owner to introduce plenty of novelty and foraging toys so that their bird is well adapted to enjoying these forms of enrichment throughout life.

Wing clipping

Wing clipping is a very common procedure that is carried out in captive parrots. There are arguments

Fig. 11.9. (a) Using a pump spray replicates rainfall for parrots who are confined indoors. (Photo: author's own.) **(b)** Showering is important for feather condition but can also provide much-needed enrichment. (Photo credit: E. Stapleton.)

for and against this measure but the author strongly believes that disabling a creature such that it is incapable of expressing a natural physical need should be avoided unless absolutely necessary. Flight is important for physical fitness as well as emotional well-being, particularly enabling anxious birds to show avoidance behaviour. Provided that owners are careful to close external windows and doors when their parrot is not caged, there is usually no safety issue with parrots remaining fully flighted. The exception to this is when a hormonal bird is in danger of attacking rival family members in defence of their territory, or a bonded human 'mate' during the breeding season. In these cases, wing clipping may be required as a short-term management strategy (see Chapter 20). If owners do wish to clip their bird it is crucial that this is done in a manner that will still allow some flight, so that the bird is able to glide safely to the ground or a perch but lacks the power for gaining height. It is also crucial that the clip is symmetrical, to allow controlled flight. Birds who are severely or asymmetrically clipped will be at risk of physical and emotional damage should they land abruptly or be unable to steer themselves away from danger. Young birds should be allowed to learn flying skills before being clipped; then as they become more capable and their flight stronger, they can be gently clipped if necessary.

Social requirements

Parrots are highly social creatures and in an ideal world they should never be in social isolation. Solo-housed birds are at greater risk of developing behaviour problems such as stereotypies and screaming compared with paired or group-housed birds (Rodriguez-Lopez, 2016; Williams *et al.*, 2017). Social stimulation is the most important and significant form of enrichment for them. Social interaction is flexible and unpredictable in a way that inanimate enrichment cannot be. However, many people do choose to keep only one bird and these owners still need to be able to leave their house for periods of time or be in areas of the house away from the parrot's living quarters. Parrots are not naturally predisposed to cope with being alone and active measures to teach the parrot to enjoy its own company should be taken. For example, once the bird is occupied with a favourite foraging toy, the owner can briefly leave the room and then return. This is a very similar process to teaching a puppy to accept being separated from owners (see Chapter 12). However, an avian companion is often an ideal solution.

Parrots of different species may learn to treat each other as social stimulation but ideally a companion should be of the same species. Wild parrots will come together in mixed-species flocks but these birds would not normally show social behaviour to those of a different species (see Chapter 5). There are many reports of cross-species friendships (Fig 11.10) but there is higher likelihood of compatibility within the same species. It is best to keep pairs of the same sex, unless owners are hoping to breed, and it depends on the species whether it is most appropriate to keep two males or two females together.

Fig. 11.10. These two Amazon parrots have bonded despite being different species. (Photo: author's own.)

Having two parrots can initially be double the commitment whilst the birds are introduced and trained, particularly if they are not siblings. Each bird will need taming and training as an individual as well as working with both birds together. Research on Orange-winged Amazons showed that same-sex pairs housed together were actually more confident to interact with their human handlers than birds living alone and were significantly less likely to show behavioural problems such as stereotypies or excessive vocalization (Meehan *et al.*, 2003).

Training

Training is not just about teaching tricks but about educating a parrot to be a pleasurable household companion. Having a selection of behaviours on cue improves the owner–parrot bond and hugely improves welfare, so that the parrot is allowed plenty of freedom out of its cage, and it also reduces the risk of having to use force for procedures such as nail trimming. Basic obedience training in young birds is particularly important to ensure that they have established behaviours before reaching full social maturity, when they are naturally independent of carers. It is often at social maturity that problems develop, so having a framework of household rules and general obedience cues before this age can make this transition into adult life smoother for the bird and the owner. Parrots will at some point require restraint, either for a procedure such as nail trimming, or blood sampling or wing clipping. Being a prey species, they are naturally very scared of being restrained and so teaching them to enjoy being held through positive reinforcement can be highly beneficial to both the parrot and the handler (Speer *et al.*, 2018). Parrots can be taught to be lightly restrained in a towel by asking them to step on to the handler's hand and then bringing the towel to the front of the bird and gradually teaching it to accept being wrapped whilst rewarding it with praise, head scratches or food. The bird must see what is happening and the handler must observe the bird's body language to ensure the bird remains relaxed. Grabbing a bird from behind is indicative of predation; it will almost certainly induce a fear response and must be avoided at all costs.

There is still considerable debate amongst those in the world of avian behaviour about the relevance of 'dominance' to explain patterns of behaviour. Wild parrots show very low levels of serious aggression and wounds are very rare. All social species have communication strategies to enable them to live in harmony with their conspecifics, because conflict uses up vital energy reserves and is not in the best interests of any individual. In the real world, the debate about dominance is relatively academic and is not nearly as relevant as the actual behaviours that are seen in parrots and how it is best to respond to these. The term 'dominance' is avoided here as it engenders a sense of power and command over another being, which is incompatible with the idea of sentient and emotionally sensitive creatures being a companion and friend to their caregiver. Instead, behaviours will be discussed in terms of what events precede the behaviour (antecedents) and what actions follow the behaviour (consequences). This is a more appropriate manner in which to examine interactions between parrots and their owners and also parrot–parrot interactions. Parrots are highly intelligent creatures and, as all animals (including humans), their prime aim in life is to maintain emotional stability through obtaining and keeping resources that are important to them for survival and for enjoyment. If a parrot shows a particular behaviour, for example opening its beak wide as if to bite, and the consequence of this is beneficial to the parrot, it will learn to continue to use that strategy in the future. If a parrot steps up on to its owner's hand and is rewarded for this with a head scratch or a nut, it is likely to do it again in the future.

Parrot training is generally similar to that in other species in terms of using reward-based techniques being most beneficial to the owner–parrot relationship (Fig. 11.11) (Martin, 2007). Aversive techniques can be particularly damaging in a prey animal: parrots will be unwilling to work with someone they feel threatened by. It is also the case that a fully flighted bird that does not feel comfortable with the techniques used will simply fly away and refuse to participate. Wing clipping must not be seen as a means of getting a parrot to cooperate. Some texts recommend aversive techniques such as a hand wobble or a sudden hand drop if a bird on the hand bites. This can be extremely damaging to the trust

Fig. 11.11. Teaching a bird to enter its head into a harness for a food reward ensures its willingness to be engaged with this process. (Photo: author's own.)

between the bird and its handler and must be avoided. The cage should also not be used as a punishment, as the bird must feel safe and relaxed in its own home. The one aversive technique that can be used very effectively as a signal to the parrot that the owner is not happy with its behaviour is brief social isolation. So, for example, if the bird is showing inappropriate chewing and cannot be redirected to an appropriate object, the owner can walk out of the room for a short time and return when the bird has stopped chewing. Some texts advise social isolation for long periods of time but this is inhumane and not necessary and also puts the bird at risk of developing serious behaviour problems such as feather picking or screaming. Parrots are intelligent and will learn very quickly which behaviour has resulted in the owner walking out and which behaviour resulted in the owner coming back.

As with all reward-based training, one must identify the rewards that are potent for each individual bird. Parrots are unable to preen the feathers on their own heads and therefore allopreening not only serves to cement and maintain bonds but also serves a functional purpose. This can be a very useful tool to use as a reward in some parrots, as many will thoroughly enjoy a good head rub (Fig. 11.11). Many will 'work' for praise, food or access to toys. Different foods and toys will have different values, with more demanding training requiring the use of higher-value rewards. For example, a bird who is scared of hands is likely to require a higher-value food reward for learning a step-up cue than a newly arrived, well-handled confident bird.

Many of the larger species are capable of complex thought and learning and this means that alternative techniques such as model-rival system used by Irene Pepperberg during her work with African Greys can be very useful (Pepperberg, 2006). This can be an excellent way of teaching parrots to try novel food items or to show them how to use new foraging enrichments. Research has shown that African Greys are more likely to share with a sharing person and more likely to be selfish with a selfish person (Péron *et al*., 2013, 2014). Therefore, the behaviour of the humans in the household is potentially influential on the behaviour of the parrot. However, it is important to note that birds will vary in their cognitive abilities – not only between species but also with different early experiences. Limited environmental and social complexity during early development will reduce their capacity for complex learning.

Parrots have a lifelong ability to learn new vocal sounds and this can be used as a tool for enhancing human–parrot communication. The parrot does not necessarily need to learn to mimic sounds for it to recognize and understand them when used by humans. For example, the parrot who learns 'bye bye' when the owner is heading out to work and learns the context of that word will be less likely to be stressed at the departure than a bird that has no contextual cues that it is to be left alone. Teaching human language to parrots is done in a similar way as teaching dogs or children: by consistently using words in a particular context, the bird associates the word with the context. This can be extremely useful for creating predictability, which is important for welfare.

Some texts suggest that parrots should not be allowed to perch in high places, as this can encourage them to be 'dominant' over their owners. As explained earlier, the natural behaviour of parrots is not compatible with the type of fixed hierarchical structures that underlie this interpretation of behaviour (Chapter 5). All birds prefer to perch in higher places for safety reasons. They are very vulnerable to predation when down low on the ground. Being high not only ensures that they are well out of the reach of ground predators and the heavier climbing predators, but also allows them an excellent view of their surrounding area to keep a lookout for potential threats. Parrots will alarm call to alert the rest of the flock if they spot a potential hazard such as an overhead bird of prey and this may result in the flock taking flight (Chapter 5). It is therefore a serious welfare issue to force parrots to stay low down and it is an essential requirement for them to have high perches in their cages and also when out and about in the house, to ensure that they feel safe. From the owner's perspective the advice about not allowing a bird on to one's shoulder is sensible. This is not to do with the parrot's perception of 'dominance', but more to do with the vulnerability of the owner's face and head to physical damage should the parrot show any aggression in this location. The bond with the owner may be strong and trusting and the owner may feel the bird would not bite them in this context. However, there is a risk of re-directed aggression; for example, if a potential rival (such as the spouse of the owner) approaches, the parrot may attack the owner as they are unable to easily reach the rival. In order for high perches to be practically implemented it is important that owners train their birds to recall on cue or to come down from the perch on to a hand or a stick. Those who argue for the 'dominance' explanations often use the argument that the bird refuses to come down from a high perch. However, birds can easily be trained to come down from high perches if there is an appropriate incentive or reward and they feel safe at the lower level.

The use of the beak

Parrots use their beak for exploring textures and tastes, manipulating food, chewing, biting and as an extra limb when climbing. Parrots can use their beak very skilfully to manipulate objects. Carducci *et al.* (2018) found that providing keas with tactile as well as visual information facilitated success in visual discrimination tasks. Young birds would usually learn the bite strength of their beak through interaction with their siblings and parents. This is a component of learning other social communicative skills, as discussed earlier. Birds will only use a severe, hard bite if seriously threatened. They therefore need to learn the strength of their bite so that they can use a milder warning bite, which may just involve a grab with a beak with very little pressure. Hand-reared parrots are often lacking in this bite inhibition training and owners should carefully teach their bird what is acceptable. The response to being bitten will depend on the exact circumstances and on the temperament of the individual bird. The owner must be careful not to scare the bird by shouting or arouse it further by squealing. Brief withdrawal of social interaction can be very effective. If the owner is very vocal this might excite the bird, who may learn to bite in order to get such a response. If the bird has bitten through fear, then a vocal reaction from the owner may further frighten the bird and make it more likely to bite again in similar circumstances in the future. If the bird has become overexcited and bites, then allowing it to calm down before interacting again would be wise.

Introduction to other pets in the household

This is an area where specific advice needs to be given by a parrot behaviourist who has visited the home environment. Assessment is needed of the parrot's temperament, the response of the other animals to a flighted, noisy creature and the level of training achieved by the owner of both the parrot and the other pet. Owners must be aware that parrots are prey species and therefore particularly vulnerable to feelings of stress from potential predators such as dogs, cats and young children. When the parrot is first introduced to the home it is best to give the bird some quiet time in its new surroundings to settle in and feel comfortable before any introductions are made. Parrots should never be left out of their cage unsupervised with dogs or cats.

Breeding seasons

As birds cannot be routinely neutered, parrot owners must be prepared for seasonal variations in their bird's temperament due to hormonal changes during the breeding season. Hormonal problems will be discussed in more detail in Chapter 20. General advice to reduce the likelihood and intensity of any problems include: (i) making sure all household members are involved in the care of the parrot, to avoid pair bonding with one individual; and (ii) using the two-cage system described above to reduce territorial behaviour around the cage. Nutrition is also important, as a high-fat diet can trigger breeding condition.

References

Aengus, W.L. and Millam, J.R. (1999) Taming parent-reared Orange-winged Amazon parrots by neonatal handling. *Zoo Biology* 18, 177–187.

Carducci, P., Schwing, R., Huber, L. and Truppa, V. (2018) Tactile information improves visual object discrimination in kea, *Nestor notabilis*, and capuchin monkeys, *Sapajus* spp. *Animal Behaviour* 135, 199–207.

Collette, J.C., Millam, J.R., Klasing, K.C. and Wakenell, P.S. (2000) Neonatal handling of Amazon parrots alters the stress response and immune function. *Applied Animal Behaviour Science* 66(4), 335–349.

Fox, R. (2006) Hand-rearing: behavioural impacts and implications for captive parrot welfare. In: Luescher, A. (ed.) *Manual of Parrot Behavior*. Blackwell Publishing, Ames, Iowa, pp. 86–91.

Garner, J.P., Meehan, C.L., Famula, T.R. and Mench, J.A. (2006) Genetic, environmental, and neighbor effects on the severity of stereotypies and feather picking in Orange-winged Amazon parrots (*Amazona amazonica*): an epidemiological study. *Applied Animal Behaviour Science* 96, 153–168.

Martin, S. (2007) The art of training parrots. *Journal of Exotic Pet Medicine* 16(1), 11–18.

Meehan, C.L., Garner, J.P. and Mench, J.A. (2003) Isosexual housing improves welfare of young Amazon parrots. *Applied Animal Behaviour Science* 81, 73–88.

Pepperberg, I.M. (2006) Cognitive and communicative abilities of Grey parrots. *Applied Animal Behaviour Science* 100, 77–86.

Péron, F., John, M., Sapowicz, S., Bovet, D. and Pepperberg, I.M. (2013) A study of sharing and reciprocity in grey parrots (*Psittacus erithacus*). *Animal Cognition* 16: 197.

Péron, F., Thornberg, L., Gross, B., Gray, S. and Pepperberg, I.M. (2014) Human–Grey parrot (*Psittacus erithacus*) reciprocity: a follow-up study. *Animal Cognition* 17, 937.

Rodriguez-Lopez, R. (2016). Environmental enrichment for parrot species: are we squawking up the wrong tree? *Applied Animal Behaviour Science* 180, 1–10.

Schmid, R., Doherr, M.G. and Steiger, A. (2006) The influence of the breeding method on the behaviour of adult African grey parrots (*Psittacus erithacus*). *Applied Animal Behaviour Science* 98(3-4), 293–307.

Serpell, J.A. (1982) Factors influencing fighting and threat in the parrot genus *Trichoglossus*. *Animal Behaviour* 30, 1244–1251.

Speer, B.L., Hennigh, M. and van Zeeland, Y.R.A. (2018) Low-stress medication techniques in birds and small mammals. *Veterinary Clinics or North America: Exotic Animal Medicine* 21, 261–285.

Suchodolski, J.S. (2018) Gut brain axis and its microbiota regulation in mammals and birds. *Veterinary Clinics of North America: Exotic Animal Medicine* 21, 159–167.

Williams, I., Hoppitt, W. and Grant, R. (2017) The auditory enrichment, rearing method and social environment on the behaviour of zoo-housed psittacines (Aves: Psittaciformes); implications for welfare. *Applied Animal Behaviour Science* 186, 85–92.

12 Running Puppy Parties and Kitten Evenings

Emily Blackwell

Puppy parties and kitten evenings provide an opportunity to supply information for owners to ensure that their new puppy or kitten grows up into an emotionally resilient adult dog or cat, with the capacity to cope with the challenges of daily life and able to make decisions without being overly dependent on their owner. Providing this service fulfils the veterinary practice's duty of care to clients and their pets, but is also important in the prevention of physical disease and problematic behaviours. Using a combination of advice and practical assistance, new pet owners can be supported during this important, but sometimes challenging, period of their pets' lives.

Puppy parties and kitten evenings also offer an invaluable public relations exercise, providing a beneficial addition to the services of the practice. As they are often the first contact that owners have with the practice when they obtain their pet, they are a useful opportunity to establish a good relationship with the new client and influence their perception of the practice as a whole. In addition, these events function to promote other practice services comprising overall healthcare, including microchipping, nutrition, dental care, pet insurance, travelling, neutering and husbandry, rather than simply being seen as somewhere to seek advice when a pet is unwell. However, a practice should not see these services as something that they are obliged to do. Unless there are enthusiastic, well-informed staff who are keen to run them, it is better for practices to collaborate with external trainers or behaviourists, as badly run puppy parties or kitten evenings, or the provision of poor advice, can do more harm than good.

Throughout the main text of this chapter, superscript numbers refer to Online Resources, detailed in the endnotes before the chapter's References section.

Puppy Parties: a Suitable Environment

The venue

The venue is important, as a very large room may not allow close monitoring of puppies and too little space will not allow puppies enough room. Puppy parties are often run in a consulting room or other larger area such as a waiting room. Whatever the location, the room needs to be large enough to allow puppies to distance themselves from other puppies, people, sounds or objects should they wish to. Comfortable seating should be provided for owners, so that they feel relaxed.

The room should be well ventilated and the floor should be non-slip, so that puppies feel secure and easily able to move towards or away from things and to avoid young puppies suffering injuries. Allowing less confident puppies to

DOI: 10.1079/9781800621312.0012

hide is also important and physical barriers to block visual access can be useful. Puppies will chew anything, so it is vital that wires from any electrical equipment are inaccessible and the room is cleared of any small items that puppies may ingest. A safe enclosed outdoor area for training is also useful, as puppies can be taken out for training exercises during the lesson.

Equipment

There should be a range of toys and safe objects for puppies to explore with their teeth. A variety of textures and sizes should be made available to allow the puppies to choose and there should be plenty for all of the puppies, to avoid any competition. There are a number of products on the market that have been designed specifically for puppies to chew on.

More than one water bowl is essential (playful puppies are guaranteed to climb into the bowl and upturn it). Cleaning materials are essential to mop up 'accidents' quickly, but the materials should be stored safely out of reach of the puppies and any children in attendance. It is helpful to have some examples of the training devices that you intend to discuss with owners, such as harnesses and clickers, as they may not be familiar with these. It is also useful to collect a number of 'props' for use in practical habituation sessions, such as a motorcycle crash helmet, umbrella, backpack, hat and fluorescent jacket. A CD player and a CD containing everyday indoor and outdoor noises (e.g. Sound Therapy for Pets[1]) is also useful for demonstrating how to habituate puppies to new sounds.

Handouts

Puppy parties provide an excellent opportunity to inform owners, using literature and resources from reputable sources. This will not only encourage clients to be responsible pet owners from the earliest possible point, but will also avoid them turning to the many sources of outdated advice and misinformation available on the internet.

Important information that is discussed during the puppy party should also be provided via written handouts so that owners can refer to them between classes. Information covering basic care of puppies (for example vaccinations, worming, fleas and feeding) should be provided, alongside training articles, advice on choosing appropriate dog training classes, doggy daycare and grooming services. Owners should also receive information about the role of early development in shaping later adult behaviour, along with written practical advice about how to socialize and habituate their new pet.

Staff

Although this type of socialization class is commonly described as a 'puppy party', it is important that the classes are not seen as unstructured 'anything goes' play sessions. It is vital that puppy parties do not descend into a 'free for all', where some puppies learn to 'bully' others, or become overwhelmed by the boisterous play of other puppies. The aim should be calm controlled interactions and so a good level of supervision is essential. Playtime is fine, as long as there are also periods where puppies are restrained and rewarded for remaining calm, learning that they do not always get to rush up to and play with other dogs.

Watching puppies to ensure that interactions between them are calm and controlled whilst being distracted by questions from owners is not easy, so *at least* two members of staff should be present in a class containing four to six puppies. This will allow at least one person to intervene where inappropriate interactions are occurring. Highly aroused puppies should be given quiet time, away from other puppies, if they become frustrated.

Staff need to have a good knowledge of the behavioural development of puppies and a good understanding of applied learning theory and training techniques. Ideally, they should have specific training in how to run puppy classes and there are a number of continuing professional development (CPD) courses available for people wishing to obtain this. Ideally course tutors should have a higher level of qualification to be able to deal with the early signs of any undesired or problem behaviours in the puppies. This includes veterinary behaviour specialists[2], or clinical animal behaviourists (CAB[3], ABTC[4]). Where classes are run by individuals without advanced behavioural qualifications, the use of force-free positive reinforcement-based techniques should be a prerequisite, along with a minimum standard of dog training qualification, for example Animal Training Instructor Level (ABTC).[4]

Who should attend?

Practice policy will vary in terms of a minimum age but generally speaking the earlier that new puppy owners receive information about socialization and habituation, the better. Most practices allow puppies between 7 and 14 weeks of age to attend, depending on assessment of local risk of infectious diseases, with older puppies being offered one-to-one sessions if they are already over 14 weeks. Puppies must all be in good health and an email in advance, plus a sign on the entrance to the puppy party room, is a good idea to remind owners not to bring their puppy in if it is unwell.

Some practices offer a first meeting with no puppies present, so that owners have a chance to ask questions and can concentrate on the information being given, without having to supervise their puppy at the same time. As many members of the family as possible should be encouraged to attend, at least the first puppy party. Well-behaved children should be made welcome, as they can actively participate in some of the practical exercises, but if they do not behave as requested then it is better to ask them not to attend further puppy parties, as children who are rough with puppies, too noisy, or unable to sit quietly when asked can disrupt the party and even cause puppies to become fearful. Managing puppies and people during classes is a skill that will improve with experience, but the aim is to offer a friendly but professionally run class, where puppy owners feel supported and confident that their precious new puppy is safe in your hands.

Topics to cover

Each class should begin with a question-and-answer session including introductions (week 1) and a discussion of progress on subsequent weeks. Small groups and a welcoming atmosphere should encourage owners to discuss their own personal experiences and ask questions. If an owner is particularly quiet, it is a good idea to take them aside for a brief one-to-one chat during a break in the class to check there are no problems.

The most important behavioural topics are outlined below. Whilst the initial information on recognizing when a puppy is worried and how to carry out socialization and habituation should always be covered during the first week, the remaining topics can be covered on an ad hoc basis. Having a rigid structure and timetable of topics for each week can lead to owners being forced to wait for advice about a specific topic. This can be detrimental, as it is likely to lead to a worsening of the behaviour and to a problem that is less easy to rectify.

Recognizing signs that a puppy is worried

Throughout each session all puppies should be making wonderfully positive associations with the veterinary environment, other puppies and with the people attending. However, it is very easy for the enthusiastic puppy owner, or practice staff member, to overwhelm a puppy in their enthusiasm to teach the puppy about new things, so making sure that both owners and staff can recognize when a puppy is worried should be a priority (Box 12.1).

A relaxed puppy will show a loose, happy expression (often with mouth open loosely in a 'smile'). Its body will not appear tense and it will loosely wag its tail. The puppy may put its front end down and rear in the air – this is a play bow, which is an invitation to play.

The use of photographs, diagrams and videos is recommended to explain these signs clearly to puppy owners.[5]

Socialization

The first months of life are an essential period for puppies to learn about social contacts and start to develop social skills (Chapter 8). Puppy parties provide an excellent opportunity for puppies to meet and interact with puppies of different breeds, adult dogs, unfamiliar people and other species, under expert supervision. They also

Box 12.1. Signs that a puppy is not comfortable with a situation.

- Cowering – lowering head or whole body
- Tucking tail between legs (sometimes whilst wagging)
- Ears sideways or held back against the head
- Lifting a front paw
- Whale or half-moon eye (this is when the white of a dog's eye is visible at the corners and/or rim)
- Yawning when not tired
- Lip licking when there is no food around
- Avoidance behaviours – moving away or reluctance to approach, hiding, turning head away
- Shaking
- Freezing/tension in muscles
- Panting
- Rolling over on to back and exposing belly
- Toileting
- Displacement behaviours*
- Vocalizing (e.g. barking or whining), sometimes whilst retreating

If a puppy shows any combination of these signs then you should intervene to remove the puppy, or whatever it is finding threatening.

(*Displacement behaviours are normal behaviours displayed out of context. They indicate conflict and anxiety. The dog wants to do something, but is unsure of the consequences and therefore supresses the urge to perform the behaviour and does something else instead. Examples of common displacement behaviours include: scratching; licking or biting at paws or other body part; sniffing the ground or other object. These are all things that dogs do in other situations too, so remember to look at the context to determine whether the dog is feeling anxious.)

allow practice staff to make sure that owners have a good understanding of how social behaviour develops in dogs (see Chapters 2 and 8). This should include an explanation of normal canine social behaviour, dispelling any misunderstandings about the social structure and behavioural motivations of dogs (such as misinterpretations of the term 'dominance'[6]) and explaining that dogs are an obligate social species that live in groups (Chapter 2).

During both socialization and habituation exercises it is important that the puppy is relaxed and calm in order to avoid sensitization. Interactions should therefore be under continuous close supervision, to ensure that they remain calm and controlled, whilst still allowing puppies to learn how to use, read and respond appropriately to canine signals. Practical socialization exercises should include opportunities for controlled interactions with other puppies and well-socialized adult dogs. This should include off-lead play and on-lead interactions, as well as periods when the puppy is expected to remain still and quiet and not approach other puppies.

Play allows puppies to develop and practise social skills, refine motor abilities, form attachments and may even facilitate learning. However, play sessions must be enjoyed by all parties involved, as inappropriate play can lead to a dislike of interactions with other dogs (Box 12.2).

Very confident adult cats attend some puppy parties, although this should be carried out with extreme caution, as the cat must be completely happy in the presence of puppies and the puppies must be kept under strict control. Puppy parties are also a great place for the puppies to meet and be gently handled and rewarded by unfamiliar people.

Habituation

The first three months or so of a puppy's life are very important, as they are rapidly learning about their environment and experiences: things they learn during this period can have impact throughout their lives (Chapter 8). Gently introducing them to all of the sights, sounds, smells and experiences that we would like them to be happy with as an adult dog is therefore vital.

Habituation is a simple form of learning whereby the response to an inconsequential novel stimulus decreases with repeated exposure. Habituation is important in filtering the large amounts of information that all animals receive

Box 12.2. Signs of appropriate and inappropriate play.

Signs of appropriate play between dogs:

- Play signals such as a play-bow posture and play face
- Bouncy exaggerated movements
- Regular breaks in wrestling, biting or chasing, with both dogs keen to return to play
- Self-handicapping: the bigger, stronger dog allows the smaller, weaker dog to knock or pull them over, or 'win' at tug games
- Role reversals:
  - Chaser becomes chased
  - Positions change – dog on top becomes dog underneath
- Biter becomes the one being bitten

Signs that could indicate 'inappropriate' play:

- Fast chasing with low, flat posture rather than bouncy pursuit
- Two or more dogs chasing one dog
- No role reversals
- Big dog being uninhibited (rough), body slamming, or running very quickly towards a smaller dog (risk of injury)
- Signs of fear in either dog
- Yelping

from their surrounding environment. By habituating to less important signals, an animal can focus attention on the features of its environment that are likely to be of importance in terms of survival.

As well as teaching puppies that these everyday stimuli are nothing to worry about, we also want to build positive associations with them to 'proof' the puppy against becoming fearful if it does have an unpleasant experience in the future in these contexts. In order to do this the puppy should be rewarded with treats to build positive associations with each new stimulus.

There are a number of 'checklists' available for owners to use as a guide to some of the things that a pet might encounter. Any list should include a wide variety of noises, such as traffic, loud voices, fireworks, roadworks, washing machines, vacuum cleaners, airplanes and children playing. Encountering different surfaces, such as sand, gravel, stairs, grass and puddles, as well as different places, such as the veterinary surgery, pet shops, other people's houses, grooming parlour and kennels, will allow the puppy to habituate to the sounds, scents and sights associated with these varied environments. Experiences such as travelling in vehicles, encountering livestock, people of all ages and ethnicity, people wearing uniforms, people carrying things, people in wheelchairs or with beards or wearing helmets or hats are also an important part of a puppy's education. Sound effects for use in habituating puppies can be downloaded free of charge at Sound Therapy for Pets.[1]

When attempting to introduce a puppy to something for the first time, the most important thing to remember is that puppies must be comfortable, happy and relaxed and should be rewarded with something nice (food, praise or play) in order to make positive associations with them. Puppies that attend puppy parties should be keen to come in through the surgery door in search of treats well into adulthood because of the good associations made during their early months of life.

As all puppies are unique and will have had different previous experiences, there is no prescriptive programme for habituation, but an example of the process for habituating a puppy to being left alone is included in Box 12.3. Controlled exposure to novelty requires close supervision, as each puppy will have differing requirements. Sensitization becomes more likely if the puppy is emotionally aroused, the stimulus is presented at high intensity, or is particularly salient. To begin with the intensity of stimulus should therefore be minimized, for example low volume or far distance, and the rate at which this is increased will depend upon the puppy's response. If the puppy remains completely relaxed, then the intensity can be increased, for example the volume can be turned up slightly or the distance decreased. Puppy classes provide an opportunity to expose puppies to everyday sounds at low level and demonstrate best practice to owners. Bear in mind that if you are attempting to do this in a class situation, you will need to proceed at the rate suitable for the least confident puppy in the group.

Handling

Prior to starting habituation to handling, owners should be made aware of what constitutes appropriate handling and why, including appropriate grooming techniques and equipment. Habituation to being handled involves gently getting the puppy used to being examined all over and is best done at home, where the puppy is likely to be most relaxed. Puppies need to be habituated to the types of handling that they might experience during a veterinary examination; they should go into consultation rooms, meet people in veterinary scrubs, see, smell and investigate veterinary equipment such as stethoscopes and at the same time make positive associations using treats and play.

Box 12.3. Example owner handout detailing how to habituate a puppy to being left alone.

Many owners spend time introducing their puppy to other dogs, unfamiliar people, household objects and noises, but overlook one of the most important things that we need puppies to see as a normal part of life – being left alone by their owner.

The best way to prevent a puppy from developing separation problems is to teach it right from the start that being alone is a normal part of life.

- To do this you need to begin to leave your puppy alone for very short periods.
- Start at a time when your puppy is relaxed and ready for a snooze.
- First make sure that your puppy has been to the toilet and has had some exercise (a play session is ideal, but allow the puppy a few minutes to calm down afterwards before you begin the leaving process).
- Settle the puppy down on a comfy bed with something tasty to chew on (make sure that it's something that is safe to leave with the puppy), then say a new word such as 'Relax' or 'Settle' and move away from the puppy for a few seconds, before going back to it and rewarding with a toy, treat or praise.
- The idea is to teach your puppy that being alone is actually fun!
- Once you've done this 3 or 4 times and are certain that your puppy is relaxed (i.e. continues to rest on the bed chewing the treat), move progressively further away and for longer. Increasing the distance depending upon the puppy's reaction – if your puppy stops chewing and cries or moves, then go back to the previous stage and try again.
- Never leave your puppy so long that it starts to become distressed.
- If at any stage your puppy does become distressed (for example, cries or scratches the door/crate if you have progressed to the point where you have left it behind such a barrier) wait until the puppy is quiet for a moment, then return to it. (If it doesn't naturally stop after a few seconds, then you can make a distracting noise, such as bag rustling, to allow you an opportunity to return whilst the puppy is quiet).
- Now go back a stage to where your puppy was relaxed and try taking even smaller steps.
- Continue to gradually increase the time that you are away until you can leave your puppy for 30 minutes whilst you are in a different part of the house.
- The speed at which you can do this varies a lot between puppies – for some it may be fairly quick, whilst others take longer.
- Then you can begin to leave the house. Start going out through the door before returning, then going out and shutting the door, then going out for longer periods of time. When you reach this point, start to vary the length of time that you are out.
- Puppies should never be left alone for long periods, as they will need to eat and toilet frequently. If you do have to go out and leave your puppy for longer periods, arrange for a dog sitter or friend to help out for a short while.
- If your puppy becomes distressed even when you try to take a few steps away, please contact a clinical animal behaviourist (CAB[3], ABTC[4]), as these things are far easier to resolve early on, rather than after the behaviour has become well established.
- Just ignoring the puppy in the hope that it will learn to cope can actually make things worse!
- Although it is tempting, don't allow your new puppy to spend the whole time with you during these early weeks, as you may be storing up problems for the future. Create periods of separation in your daily routine.

Owners should also be shown how to administer medication to their puppy, clip its nails and clean its eyes and ears.

Training a puppy

Dog training is not just about teaching traditional 'obedience commands' but should instead be thought of as a means of ensuring that puppies know what is expected of them and have realistic expectations about social interactions and how they respond to their environment. This avoids the risk of either anxiety or frustration because puppies do not know how to cope with new situations, or have misplaced expectations. For example, puppies that are always permitted to rush over and boisterously greet any other dog or person that they encounter may become frustrated if they are later on a lead and unable to do so. Teaching puppies that there are times when they have to sit quietly apart from other puppies in the class is therefore an essential part of their early learning.

Puppies do need house rules, but these are not to 'show the puppy who is boss', or produce a subservient companion, but rather to make the puppy's environment consistent and predictable, making it less prone to confusion and anxiety. In many cases, 'house rules' vary between each household depending on owner preferences (for example, whether puppies are allowed on sofas and beds, have access to particular rooms, etc.). What is essential is that the 'rules' are applied consistently by everyone. This means that each household will have its own set of rules, agreed by the whole family, that are positively and consistently applied. Where there is disagreement, or different rules for different contexts, for example the dog is allowed on the sofa only at certain times, then the puppy will need to be given clear signals (known sometimes as 'discriminative stimuli') that make the context clear, such as being verbally invited to get up, or the presence of a cover or blanket on the sofa. This enables the puppy to learn not only what behaviour is expected, but also when and where the different rules apply. Puppies are likely to learn at different rates and will be motivated by different things (such as praise, food, play) and it is important that owners do not compare progress and feel that they have failed if their puppy does not achieve as much during practical training sessions as some of the others. Owners should also be made aware that a class situation may not be suitable for all puppies and whilst most do gain in confidence if allowed to proceed at their own pace (some puppies remain on their owner's lap throughout the first session and only go down on the floor to explore once the other puppies have left the room), others may be more suited to one-to-one sessions.

Whilst training is a really important part of puppy education, over-training can limit self-confidence. Owners need a good level of control over their pet in order to avoid antisocial behaviour and for safety purposes, but this does not require the dog to be dependent upon instruction from the owner for its every move. A highly responsive, self-confident, 'well-trained' dog can be attained through consistently applied positive training methods, where instructions should be thought of as 'cues' rather than 'commands'.

Although there are an increasing number of studies supporting the notion that inappropriate training[7] using aversive methods induces fear and limits emotional stability, there is still a great deal of 'traditional' advice online and in the popular media that recommends punitive methods.

It is therefore essential that owners attending puppy parties are provided with information about the most humane and effective training methods and how to apply them. All training should be positive and 'force free' and owners should be guided in choosing a follow-on training class. Things to look out for include classes containing relaxed dogs, instructors who are able to recognize and deal appropriately with anxious dogs, for example by removing them from the situation that is causing them to be worried. Owners should be cautioned to avoid classes where lots of dogs appear to be cowering, have their tail between their legs, or do not make eye contact with their owner or training instructor. They should find out in advance what training equipment and methods are used and avoid any that recommend techniques that rely on inducing fear or pain, such as aerosol sprays or check chains. Advise owners to avoid classes where shouting at dogs or any form of physical punishment is advocated. Instead, advise them to seek classes in which dogs are motivated to show the desired behaviours through the use of rewards such as food treats or playing with a toy, and not through fear of the consequences of showing the wrong behaviour. Competent, qualified instructors will welcome scrutiny and are more than happy for an owner to attend to observe without their puppy prior to formally enrolling for the classes.

Training classes, although generally larger than puppy parties, should still be limited to an appropriate number of dogs and owners for the situation. For example, the Association of Pet Dog Trainers UK (APDT (UK)) recommends no more than eight puppies in a class with an instructor and one assistant. You should also check that the instructor is not giving advice beyond their level of knowledge or qualification. For example, you should be wary of instructors who give advice about serious behavioural disorders, such as aggression, or severe fears. Owners should be advised to seek referral by their veterinary surgeon to a qualified clinician[3] in such cases.

Common puppy problems

Puppy classes can be used to help the owners of new puppies to deal with the common behavioural issues that arise with their new pets (Box 12.4). A note should be made on the puppy's clinical record of any issues raised and these should be followed up with owners in the following week, or by telephone if raised in the final session. More serious issues such as aggression towards people or other animals, separation-related behaviour or fearful responses should be directed towards the veterinary surgeon, who can arrange appropriate referral.

Box 12.4. Puppy party training topics.

Training topics that could be covered in a puppy party setting include:

- Training equipment available and how to use it
- Training techniques, e.g. clicker training
- Recommended reading (e.g. Bailey, 2008; Wilde, 2012; Zulch and Mills, 2012)
- Practical training exercises:
  - Recall practice
  - Walking on a loose lead
  - Learning to sit
  - Learning to lie down
  - Learning to stay.

Play-biting and mouthing

As soon as puppies become active, they begin to play physical games with their littermates. In the very early days this behaviour is allowed, but as they grow stronger their sharp teeth are no longer tolerated by adults or littermates. Puppies soon learn that hard biting or mouthing is not acceptable and will result in an end to the fun and by 18 weeks of age puppies have usually learnt to inhibit their biting. Puppies sold as pets usually go to their new home before this process is complete and as a result may continue to mouth or bite hard in their new home. The simplest way to teach a puppy not to do this is simply to stop playing when you feel your puppy's teeth and, once the puppy has calmed down, redirect the behaviour to an alternative object, e.g. a chew.

Toilet training

For many puppy owners, toilet training is a simple process, but for others it can be a source of much frustration. Puppies have limited bladder control and need to urinate at least every hour or two. They will urinate soon after waking up and shortly after eating, exploring or playing. Basically, a puppy may need to toilet after any form of activity, so their toileting needs should be accommodated without delay. In order to teach a puppy to toilet on cue, a word or phrase should be repeated whenever the puppy starts to toilet. Puppies should never be punished for accidents indoors as this simply teaches them not to toilet in front of the owner. Every opportunity possible should be taken to reward them when they toilet in the correct place. Every toileting episode is a potential learning opportunity and the more often the owner manages to get the puppy to the correct toileting location, the quicker the learning will occur.

Inappropriate greeting behaviour: jumping up

Adult wolves returning to the cubs after a hunt are ceremoniously greeted by them rushing up and licking their muzzles, stimulating them to regurgitate food. Adult wolves also show this muzzle-licking behaviour as a form of greeting towards other adults to inhibit aggression. A puppy in a domestic environment will also try to do this, but in the puppy–human interaction the puppy must jump up in order to reach the faces of returning humans. Most owners greet their puppy with enthusiasm and encourage the puppy to jump up. Even looking at the puppy, or telling it to stop, can be enough to encourage it to continue. This results in the puppy learning that greeting people in this way is rewarded. As the puppy grows bigger, jumping up at people can become a problem and often leads to confusion and anxiety when owners begin to show a negative reaction to the puppy's greeting.

The most effective way to combat this is to teach the puppy an alternative greeting behaviour, such as sitting down, and consistently reward it for not jumping up.

Stealing

Puppies will naturally pick things up to investigate them. In addition, stealing things is often an attempt to gain human attention. Puppies learn very quickly from an owner's immediate reaction that picking up something and running off with it is a very good way of getting their owner to laugh or jump up and chase them. The object may be something precious, like one of a pair of favorite shoes, or something dangerous, like a knife. Hence the response from the owner to immediately retrieve the object is totally understandable. Unfortunately, it can also be very entertaining for the puppy, and teaches it a neat trick to guarantee people's attention.

To avoid this situation, owners should be advised to try to keep all dangerous or precious things out of their puppy's reach. Puppies will initially pick up objects to play with, or to chew, both of which are natural behaviours; hence ensure that the puppy also has plenty of appropriate objects available (e.g. safe dog toys, chew 'toys'). When the puppy picks up an appropriate object the owner should reward them with praise or a game.

If the puppy picks up something inappropriate the owner should not tell the puppy off, as punishing this behaviour may either been seen as reward (often even 'bad' attention is better than no attention) and encourage them to do it again, or as threatening, which may lead to a defensive response from the puppy in the future. If the puppy does take something and is likely to harm itself, the owner should distract the puppy with an appropriate object and then reward it when it changes its attention to it. Puppies do not know instinctively what are their own toys/chews and what is inappropriate to pick up, so be ready for them to make mistakes and respond accordingly. It can also help to train the puppy to 'give' or 'leave' objects, so that the owner can quietly and calmly ask them to give it up in return for a reward.

Attention seeking

Most puppies highly value human attention and will learn very quickly what works to obtain it. Some of these behaviours can be problematic for the owner, e.g. whining or nudging. However, once made aware of why their puppy is showing this behaviour and given suggestions for alternatives to reward, such as lying quietly on their bed, this behaviour can be modified fairly quickly. If the behaviour worsens, owners should be referred to a qualified clinical animal behaviourist.[3]

Humping

Owners frequently assume that this behaviour is hormonally driven, but it can also be emotionally driven. Frustration, fear and anxiety are other possible underlying motivations for humping behaviour. Owners should attempt to identify the specific situations in which the puppy shows the behaviour. If it appears to be situations of high emotion, they should seek one-to-one help from a clinical behaviourist.[3]

Kitten Evenings: a Suitable Environment

Kitten evenings are a useful way of distributing important information about kitten development and normal feline behaviour, an area where knowledge is generally poor. There are differing opinions about the value of having kittens attending these events with owners, and practices may run either kitten education evenings for owners alone, or kitten kindergartens ('kitty kindies') where owners also bring their kittens along. When owners are invited to bring their kittens along, the aims of kitten evenings are quite different from those of a puppy party, where one of the primary aims is to allow puppies to develop social skills. Due to the very different behavioural needs of kittens and cats (Chapter 9), kitten kindergartens should not aim to provide a venue for socialization with other kittens and there should be no forced social contact. Kitten kindergartens can provide an opportunity for passive exposure to novelty but, as with puppies, for habituation to occur it is vital that the kittens are in a positive emotional state throughout this process, and so avoiding social stress is essential.

The venue

The venue requirements for kitten evenings, where kittens are attending, are similar to those for puppy parties. However, providing an appropriate safe location for kittens can prove more challenging, as they require a more complex 3D environment. The size of the room will depend upon the number of kittens: allowing kittens to explore, feed, drink and toilet without necessarily coming into close proximity with another kitten means that a large room is necessary. Plenty of hiding spaces, such as tunnels and boxes, along with easily accessible elevated platforms or cat trees, are essential if several kittens are going to be in the same room. Kittens can access higher surfaces more easily than puppies, so kitten-proofing a room will need to be thorough.

Equipment

As for puppies, there should be a range of toys (fishing-rod type toys and toys with feathers are often a favourite) and safe objects for kittens to investigate. At least one water bowl and litter tray per kitten are essential, along with cleaning materials and similar props that are used for puppies (see Puppy Parties section above), with the obvious exception of vehicles and vehicle noises.

Handouts

Both kitten evenings and kindergartens provide an opportunity to inform owners using literature and resources from reputable sources. Information covering basic care of kittens (such as vaccinations, insurance, worming, finding a cattery, travelling with kittens, fleas and feeding) should also be provided alongside information about behavioural development and detailed instructions on how to habituate a kitten to everyday experiences.

Staff

As with puppy parties, a good level of supervision is essential in kitten kindergartens, as watching kittens whilst answering owners' questions is impossible for one person to do effectively. As with puppy parties, at least two members of staff should be present in a kindergarten and a maximum of six kittens (aged between 7 and 15 weeks) should be present at any one time. As with puppies, kittens must be in good health if they are attending the classes. Staff will need to have a good knowledge of the ethology and behavioural development of kittens and an understanding of applied learning theory. There are few courses available specifically for running kitten kindergartens, but a number of those covering how to run puppy parties also include information on kitten evenings and kindergartens.

Who should attend?

As many members of the family as possible should be encouraged to attend kitten information evenings. However, whilst it is good that kittens have an opportunity to habituate to other owners and have positive experiences with unfamiliar people of all ages, only extremely well-behaved children should be permitted to attend and kittens should not be forced to interact with them.

Topics to cover at kitten evenings

The topics covered at kitten evenings will vary slightly from those covered in puppy parties, due to their differing environmental needs. There should be no attempts at forced socialization.

Recognizing signs that your kitten is worried

As with puppy owners, kitten owners should be taught how to recognize signs of negative and positive emotional states in their cat (Box 12.5). If kittens are attending the class, then care must be taken to ensure that they experience positive emotions throughout the session, otherwise this will be counterproductive. Whether carried out as part of a kindergarten session or at home, habituation should follow the same rules as for puppies (see Habituation section in Puppy Parties, above). As it is relatively easy for the overenthusiastic owner to overwhelm their kitten, teaching owners how to recognize when a kitten is uncomfortable should be a priority.

Videos, diagrams and photographs are useful for disseminating this information.[8] The first sign that a cat is worried usually comes from their facial expression, as this tends to change before they alter their body position.

Signs that a cat is not worried are broadly the opposite of these. It may have a raised head and tail, half-closed eyes and a relaxed facial expression. A happy cat will have a relaxed body posture, will be confident enough not to have all four paws in contact with the ground and may rest with its front paws tucked under its body.

Socialization

Kitten information evenings provide an opportunity to explain to owners the limitations of social behaviour in the cat. Whilst socially compatible cats can live together peacefully, and come together for feeding, grooming or resting, living in close proximity with cats that are not part of the same social group can be a considerable source of stress. Probably the greatest cause of welfare problems for cats is the presence of cats that are not socially compatible.

Kitten information evenings can be used to deliver information about the social environment required by their new kitten. It is particularly important that owners understand that cats are unable to share resources in a restricted environment and need free and immediate access to essential resources. The location of resources relative to one another is also important. The main factor to consider is whether the cat is able to access the resource without encountering any form of threat. This means placing resources in quiet and private places, with easy access. Cats will often avoid eating, drinking, or toileting if they have to run the gauntlet of another cat or the family dog to get to the resource. Cats tend to eat away from the location where they drink, so require food bowls to be placed apart from (or with a visual barrier between) water bowls. Ample hiding places and easy access to elevation will allow the kitten to cope with any social tension that might arise. Information should also be provided about introductions to other species within the household.[9]

Box 12.5. Signs that a kitten is not comfortable with a situation.

- Flattened ears
- Dilated pupils; wide-open eyes; eyes pressed together (false sleep)
- Whiskers pulled back
- Immobile/freezing
- Crouched/tense body posture
- Arched back
- Head lower than body
- Rapid breathing
- Involuntary toileting
- Vocalizing (e.g. purring, growling or hissing)
- Tail close to body/swishing
- Aggression; avoidance; climbing
- Hiding

Habituation

The first months of a kitten's life are very important, as they are rapidly learning about their environment and experiences, and things they learn have impact throughout their lives (Chapter 9). It is therefore important to provide opportunities to habituate by gently introducing them to the sights, sounds, smells and experiences that we would like them to be happy with as an adult cat (see puppy parties Habituation section earlier in this chapter for explanation of habituation process and suggestions for things that they need to habituate to). This will involve exposure to things that they may encounter later in life, but these will differ from the things that puppies should be habituated to; for example, cats should not be habituated to traffic, as it is important that they are fearful and avoid it. We want kittens to build positive associations, so the kitten should be rewarded with treats (such as a small amount of cat food on a spoon) to build positive associations with the environment. As with puppies, the most important thing to remember is that the kitten must be comfortable, happy and relaxed for habituation to occur – these should be passive encounters whilst in a positive emotional state. If there are any negative consequences, this will have the opposite effect as the kitten will become sensitized to the stimulus.

Minimal restraint handling

Any form of physical restraint is potentially threatening to a kitten. Handling and social interactions should always be on the cat's terms. Unlike us, cats tend to prefer high-frequency, low-intensity interactions, so owners ought to adopt a 'hands-off' approach, allowing the cat to decide upon the nature and duration of interactions. Whilst this may be frustrating to the owner who wanted a cuddly kitten, helping them to understand that allowing their cat to choose will result in a better-quality relationship is important and in the long term to the benefit of the owner as well as the cat.

Prior to starting habituation to handling, owners should be made aware of what constitutes appropriate handling and why, including appropriate grooming techniques and equipment.[10;11] Habituation to being handled involves gently getting the kitten used to being examined all over and is best done at home, where the kitten is likely to be most relaxed. The kitten should be allowed to remain in a natural position, whilst the handling fits around this. Kittens need to be habituated to the types of handling that they might experience during a veterinary examination, such as going into consultation rooms, meeting people in veterinary scrubs, seeing, smelling and investigating veterinary equipment (e.g. stethoscopes) and at the same time making positive associations using treats and play. Owners should be specifically advised to avoid scruffing or 'clipping' techniques, used to inhibit behavioural responses, which are potentially painful and distressing. Owners should also be shown how to administer medication to their kitten, trim claws and clean eyes and ears.[12]

Box 12.6. Training topics that could be covered in a kitten information evening setting.

- Training equipment and techniques available and how to use them, e.g. clicker training.
- Jumping into a cat carrier.
- Coming in when called.
- Remaining still for grooming/examination.
- Recommended reading for owners (e.g. Bradshaw and Ellis, 2016).

Training a kitten

Contrary to popular belief, it is possible to train a kitten. Although there are no formal practical training classes available, the techniques and principles are the same as those for puppies (see puppy training section). The key is to find something that the kitten really likes, but in some cases this may take a little more creativity and patience. Whilst some kittens will undoubtedly find owner attention very rewarding, others may require highly palatable treats such as chicken or tuna to reinforce a desired behavioural response.

Common kitten problems

Providing guidance for owners about how they should deal with unwanted behaviours is essential in preventing problem behaviours later in life. As with puppies, early intervention makes treatment not only easier but more likely to completely resolve the problem. As with any problematic behaviour, owners should first speak to their vet to rule out underlying medical causes.

Litter training

Cats' strong substrate preference makes them relatively easy to litter train, as long as this is done in the correct way. Cats do not tend to like dirty litter, so regular cleaning is a must to make the box attractive to them. It is of course important that the litter tray is easy to access, particularly in multi-cat households, and ensuring that there is at least one more litter tray than there are cats in the household is a general rule of thumb to achieve this. Location is also important, as having to 'run the gauntlet' past other cats, dogs, children or anything else that might be threatening is not going to encourage use of the tray. Small kittens may struggle to get into high-sided litter boxes, so owners should ensure that they can actually climb in without difficulty.

Aggression during play

Kitten information evenings should include information about appropriate games to play with cats.[13] Play is an important part of a kitten's development and may enhance welfare. However, over-exuberant or inappropriate play is a common problem voiced by new kitten owners. Where the kitten has learnt to target human hands or other body parts, this behaviour should be redirected to toys. The owner should be advised not to punish the kitten, run away or scream, as all of these are likely to make the behaviour worse.

Scratching

Scratching is a normal and beneficial cat behaviour but this can be problematic for owners if the target is their furniture. Providing appropriate scratching posts or surfaces and encouraging the kitten to use these from the beginning is the ideal way to avoid damage elsewhere.[14]

Jumping on surfaces

It is a necessary and normal part of kitten development to explore their environment and high perches will naturally appeal to them, as they gain a good vantage point. Pair that with the odd tasty morsel that they discover there and the attraction of kitchen worktops and dining tables to an athletic kitten suddenly becomes clear. If owners do not want their new kitten to hang out in food preparation areas, then they will need to offer easily accessible alternatives. Playing games and delivering treats in the alternative locations will help to make them more appealing. Making it less easy to access worktops and ensuring that the kitten never finds food there will also help the kitten to make the desired choice.

Nocturnal disturbances

Kittens that spend a large proportion of the day sleeping may be prone to bursts of energy during the small hours. Providing them with lots of activities during the day, including things to do when the owner is absent, such as puzzle feeders, will help to provide exercise during the day. Games in the evening, before bedtime, can also help. Commonly kittens learn that if they make a noise at night their owner will get out of bed and join in the fun, so it is important that owners do not reinforce their night-time antics. There are some great sources of information about cat behaviour online.[15]

Conclusion

This chapter gives an overview of the things that should be considered when planning either puppy socialization classes or kitten information evenings. Although different, both of these services offer a valuable opportunity to establish positive relationships with new pet owners and provide owners with important information for influencing their pets' future well-being, whilst at the same time giving practical help with this process. They offer a great opportunity to prevent behavioural problems from ever arising, or, where they do appear, to deal with them before they become well established, and are therefore one of the most important roles that the veterinary practice can play in preventive veterinary medicine. However, if run badly, using outdated information or practical techniques based upon coercion, they can actually create problems. Time must therefore be devoted to finding the right evidence-based information and training staff adequately to deliver the appropriate material in the appropriate manner required to develop the behaviour and confidence of new puppy and kitten patients.

Notes: Online Resources

(All accessed 6 May 2021)

[1] Sound Therapy for Pets
https://www.dogstrust.org.uk/help-advice/dog-behaviour-health/sound-therapy-for-pets

[2] Veterinary Behaviour Specialists
http://www.rcvs.org.uk/education/specialist-status/rcvs-list-of-specialists/

[3] Association for the Study of Animal Behaviour Register of Certified Clinical Animal Behaviourists (CCAB)
http://www.asab.org/ccab-register
[4] Animal Behaviour and Training Council
www.abtc.org.uk
[5] Signs of fear in dogs
https://www.rspca.org.uk/adviceandwelfare/pets/dogs/behaviour/understanding
[6] Dog Welfare Campaign
http://www.dogwelfarecampaign.org/why-not-dominance.php
[7] Dog Welfare Campaign
http://www.dogwelfarecampaign.org/implications-of-punishment.php
[8] Recognizing signs of stress in cats
http://icatcare.org/advice/problem-behaviour/stressed-cats
[9] How to introduce a new kitten to your resident cat
https://icatcare.org/advice/adding-an-additional-cat/
[10] ISFM Feline friendly handling guides
https://icatcare.org/advice/handling-cats-videos-for-owners/
[11] Kitten handling for veterinary restraint
https://drsophiayin.com/blog/entry/kitten-socialization/
[12] International Cat Care cat care videos
http://icatcare.org/advice/videos
[13] Playing with your cat
http://icatcare.org/advice/playing-your-cat
[14] Choosing a Scratching Post
http://icatcare.org/advice/how-choose-scratching-post
[15] International Cat Care cat behaviour advice
https://icatcare.org/advice/?per_page=12&categories=problem-behaviour

References and Further Reading

Atkinson, T. (2018) *Practical Feline Behaviour: Understanding Cat Behaviour and Improving Welfare*. CAB International, Wallingford, UK.

Bailey, G. (2008) *The Perfect Puppy*, rev. edn. Hamlyn Publishing, London.

Bradshaw, J.W.S and Ellis, S. (2016) *The Trainable Cat: A Practical Guide to Making Life Happier for You and Your Cat*. Basic Books, New York.

Chin, L. (2020) *Doggie Language: A Dog Lover's Guide to Understanding Your Best Friend*. Summersdale Publishers, London.

Herron, M., Horowitz, D.F., Siracusa, C. and Dale. S. (2020) *Decoding Your Cat*. Houghton Mifflin Harcourt Publishing, New York.

McCune, S. (1995) The impact of paternity and early socialisation on the development of cats' behavior to people and novel objects. *Applied Animal Behaviour Science* 45, 109–124.

Seksel, K. (2008) Preventing behavior problems in puppies and kittens. *Veterinary Clinics of North America-Small Animal Practice* 38, 971.

Seksel, K., Mazurski, E.J. and Taylor, A. (1999) Puppy socialisation programs: short and long term behavioural effects. *Applied Animal Behaviour Science* 62(4), 335–349.

Serpell, J. and Jagoe, J.A. (1995) Early experience and the development of behaviour. In: Serpell, J. (ed.) *The Domestic Dog: Its Evolution, Behaviour and Interactions with People*. Cambridge University Press, Cambridge, pp. 79–102.

Wilde, K. (2012) *What Your Dog Wants: 7 Key Skills of a Perfect Dog Owner*. Thunder Bay Press, San Diego, California.

Zulch, H. and Mills, D. (2012) *Life Skills for Puppies: Laying the Foundation for a Loving, Lasting Relationship*, 1st edn. Veloce Publishing, Dorchester, UK.

Part III

Behavioural First Aid

13 Identifying Problems

Amanda L. Roshier

Although some owners come to the practice and directly ask for help with undesired behaviours, in many cases problems are not directly addressed or are left until they are well established and serious before discussion takes place. This chapter gives practical advice about picking up the 'clues' that may indicate underlying behavioural problems from both the behaviour of pets and the conversation of owners. An obvious example might be where a dog shows aggression on handling in the consultation room, but a less obvious clue may be where owners ask about neutering an older entire male dog.

Are Behaviour Problems a Client Concern?

Studies of veterinary consultations have identified that behaviour is frequently a topic of conversation. Roshier and McBride (2013a) video-recorded and thematically analysed dog booster vaccination appointments and turn-taking during discussions between owners and veterinarians. They found that both instigated discussions relating to behaviour equally (although there was variation between the six veterinarians), whereas in contrast veterinarians led discussions on themes relating to navigating the consultation, medically related issues (e.g. neutering, injuries), husbandry (e.g. weight, worming, diet) and the costs related to veterinary treatment. Owners were more likely to discuss weight and worming, topics that veterinarians will revisit at most consultations, and perhaps therefore owners engage in discussion because they involve an ongoing process of measurement in the case of weight and the regular administration of a given product in the case of flea treatment and worming. This is in contrast to being able to identify behaviour problems, where often the onus is on the owner to recognize and report a problem, which requires knowledge and interpretation of observations, because behaviour problems may not be apparent to the veterinarian in the clinic.

Owners also completed paper-based questionnaires on the day they were observed. This revealed that all 17 clients considered that their dog performed one or more behaviours of concern (58 problems in total), but only ten were discussed during the consultations. So, although behaviour problems were common, they were rarely discussed with the veterinarian during the consultation; and when they were mentioned, they were not fully explored and managed. Where behaviours were implied, such as the need to trim claws because the dog was jumping up, or displayed during the consultation, such as aggression towards a person, these were also rarely discussed. Interestingly, owners reported that they were satisfied with the consultation and did not have any unstated concerns. It is not known if this is because they

DOI: 10.1079/9781800621312.0013

did not recognize there was a problem, or if they were getting support at another time, or from a different resource than the veterinarian.

When 622 veterinarians responded to a questionnaire about topics they discussed during booster vaccination consultations (Robinson *et al.*, 2019), they reported that behaviour was discussed, but more frequently in canine consultations than feline. The authors suggested that this may reflect changes in legislation around the 2014 amendment to the Dangerous Dogs Act (Veterinary Record, 2014).

Although further studies are needed to fully explore the decisions of owners regarding accessing behaviour support, there are indications that the veterinary practice is not fully utilized, for reasons that are unclear, and that clients identify alternative resources to access this information (Roshier and McBride, 2013a). From the veterinarian's perspective, behaviour is often viewed as a specialist and time-consuming topic that precludes it from general practice, with veterinarians saying they do not feel confident in their ability to identify and support a behaviour problem, with some concerned about deficiencies in their veterinary training (Roshier and McBride, 2013b). Veterinarians have also reported that they would rather not discuss behaviour problems, with reasons including lack of time, poor confidence in their own knowledge and an expectation that owners would ask, if they had a problem (Belshaw *et al.*, 2018).

If owners perceive a lack of expertise or interest (particularly if support is not offered, or only given a light touch), then this may contribute to choosing not to access veterinary support. Where a full history of an individual animal's problem might not be taken and only generic solutions suggested, it would follow that if the advice given is not tailored to the individual animal and the owner's circumstances, and perhaps with no follow-up opportunities or ongoing support, then behaviour issues would not resolve and hence owners would seek advice elsewhere or choose not to discuss concerns in the future. Finally, some owners might not be aware that their pet has a behaviour problem and these issues would not be discussed. Clients should be encouraged to discuss behaviour, helped to identify if behaviour problems exist and, if so, advised on how they may be managed or referred on for additional support.

A Veterinary Team Approach to Identify Problems

The veterinary profession acknowledges that veterinarians should provide behaviour support (FVE and AVMA, 2011) either directly, or indirectly by referral. The practice team should be viewed more broadly than veterinarians and veterinary nurses providing this support, as the team may include receptionists, practice managers, behaviour specialists via referral, and services that the practice might incorporate (such as puppy classes, dog grooming and boarding facilities). All members of the practice team, veterinary and administrative, should be familiar with the behaviour support available within the practice, as this will help provide a transparent and coherent message and facilitate a clear process of operation. Ideally, a member of the team should take the role of behaviour champion to aid dissemination of a congruent message, facilitate a way for the team to work together, enable review of internal processes and ensure that the practice operates in a way that is behaviourally aware. At the very least, practices should have a clear process for referral of behaviour cases and recommendations to animal trainers; they will also need to ensure that these paraprofessionals are suitably experienced and qualified (e.g. ABTC, 2020; RCVS, 2020).

A whole-team approach will lead to increased opportunities for client interactions with behaviourally skilled staff members and this creates multiple opportunities and perspectives to help identify if problems exist. An increased engagement with clients helps bond them to the practice, which in turn helps engender rapport and trust, thus aiding disclosure of concerns about behaviour problems. This relationship with the practice can be enhanced through interactions with the team at various engagement points, where this starts with the owner identifying that they want to be a client of the practice, is followed by the process of registering as a client, then visiting reception and attending appointments and possibly also attending client events.

The use of social media and newsletters can facilitate communication of practice messages and enable continuity of the client practice relationship when clients are not attending the practice in person. Provision of supplementary information, such as articles and handouts, can provide behavioural support in their own right, but will also demonstrate the practice's interest and how the practice can offer direct support; this can lead to further discussions and appointments to focus on clients' specific needs. The following are some examples of themes that could be included in these articles and handouts.

- **Advice on selecting an appropriate pet** relevant to client expectation, experience and circumstances.
- **Information relevant to animal life stages**, including helping the young animal integrate into the human world (for example, the need for socialization, training, and how to identify appropriate sources of support, such as puppy classes and trainers).
- **Considerations for the ageing pet**, where highlighting potential changes in behaviour can help manage client expectations of their pets and provide suggestions to implement age-appropriate care. For example, the arthritic cat that can no longer comfortably access its litter tray, which could potentially result in house soiling problems, could be given a litter tray with lower sides.
- **Information specific to a behaviour problem** (e.g. house soiling, firework fears), with the aim of helping clients identify that a problem exists or avoid one developing, highlighting that support is beneficial, and encouraging discussion with their veterinarian.
- **Articles that enhance client understanding of their pets**, such as recognizing species-specific behaviours and sensory capabilities (e.g. vision, olfaction, hearing), would help manage client expectation of their animal's behaviour; owners may also then be more accommodating to their pet's needs and understanding of how their pet reacts in a given situation. Information could aim to improve interpretation of behaviour and how animals communicate; this would help clients identify problematic behaviour and address the issue with their veterinarian. For example, if an animal's body language is indicating that they are anxious and this goes unnoticed or is misinterpreted, then it can lead to a behaviour problem developing or an existing problem being overlooked.

The capacity for the veterinary practice to work as a team throughout the client's journey with them creates a powerful model to support the behavioural health of pets, which in some cases will have wide-reaching impact and support public health.

Recognizing Opportunities to Identify Behaviour Problems

Where do discussions about behaviour happen?

The content of discussions is influenced by the purpose of the consultation, where veterinary consultations typically fall into one of the following categories.

1. Presentation of an animal with a health problem.
2. An animal returning for follow-up treatment.
3. Wellness appointments (a routine appointment, e.g. vaccination, health check).

Analysis of discussions during a health problem consultation compared with a wellness appointment found 90% of conversation in a health problem consultation is the veterinarian gathering biomedical information, whereas in wellness appointments 50% of the conversation is related to information gathering and 27% to client education (Shaw *et al.*, 2008). These findings indicate that there may be more opportunity to discuss welfare during wellness appointments and this concurs with the finding that behavioural problems are more frequently discussed during preventive healthcare consultations than during specific health problem consultations (Robinson *et al.*, 2016). However, clients will typically visit veterinarians when their animal is unwell and therefore the practice team should use strategies to initiate discussion of behaviour at each interaction they have with the client, including outside the consultation. For example, information and posters can be displayed in the practice, practice social media and websites can be utilized, various members of the practice team can engage in discussion, and questionnaires used to gather further information about pets can be completed in the waiting room or at home. Once behaviour has been mentioned, further appointments can be arranged or plans made to explore issues further if required.

Life stages and events relating to behaviour problems

Although it is essential to collect a complete case history, which will provide information to help inform assessment of the case and keep behavioural differential lists broad, an awareness of typical life stages and events can be useful to appreciate circumstances when problem behaviours may arise. For dogs, most problem behaviours are reported between the ages of 12–42 months (Lund *et al.*, 1996), although problems are likely to be developing before the stage they get reported. However, medical conditions, such as those that cause pain, anxiety and altered perception, changes in hormones and biochemistry and cognitive decline and ageing, can all be accompanied by changes in behaviour (see Chapter 15). For example, musculoskeletal conditions occur predominantly after 7 years of age, and where an animal is in pain and protective of themselves they may show signs of aggressive behaviour or changes in temperament. Other behaviour changes may also be noticed; for example, toileting behaviour can be affected. An arthritic cat may associate pain with using the litter tray and develop an aversion, leading to toileting elsewhere.

Calendar events may predict the likelihood of problem manifestation. For example, fireworks can exacerbate noise fear; and periods when the family dynamics change such as holidays, changes in owner work patterns, changes in the composition of the family unit (people and pets), moving house or building work being done may create a problem, or make behaviour problems evident. When households acquire new pets, the household dynamic changes but behaviour changes are not always immediately evident. For example, spraying and elimination behaviours in cats may occur as multi-cat households reach various stages of maturation which change the dynamics between pets. It is important to note that, even when an owner may predict a problem, such as in a pet that is noise fearful, they typically only seek help during this period, leaving it too late to implement long-term behaviour support (Bowen, 2008).

Another frequently predictable calendar event would be elective surgery. Where such a surgery is being considered, discussion with clients should include plans for the management of the animal during rest and recovery, and finding out how both animal and client will cope with this situation. These discussions may identify existing behaviour problems and help when considering treatment options: perhaps treatment will need to be postponed and a behaviour problem addressed to aid the recovery process (Ryan *et al.*, 2014).

Bearing all of the above in mind can help the practice to offer behavioural support proactively to their client base, especially if the practice has made it clear to clients that they are available as a source of support.

Clues indicative of behaviour problems

Underlying behaviour problems can impact on how frequently a client attends the practice. For example, the client with a cat that soils when travelling in the car may decide to attend only for emergencies, and may even forego vaccinations. Clients may feel that their pet becomes too distressed in the practice environment; they might worry about interactions in the waiting room and that the veterinarian or others could be injured, or perhaps their animal taken away if it appears aggressive. Consideration should be given to tailoring clients' visits to ease their concerns and mitigate situations that they and their pets find distressing. It is vital that every effort is made to facilitate client attendance at the practice to ensure that timely behaviour support is provided and also to ensure that other aspects of health and welfare are not compromised, such as missing out on health checks and vaccinations.

Communicating with clients to identify if there are any areas of concern in the first instance is helpful, either through direct conversations or utilizing various media channels to acknowledge that some clients might feel restricted visiting the practice, and provide information about what measures the practice takes to support them. These measures could include advice for travelling to the practice, arranging home visits, or referral for support to help with travel. The layout of the waiting room, kennelling and treatment rooms should be reviewed; with an aim to create separate species areas, or arrange species-specific clinics (see Chapter 6). Schemes such as the cat-friendly clinic (ICC, 2020) provide guidelines to improve behaviour husbandry or behaviour wellness in practice, and practices can be evaluated and certified if they meet the organisation's standards. Clients could be given an option to arrive before their appointment and allow the animal to become accustomed to the consultation room before the veterinarian arrives; the owner can help create positive associations with the practice environment (e.g. treats, grooming, play) and the practice could encourage doing this at times when there is not an appointment scheduled. This approach can be beneficial for animals that are anxious, young, or have no experience of the practice environment and can help mitigate the fact that the nature of practice visits often means that pets experience procedures that cause anxiety, through restraint handling, pain associations and novel situations.

Less obvious clues indicating that behaviour support is needed may be when a client asks about neutering an older entire male dog: potentially they are looking to resolve a behaviour issue and are under the impression that this can be achieved with surgical intervention. Discussions are needed to ascertain if neutering will provide a solution or escalate the problem. The behaviours displayed may be under hormonal influence, but may also be learned. Injuries or rub marks caused by the accoutrements that are used for animal handling may indicate a problem; for example, a dog with a cough or neck injury (that may also present as a sore or rub mark) may indicate that they pull on the lead or lunge at other dogs (Carter *et al.*, 2020).

The veterinary team should also consider the population that is not their client base and the potential barriers, or behaviour problems, that may be preventing them from joining the practice. Efforts to engage with new clients could be via social media, local press and practice events; these could help open channels of communication, or at least provide advice to encourage owners to seek support. Events can be used to promote the profile of the practice as being behaviourally aware, both to encourage owners to become clients of the practice and to encourage existing clients to address their behaviour concerns. This can be achieved by organising timely behaviour-focused client events throughout the year. These events could focus on specific behaviour problems, or broad concepts such as species-specific behaviours and communication. All these initiatives have the potential to show clients that the practice has a genuine concern for their animal's behavioural health, which enhances client bonding to the practice and encourages discussions and disclosure of behavioural concerns.

Discussion Strategies for Illuminating Behaviour Concerns

This section considers what might be discussed and what might be observed that may provide the veterinary team with an indication that a behaviour problem may exist. In reality the veterinarian will have to discern this intertwined information carefully. As with all aspects of veterinary medicine, the veterinarian will need to employ careful questioning and clarifying techniques to gain understanding of the case. Encouraging a client's narrative will provide insights into their interpretation of the problem, how much of an issue it is to them, how they are managing it, and what outcomes they would like. In addition to gaining an understanding of the client's interpretation, it is important to ask them to describe the behaviours and the circumstances in which they occur, as without appropriate probing it is possible for the client to misinterpret what is going on and thus for the veterinary professional to evaluate the problem incorrectly.

Encouraging disclosure of a problem

Roshier and McBride (2013a) proposed reasons for behaviour concerns identified by owners not being discussed during the consultation:

1. Discussed elsewhere in the practice. It is possible that concerns were already being addressed at another time or with another member of the veterinary team.
2. Timeliness. Some behaviour concerns may be intermittent, such as firework phobias that may only be an issue at certain times of the year (Sheppard and Mills, 2003), or when the problem has escalated.
3. Not realizing support is available. Clients may not be aware that behaviour support is available; for example, Blackwell *et al.* (2005) found that owners did not realize they could speak to their veterinarian about noise phobias.
4. Not confident in the support offered. Do not discuss issues with the veterinarian.
5. Access support from elsewhere. Clients recognize support services external to the veterinary practice and have indicated that these resources would be used.

Further studies are needed to gain an understanding of behaviour support seeking, exploring questions such as under what circumstances support is sought and what influences the type of resources accessed for this help. However, there are approaches that a veterinarian can take to try to encourage discussion of behavioural challenges that an owner may be experiencing.

Medical professions recognize adopting a client-centred approach aids discussion, as the focus is on enabling the client to tell their own narrative and to express their concerns – an approach that has also been found to increase client satisfaction (Kinnersley *et al.*, 1999). Robinson (2001) identified two main barriers that might preclude discussion: (i) *psychological barriers* where the client does not feel comfortable talking about the issue, as they think it might be deemed socially unacceptable, embarrassing or too trivial to mention; and (ii) *interactional barriers* where the conversation needs to flow to enable a new topic to be introduced. Roshier and McBride (2013a) provided suggestions on how these barriers might be addressed. Developing trust and rapport and creating a 'safe' environment to encourage disclosure were deemed important to overcome psychological barriers. The challenge of achieving this when a client may visit infrequently for a short consultation, with various veterinarians, was highlighted. A study by Belshaw *et al.* (2018) supported this notion, where clients discussed the importance of a familiar veterinarian that they know and trust for the vaccination consultation. Then, when interactions take place, the requirements for good questioning and listening skills to foster trust and rapport are needed.

Rapport may be developed more readily with reception staff, particularly if clients do not see a designated veterinarian, and interactions with individual staff are more frequent here. Conversations may take place more readily at reception as a longer time may be spent in the waiting room, giving opportunity for rapport-building conversations to take place, including disclosure of concerns. Additionally, reception may be perceived as less judgemental, and there may not be any expectation for the behaviour to be assessed. Reception staff are in a position to offer support to clients and their pets prior to a veterinary appointment, both through telephone conversations and observations in the clinic. Staff can be trained to recognize comments or encourage discussion that aids disclosure of behaviour concerns; it can then be followed up by encouraging the client to discuss this with the veterinarian and, with permission, either noted on the animal's appointment, or arrangements made to speak to another member of the team. Additionally, reception staff can facilitate a smoother visit as they can check where clients would like to wait for their appointment: they may prefer to wait outside or in their car if their animal is more relaxed here, and reception can offer to collect them when it is time for their appointment. Minimizing arousal and stress in the waiting room will also benefit other animals and clients by keeping everyone calm; this will also facilitate any subsequent interactions with the veterinary team. The potential to identify a behaviour problem in the waiting room should not be underestimated, but care must be taken that reception staff do not offer unsolicited advice, including commenting on an animal's behaviour; hence the need for a whole veterinary team approach.

The second barrier highlighted, interactional barriers, is where a client does not find an opportunity to discuss their concern and it requires effort to create opportunities to do this. Bergman *et al.* (2002) found that some clients did not think they should mention their pet's behaviour in a medical appointment. Roshier and McBride (2013a) noted from transcriptions of the consultations that all veterinarians had a phrase to encourage discussion of issues; for example: 'What can we do for you today?' or 'Do you have any concerns?', and where discussion looked to reach an end point, a common phrase 'any lumps or bumps?' would often encourage a 'now you mention it!' moment. For these reasons, it is recommended that veterinarians inquire about behaviour directly, ideally with open-ended questions to encourage discussion. For example: 'Do you have any concerns about their behaviour?', 'What does your pet typically do during the day?', 'How well do they get on with your other pets/people?', 'How is their training going?', 'What commands do they know/tricks can they do?', or 'What is the animal's role within the family?'. Such questions provide insight into the client's experience and perceptions of animal management and may help lead to discussion about behavioural concerns. The team should have an awareness of key behaviour problem triggers (such as fireworks or the animal's age) to create timely practice initiatives and discussions. It would also be prudent to inquire if the owner has any behaviour concerns with any of their other pets not attending the consultation, in case they have a problem preventing them from attending the practice. Not only can opportunities be created by asking these direct questions but also the veterinarian can follow up on remarks made by the client, such as the client who mentions during claw trimming that the dog bruises them terribly when they jump up. The veterinarian can use these comments as a starting point to probe further.

Owners may recognize that behaviours are performed but may not realize that behaviour support would be beneficial; and without probing further the veterinarian may miss an opportunity to discuss behaviour or training support before a problem escalates.

To give another example, when dogs are muzzled during a consultation, not only does this restrict performance of the clinical examination, but also it is evidence of the presence of a behaviour that could have implications if it also takes place outside the consultation. In the Roshier and McBride (2013a) study, only one veterinarian inquired about a muzzled dog's behaviour with other people and the client acknowledged that the dog could be aggressive towards his wife. Where a veterinarian overlooks aggressive behaviour displayed in the consultation, and does not explore the contexts where aggressive behaviour occurs, not only could this miss potential health and safety issues for the owners, their friends and family, but also the wider ramifications are that outside of this 'family' unit, a member of the public (and other animals) could feel threatened or be injured. The other concern is that where a veterinarian does not discuss the behaviour, this may convey acceptance of the behaviour on their part, which in turn may lead to an owner not attempting to address the behaviour.

An outline of key behaviour points for discussion could be created by the practice for veterinarians to cover during consultations; this creates consistency across consultations and assists veterinarians not confident in this topic. The practice team should be prepared to direct clients to the practice's supporting resources, both information resources and personnel. This can then lead to further discussions about the behaviour problem and referral to suitably qualified behaviourists and trainers (either internally or externally).

In addition to the above-mentioned barriers, other studies indicate that veterinarians do not want to discuss a topic where they do not feel qualified to provide support (Belshaw *et al.*, 2018), which raises the issue of how behaviour medicine is taught to students. Where veterinarians felt behaviour required specialist input and they did not have training, it was interesting to see that these veterinarians had not undertaken behaviour related continuing professional development training (Roshier and McBride, 2013b). Another issue raised is not having time to discuss the problem. While it is encouraging that they recognize that behaviour support is often an involved process, they do need to be in a position to identify that a problem exists and provide behaviour first aid; and if they are unable to arrange a further appointment to discuss the issue, then they should refer to a suitably qualified person either in-house or via external referral. Feeling unqualified to discuss behaviour is also a concern where behaviour specialists are often working on referral from a veterinarian and they need the support of each other to understand any medical concerns and for prescribing psychopharmacological medication and supporting its intervention (see Chapter 22). This highlights again the importance of a team approach for providing behaviour support, where the team has a responsibility to provide a service to clients (including referral when something is beyond their capabilities, or resources are limited); and also to help each other's professional development through identifying training needs and providing in-house training, or reporting back to the team on any continuing professional development that is undertaken, and incorporating this into practice processes. Increasingly efforts have been made by organizations involved in veterinary medicine, behaviour counselling, animal training and animal welfare to provide support for veterinarians in incorporating behaviour awareness into their practice, either through training days or featuring talks in their annual conferences (e.g. British Small Animal Veterinary Association, British Veterinary Behaviour Association, Behaviour Practitioner Organizations).

Not all problems will be evident in the clinic and these will go undetected unless the client is questioned. It can make all the difference to detect and address problems at an early stage, as it tends to follow that the longer behaviour problems exist, the more established they become and the more difficult they are to change; this is often the case for those involving fear and/or aggression (Horwitz *et al.*, 2002). Another outcome for problems that have persisted is that the owner's bond with the animal is likely to be damaged to the point that relinquishment or euthanasia is the favoured option (Salman *et al.*, 2000). For all of these reasons, ensuring that a practice is behaviourally aware can make a great difference to both pets and their owners.

Animal Interactions as Indicators of Behaviour Problems

Observing and interpreting behaviour

Interactions with the client and pet take place throughout the practice environment, including the car park, waiting room, consultation room, back of house, and with any ancillary services provided (such as kennelling and dog grooming). Being mindful to observe how an animal interacts with other animals and humans, and if possible recognizing how this varies in different environments, can provide useful information, or provide a conversation starter with an owner. These interactions are likely to be snapshot events, but clues can be gleaned from recognizing a behaviour and this may become more meaningful if it is possible to see what happens before, during and after a behaviour is performed.

When a veterinarian calls a client and pet in from the waiting room, they have an opportunity to see how the animal interacts with other people and animals and how the owner manages these interactions. Where anxious or aggressive behaviour is observed, the veterinarian can engage the client for further discussion about their pet's behaviour. Even acknowledging positive interactions or relaxed behaviour can lead to further discussion and possibly identification of other concerns. Where animals are not overtly aggressive, but may be passive or still, this should also be explored as the animal may be fearful but using this behavioural approach as a coping strategy.

The use of food treats during visits to the practice is encouraged as it can help form positive associations with the veterinary experience and help manage interactions during veterinary procedures. The use of treats also provides a tool to help make observations. Treats can be given to reward desired behaviours, used as a distraction or lure, or given to create a pleasant association with the environment. How an animal reacts to treats can provide useful insights into their training, and perception of their current situation. If the owner provides treats (or does not) this can demonstrate their animal handling experience and approach, such as whether they use reward-based training methods, and provides a conversation starter about training and behaviour. If an animal does not want a treat, it may be that they are too stressed, although it may also mean that tastier treats are needed or that the owner needs to consider using something else that they may find rewarding, such as a toy, or stroking them. The owner should always be consulted about what rewards they think are appropriate for their pet; they may have opinions about the use of rewards (another potential conversation starter), or there could be dietary restrictions to be aware of.

The ability to interpret body language and behaviour for a given species is an important skill to develop for anyone interacting with animals. Recognizing the animal's intent or emotional state can be used to inform how to proceed with interactions, or interpret possible meanings behind an animal's behaviour, and help identify behaviour problems. However, it is important to remember that there is a paucity of evidence-based research relating to interpreting animal communication and much of the literature has been developed by experts in the field and their opinions supported by their peers. An example where expert opinion has been further investigated in a scientific manner is the 'ladder of aggression', which is a diagrammatic representation of how a dog might react to remove a threat. The behaviours range from subtle signals (e.g. yawning, turning head away) to biting (Shepherd, 2009). The use of such signals was investigated in a study that found that behaviours sometimes identified as 'calming signals' (e.g. yawning, turning head away) may have a role in social facilitation and preventing further aggressive behaviours (Mariti *et al.*, 2017).

Another method to help classify behaviour uses a traffic light system which looks at a range of behaviours, including posture, eye contact, movement and vocalization. In this system, red indicates that the animal is not safe to handle, the animal is ready to use aggression if handling continues; yellow indicates that the animal perceives danger and is alert, it may proceed to aggression if provoked further, so use handling techniques that minimize threat; and green indicates that the patient feels safe, is relaxed in the environment, so it is safe to proceed with handling (Herron and Shreyer, 2014). Again, the aim is to help clinicians without specific behaviour training to make appropriate decisions when dealing with a pet. This body language literature raises awareness of a need to interpret behaviour, and indicates how to do this, but a critical eye is needed and care taken that this is not adoption of the literature by consensus.

What is observed in the clinic is a snapshot in time, and, according to the statistics on the average length of a consultation, this could be a very short time. The average time available in 17 booster consultation appointments was 9 minutes (range: 5–15 minutes) (Roshier and McBride, 2013a). The veterinarian may miss making observations if their focus is on a task such as the physical examination, reviewing notes, or giving a treatment. Therefore, together with asking questions of the client, reviewing video footage of a behaviour of concern can be a very useful tool to provide helpful insights to the client. Additionally, video will capture behaviour in context and video recordings enable multiple reviewing opportunities of the behaviour and discussions about it. There is no time limit on making these observations and the animal does not need to be present. This is helpful, as it may not be possible to replicate the behaviour, for example if it takes place in a specific context, or it would not be safe to repeat, or on welfare grounds it would be distressing to the animal to replicate the situation. A video can be shared to garner the opinions of multiple people, at various locations. It may also be safer to review footage than interact with the animal; this is pertinent where the veterinarian is providing first-opinion advice and behaviour first aid and would not be directly involved but is referring the case to a person with the necessary expertise. Therefore, requesting clients to video behaviours of concern, when it is safe and in the pet's interest to do so, can provide a great deal of information that would otherwise not be possible to assess.

Role modelling

During interactions with pets and clients, the veterinary team, including reception staff, are in a position to be role models. An evidence-based approach should be adopted to guide best practice for interactions with animals. Currently this recommends using reward-based methods and has identified risks associated with positive punishment-based training techniques to both animals and owners. Low-stress handling techniques aim to increase the safety of staff and welfare of the patient (Yin, 2009; Herron and Shreyer, 2014) and additionally convey the message to owners that this is the correct and appropriate approach to take to support their pet's emotional and psychological welfare.

An animal's experience of the veterinary practice is often stressful. Adopting low-stress handling techniques and making efforts to respond appropriately to an animal's emotional state will improve the animal's experience.

Additionally, by explaining approaches to handling and restraint, or commentating on how the animal is interacting, this could help overcome the interactional barrier highlighted previously, where opportunities to discuss problems must be created; therefore, explaining interactions could facilitate these conversations.

Staff professional development

To ensure that low-stress handling techniques and behaviour first aid can be implemented by the veterinary practice, it is imperative that the practice identifies any gaps in expertise and invests in supporting staff professional development. Where veterinarians refer clients to trainers or behaviour specialists, they should be confident of the techniques and approaches these paraprofessionals are providing their clients, as ultimately the veterinary team are responsible for endorsing these paraprofessionals. If the veterinary team feel they are unable to assess the standard of animal training and behaviour support they are referring clients to, then there are organizations providing directories of paraprofessionals with recognized experience and qualifications. In summary, all veterinarians should be equipped with the day-one competences that demonstrate behaviour awareness when interacting with animals to minimize fear and anxiety, that they will 'do no harm', be able to identify problem behaviour (through both observation and information gathering), 'apply behaviour first aid' and refer to a suitably qualified person where necessary. It is vital that every effort is made to ensure animal welfare and mitigate the veterinary practice experience creating issues (fear or anxiety of the practice or procedures): you would not expect an animal to come home from the practice with a physical injury; equally you do not want them to come home with a behavioural issue through inappropriate interactions and you would not want them to return home with key problems, including behavioural, unaddressed.

Summary

Studies indicate that owners are not fully utilizing the veterinary practice as a resource for behaviour support, despite owners identifying behaviours considered to be problematic. It is a concern if support is not accessed, as this may impact on a range of factors such as the animal's welfare, the owner's health and safety, the pet–owner relationship and wider society. Guidance has been provided on how the whole practice team can engage with clients to stimulate discussion about behaviour, and opportunities have been highlighted as to how the practice can provide this support and work to identify behaviour problems. Further studies are needed to comprehend clients' perspectives more fully and therefore develop strategies to optimize support for a variety of species. As a minimum, the veterinary team should 'do no harm' during interactions with their clients' pets, identify if a behaviour problem exists, 'apply behaviour first aid' when necessary, and make appropriate referral if expertise is not available within the practice (Roshier and McBride, 2013b). The veterinary team are role models to clients and their actions and interactions may be emulated. In addition, explanations by the team will aid client understanding of appropriate animal handling, which will also open conversation for behaviour discussion. Identifying behaviour problems involves observations in practice to spot behaviour issues, listening for clues during conversations with clients, probing for further information, and asking behaviour-related questions directly. When the practice is able to identify if a pet has a behaviour problem, they are in a stronger position to provide comprehensive welfare support.

References

ABTC (2020) *Practitioner Directory*. Animal Behaviour & Training Council, Taunton, UK. Available at: www.abtcouncil.org.uk/practitioners (accessed 31 July 2020).

Belshaw, Z., Robinson, N.J., Dean, R.S. and Brennan, M.L. (2018) Owners and veterinary surgeons in the United Kingdom disagree about what should happen during a small animal vaccination consultation. *Veterinary Sciences* 5(1), 7.

Bergman, L., Hart, B.L., Bain, M. and Cliff, K. (2002) Evaluation of urine marking by cats as a model for understanding veterinary diagnostic and treatment approaches and client attitudes. *Journal of the American Veterinary Medical Association* 221, 1282–1286.

Blackwell, E., Casey, R. and Bradshaw, J. (2005) *Firework Fears and Phobias in the Domestic Dog.* Scientific report for the RSPCA. University of Bristol, Bristol, UK.

Bowen, J. (2008) Firework fears and phobias. *Companion Animal* 13(8), 59–63.

Carter, A.J., Roshier, A.L. and McNally, D.S. (2020) Canine collars: an investigation of collar type and the forces applied to a simulated neck model. *Vet Record* 187(7), e52. doi: 10.1136/vr.105681

FVE and AVMA (2011) Pre-eminent veterinary groups in Europe and U.S. issue joint statements on animal care, treatment and veterinary education. *The Federation of Veterinarians of Europe and the American Veterinary Medical Association*, 14 September 2011. Available at: https://www.anmvioggi.it/media/files/FVE%20AVMA%20PRESS%20RELEASE.pdf (accessed 2 May 2022).

Herron, M.E. and Shreyer, T. (2014) The Pet-friendly Veterinary Practice: A Guide for Practitioners. *Veterinary Clinics of North America: Small Animal Practice* 44(3), 451-81. doi: 10.1016/j.cvsm.2014.01.010

Horwitz, D., Mills, D. and Heath, S. (2002) *BSAVA Manual of Canine and Feline Behavioural Medicine.* BSAVA, Quedgeley, UK.

ICC (2020) Cat Friendly Clinic is a worldwide programme from the International Society of Feline Medicine, the veterinary division of the charity International Cat Care. Available at: https://catfriendlyclinic.org (accessed 31 July 2020).

Kinnersley, P., Stott, N., Peters, T.J. and Harvey, I. (1999) The patient-centredness of consultations and outcome in primary care. *British Journal of General Practice* 49, 711–716.

Lund, J.D., Agger, J.F. and Vestergaard, K.S. (1996) Reported behaviour problems in pet dogs in Denmark: age distribution and influence of breed and gender. *Preventive Veterinary Medicine* 28, 33–48.

Mariti, C., Falaschi, C., Zilocchi, M., Fatjó, J., Sighieri, C. *et al.* (2017) Analysis of the intraspecific visual communication in the domestic dog (*Canis familiaris*): a pilot study on the case of calming signals. *Journal of Veterinary Behavior: Clinical Applications and Research* 18, 49–55.

RCVS (2020) *List of RCVS Advanced Practitioners*. Available at: https://findavet.rcvs.org.uk/find-a-vet-surgeon/by-advanced-practitioner/companion-animal-behaviour/ (accessed 31 July 2020).

Robinson, J.D. (2001) Closing medical encounters: two physician practices and their implications for the expression of patients' unstated concerns. *Social Science and Medicine* 53, 639–656.

Robinson, N.J., Brennan, M.L., Cobb, M. and Dean, R.S. (2016) Investigating preventive-medicine consultations in first-opinion small-animal practice in the United Kingdom using direct observation. *Preventative Veterinary Medicine* 124, 69–77. doi: 10.1016/j.prevetmed.2015.12.01

Robinson, N.J., Belshaw, Z., Brennan, M.L. and Dean, R.S. (2019) Topics discussed, examinations performed and strategies implemented during canine and feline booster vaccination consultations. *Veterinary Record* 184, 252.

Roshier, A.L. and McBride, E.A. (2013a) Canine behaviour problems: discussions between veterinarians and dog owners during annual booster consultations. *Vet Record* 172(9), 235. doi: 10.1136/vr.101125. Epub 28 December 2012.

Roshier, A.L. and McBride, E.A. (2013b) Veterinarians' perceptions of behaviour support in small-animal practice. *Vet Record* 172(10), 267. doi: 10.1136/vr.101124. Epub 28 December 2012.

Ryan, S., Zulch, H. and Baumber, P. (2014) *No walks? No worries! Maintaining Wellbeing in Dogs on Restricted Exercise*. Velcoe Publishing, Dorchester, UK.

Salman, M.D., Hutchison, J., Ruch-gallie, R., Kogan, L., New, J.C. Jr *et al.* (2000) Behavioral reasons for relinquishment of dogs and cats to 12 shelters. *Journal of Applied Animal Welfare Science* 3, 93–106.

Shaw, J.R., Adams, C.L., Bonnett, B.N., Larson, S. and Roter, D.L. (2008) Veterinarian-client-patient communication during wellness appointments versus appointments related to a health problem in companion animal practice. *Journal of the American Veterinary Medical Association* 233, 1576–1586.

Shepherd, K. (2009) Ladder of aggression. In: Horwitz, D. and Mills, D. (eds) *BSAVA Manual of Canine and Feline Behavioural Medicine*, 2nd edn. BSAVA, Quedgeley, UK.

Sheppard, G. and Mills, D.S. (2003) Evaluation of a dog-appeasing pheromone as a potential treatment for dogs fearful of fireworks. *Veterinary Record* 152, 432–436.
Veterinary Record (2014) Comment: Missed opportunity. *Veterinary Record* 174, 514.
Yin, S. (2009) *Low Stress Handling Restraint and Behavior Modification of Dogs and Cats: Techniques for Developing Patients Who Love Their Visits*. CattleDog Publishing, Davis, California.

14 Appropriate Handling of Challenging Patients

Stephanie Hedges

Challenging Behaviour in Practice

Veterinary examination and treatment require an element of cooperation on the part of our patients. However, whilst domesticated species such as dogs and cats generally enjoy the company of people, this does not necessarily mean they enjoy or will tolerate human handling, especially by a stranger or if this is uncomfortable. Young, fidgety, or excitable animals can also be difficult to keep still long enough to perform the required interventions and many veterinary patients are fearful of visiting the vet. Studies suggest that at least half of dogs show reluctance to enter the practice and fear during examination and as many as 80% show fear when placed on a table (Edwards *et al.*, 2019). Similarly, 73% of cat owners said their cats show signs of distress when in the waiting room and 87% reported signs of distress on the examination table (Mariti *et al.*, 2016). Data is less freely available for other species. However, it is reasonable to assume that this will be equal or even higher in prey species or those that have undergone less adaptive change to the presence of people.

Traditionally, uncompliant patients were managed using physical restraint. However, this carries many risks. The most immediate of these are the risk of injury to both staff and patient from a struggling or defensively aggressive animal, and compromise of the patient's emotional welfare. The inevitable patient stress associated with restraint can also negatively impact general health, especially the immune system and so recovery, and can affect key parameters used for examination and diagnosis, such as heart/respiratory rates or blood values. Clients are also known to avoid or delay bringing their pet to the practice if they feel they find this distressing and to be less likely to return to a practice at which they feel their pet has been handled in a way that worried them (Lloyd, 2017). Physically restraining a fearful animal is also time consuming, disruptive to the smooth running of the practice and invariably triggers or exacerbates fear, increasing the likelihood of further challenging behaviour during subsequent visits. Adapting established behaviour modification techniques, such as managing the environment or manipulating the animal's emotional state or preferred behaviour, to avoid the need for physical restraint therefore has many advantages (Box 14.1).

Understanding Problem Behaviour in Practice

An animal's behaviour in any given situation is influenced by a combination of their physiological state, past experience and whatever is happening around them at the time. Identifying and then managing or manipulating these can therefore motivate the animal to choose a different – more desirable – behaviour. Not all influences over

DOI: 10.1079/9781800621312.0014

Box 14.1. Advantages of avoiding physical restraint.

- Reduces fear, improving patient welfare and reducing the risk of defensive behaviour at the time and during subsequent visits.
- Minimizes the risk of injury to both staff and patient due to struggling or defensive behaviour.
- Improves reliability of parameters such as heart rate, respiratory rate and biochemistry values.
- Prevents stress negatively impacting on general health, the immune system and recovery.
- Reduces staff stress and improves morale and job satisfaction.
- Increases practice attendance and client loyalty.
- Facilitates smooth running and practice efficiency.

behaviour can realistically be managed in the practice setting, especially in the timeframe of a typical consultation. However, it is surprising how many can be. Understanding those that cannot will help the practitioner decide when to opt for alternative approaches (see 'Strategies for Managing Challenging Behaviour', below).

What influences behaviour?

Physiology

An animal's genes provide the building blocks for behaviour and are relatively constant. They determine appearance, such as ear shape, coat length and tail carriage, which play an important role in communication. They can also affect size, strength and speed, which in turn impact on whether an animal opts for 'fight', 'flight' or 'freeze' when threatened, and the innate drives that underpin the natural behaviour for the species (see Chapters 2–5). The genetic make-up of a species arises from a combination of evolutionary pressures on the wild counterpart/ancestral species and the effects of domestication and selective breeding. However, the laws of inheritance mean that there is still a great deal of genetic variation within species, breeds and even litters, and genetic expression is also affected by experience. Therefore, whilst it is important to understand the natural behaviour for the species and typical variations between wild and domesticated variants or different breeds, it is also important to keep in mind that the genetic code is only a template and that each animal needs to be assessed according to its individual behaviour without making assumptions.

Other physiological influences over behaviour, such as age or reproductive state, are dynamic. For example, young animals are generally more curious, playful and excitable and more likely to appease or use avoidance if they feel threatened. As they reach puberty, they may start to show changes in sleep patterns, activity levels and learning ability (Turcsán *et al*., 2020), or go through phases of increased willingness to wander or to challenge 'authority' (Asher *et al*., 2020). At maturity they may become more likely to behave defensively if they feel threatened, especially if they learnt that appeasing was not effective when they were younger, and intact sexually mature animals may behave differently according to the stage in their oestrous cycle (for females) or the presence of potential mates or rivals. As they then enter old age they may slow down, sleep more and lose confidence, and so revert to more appeasing behaviour. However, they are also more likely to be sore, have slowed reactions, or be confused due to cognitive decline, and so may resort to defensive behaviour if they feel unable to escape.

It is also important to keep in mind, especially in practice, the possible influences of illness (especially pain) and medication on behaviour. The evolutionary purpose of pain is to motivate the animal to prevent damage to the body and so it is to be expected that they will resist handling that causes pain, or that they have learnt from past experience may do so. Illness and medication can also affect interpretation of or responses to possible triggers. For example, an animal may react defensively if spooked, due to sensory compromise, and any disease or medication that affects the HPA (hypothalamus, pituitary gland and adrenal glands) axis can affect how an animal copes with stressors (see Chapter 15). Illness can also simply make the animal more irritable and less tolerant, increasing the risk that they may behave more defensively (also see 'Understanding and managing fearful aggression', below).

Experience

Every experience an animal has will affect its future behaviour. Learning starts even before birth and all species pass through heightened periods of sensitivity to learning during early development. The timing and flexibility of these sensitive or critical periods varies between species, but all such experiences invariably have a strong influence over lifelong behaviours and can be difficult to undo later in life. Careful management of early experiences at the practice can therefore reap dividends in later years.

Later learning does tend to be more flexible. However, it still has a powerful influence over the animal's choice of behaviour. Learning can be conscious or subconscious and, although it can be manipulated to teach preferred behaviour, the majority of learning occurs spontaneously and outside of human control. Subconscious learning involves making associations between external stimuli and internal states, especially emotions such as fear, which then affect future reactions in similar situations. For example, if an animal undergoes a painful treatment at the practice this can create a link between environmental triggers, such as the smell of disinfectant or the sound of clippers, and the pain triggered by the treatment. The animal is then likely to anticipate pain, and so behave defensively, the next time they smell a similar disinfectant or hear a pair of clippers, even in the absence of any further painful interventions. Fear alone is also intrinsically unpleasant and so the same association can be made if the animal is fearful, even where nothing physically harmful happens. Minimizing fear is therefore of equal importance as avoiding painful interventions. Animals can be helped to overcome this kind of learning by managing experiences to minimize pain and fear and to create positive emotions, such as those linked to food or play (also see 'Desensitization and counter conditioning', below). Multiple pleasurable experiences can also help reduce the effect of a later worrying one. Facilitating these is therefore a worthwhile investment, even where the animal is not currently showing any signs of fear, to minimize the effect of any later unavoidably distressing practice visits.

Conscious learning can occur through observation of others, or by making deductions based on existing knowledge. However, the most common form of conscious learning in practice comes from a process referred to as trial-and-error, in which the animal learns from the outcome of their chosen behaviour. Put simply, if an animal's behaviour results in something they want, they are much more likely to repeat the same behaviour in similar future situations. If the behaviour results in something they want to avoid (including the loss of something they want) they are likely to try a different strategy next time. To give an example, if a dog learns that the vet will stop touching his sore leg when he uses appeasing signals, he will use them again the next time he is worried by something the vet is doing. However, if the vet continues to touch his sore leg even though he is using appeasing signals, he will then try a different strategy such as growling, either at the time or the next time he finds himself in the same situation. If growling is successful, he is then likely to growl on future occasions. However, if that also fails, he may then escalate to trying to bite on this or a future occasion (Fig. 14.1). Animals that appear to bite without warning often do so because they have learnt through trial-and-error that appeasing and threat signals do not work, or may even result in their being punished, and so feel they have no alternative but to skip these and try to bite to defend themselves instead.

The current environment

Physiology and experience both exert a strong influence over an animal's choice of behaviour in a given situation, but it is the situation itself that provokes the behaviour. For example, a cat may be in pain and have learnt from past experience to anticipate a painful examination as soon as it enters the practice. However, the cat will only bite if the vet approaches and attempts to examine it despite attempts to avoid this. From the veterinary perspective this is good news, as the environment is the parameter over which we have the most control.

Understanding and managing fearful aggression

As discussed, the most common reason for challenging behaviour in the practice is fear. Fear is a

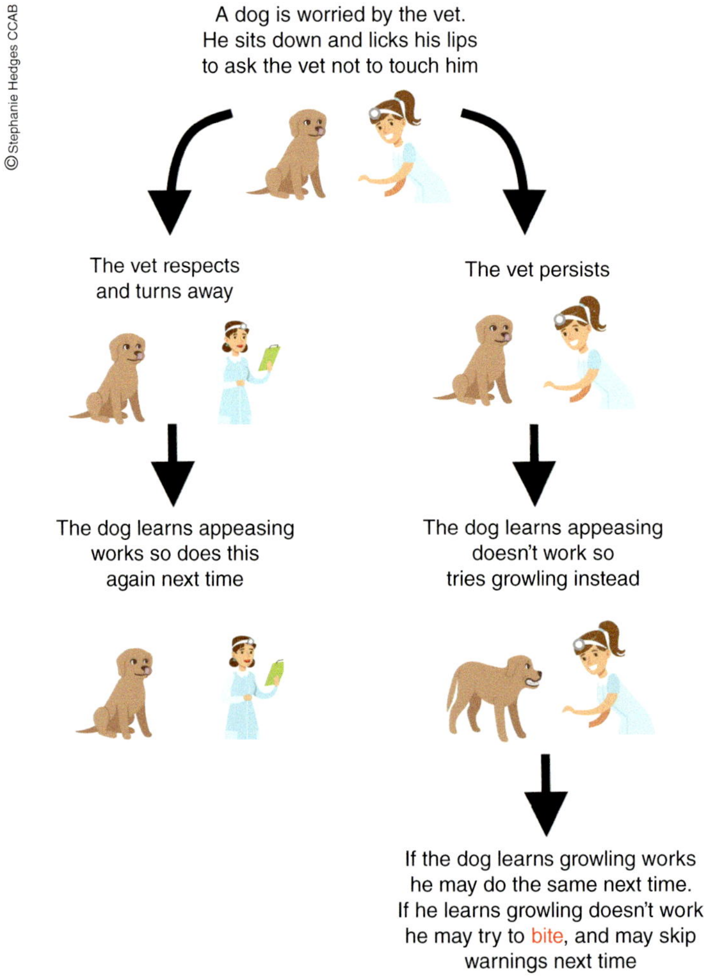

Fig. 14.1. Trial-and-error learning in the practice. If a dog's request to be left alone using appeasing signals is respected, it is likely to use them again during future visits. If they are not, the dog may try alternative ways to ask to be left alone, including the use of threat. (Figure: author's own.)

natural and necessary emotion that prompts an animal to take action when it senses something may threaten its survival. How an animal responds to a perceived threat depends on a combination of its level of fear and its options for escaping from it.

Managing an animal's level of fear

An animal's level of fear may be affected by any or all of the factors discussed above. It will also be affected by the number of those factors the animal is having to cope with at any one time. For example, a dog that is worried by loud noises, other dogs and being examined on a table may be able to cope and seem quite calm when on the examination table in a quiet, private consulting room. However, if the dog is on an examination table in a shared prep room and someone walks in with another dog and slams the door behind them, the combination may be too much for the dog to cope with. It may then try to get away by struggling, or showing threat to whoever it sees as preventing it. This is referred to as 'trigger stacking' (Fig. 14.2). It can be avoided by identifying and managing as many triggers as possible,

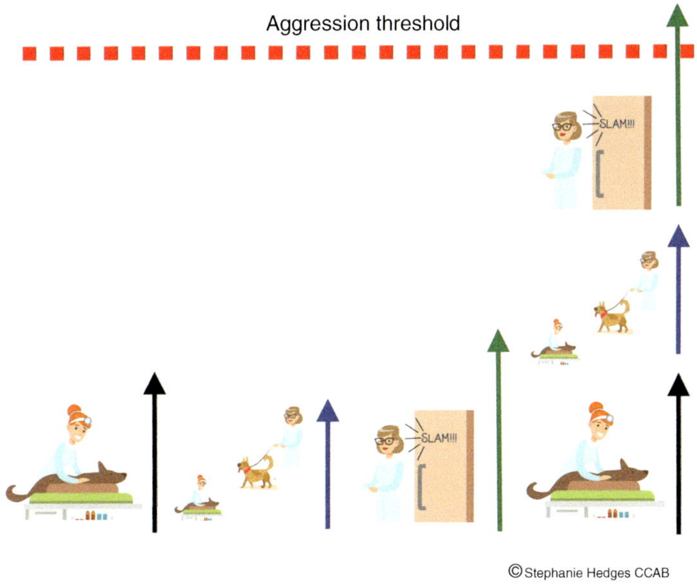

Fig 14.2. The effect of trigger stacking. A dog may cope with things that worry it individually but may become defensive if exposed to them in combination. Managing individual triggers can reduce the likelihood of this occurring. (Figure: author's own.)

even those that seem relatively minor (see 'Strategies for Managing Challenging Behaviour', below).

Managing an animal's responses when fearful

If an animal is afraid, its primary aim will be to increase the distance between itself and the (perceived) threat. Depending on species, learning and physical ability, this usually initially involves trying to avoid or move away from the trigger. Dogs that are scared will be reluctant to enter the practice and cats or exotics may seek refuge in their carrier. If the (perceived) threat is another animal (including a human) they may also use signals to try to make the threat move away, especially if escape is not an option. In strongly social species such as dogs, this initially involves appeasing signals (Fig. 14.3). Selectively social species such as cats also use appeasing signals, or may adopt displacement behaviours such as sniffing the ground or self-grooming when they are worried (Fig 14.4). Prey species typically freeze – or 'play dead' – if escape is not an option. For example, if a rabbit is held on its back, commonly referred to as 'trancing', it will typically freeze as an 'anti-predation' strategy.

If these signals are successful at repelling the threat, the situation is unlikely to escalate and the experience is unlikely to have a long-term negative effect on the animal's welfare or future behaviour. However, many of these early responses to threat can be missed or misunderstood. Lower-level appeasing signals in dogs, such as licking the lips or turning away, are very subtle and so often are not noticed, and displacement signals such as self-grooming in cats may not be recognized as a sign that they are worried. There is also a temptation to take advantage of 'freeze' behaviours in prey species, such as 'tranced' rabbits, to facilitate examination or treatment. Not only does this compromise welfare, but it may also force the animal to adopt alternative strategies. Dogs may try stronger (and more obvious) appeasing signals such as cowering or rolling on to their backs, or threat behaviours such as growling, snarling, lunging, or snapping (Fig. 14.5). Cats may also use threat behaviours such as hissing, growling or swiping with the paws, and rabbits may thump the

Copper is uncomfortable being touched on the back of the head. Note how she has lowered her body and is leaning away, the tension in her paws and around the mouth, and the furrowed brow and exposed sclera.

Toddy is uncomfortable with this examination. Note how he has turned his head and body away and is avoiding eye contact. He has also pulled his ears back and has a furrowed brow

This westie is narrowing his eyes and licking his lips to appease a threat off camera. Dogs may also yawn, sit or lay down, sniff the ground, pant or refuse to eat if they are worried.

Fig. 14.3. Appeasement signals and signs of fear seen in the dog. (Images courtesy of Laura Wyllie.)

ground or rise on to their haunches in threat before kicking or trying to bite. The precise nature of the signals used and speed with which the animal escalates from avoidance to threat will depend on a combination of its level of fear, perceived alternative options, past experiences in similar situations, and emotional and physical state as discussed above. Whatever the signal, the intention behind it is still to try to repel the perceived threat and to avoid a situation in which the animal feels forced to cause injury and so risk retaliation. Recognizing and respecting that these behaviours are a sign the animal is afraid will in most cases diffuse the situation, whereas if they are ignored the animal may feel it has no choice but to – in its mind – defend itself.

Understanding other unwanted behaviour

Aggression is not the only challenging behaviour seen in the practice. Stressed or fearful animals can show a range of behaviours in an attempt to cope, including 'displacement' behaviours or compulsions such as pacing, vocalizing or self-trauma. Young animals are also often playful and can find it hard to keep still for long, and both immature and mature animals may not

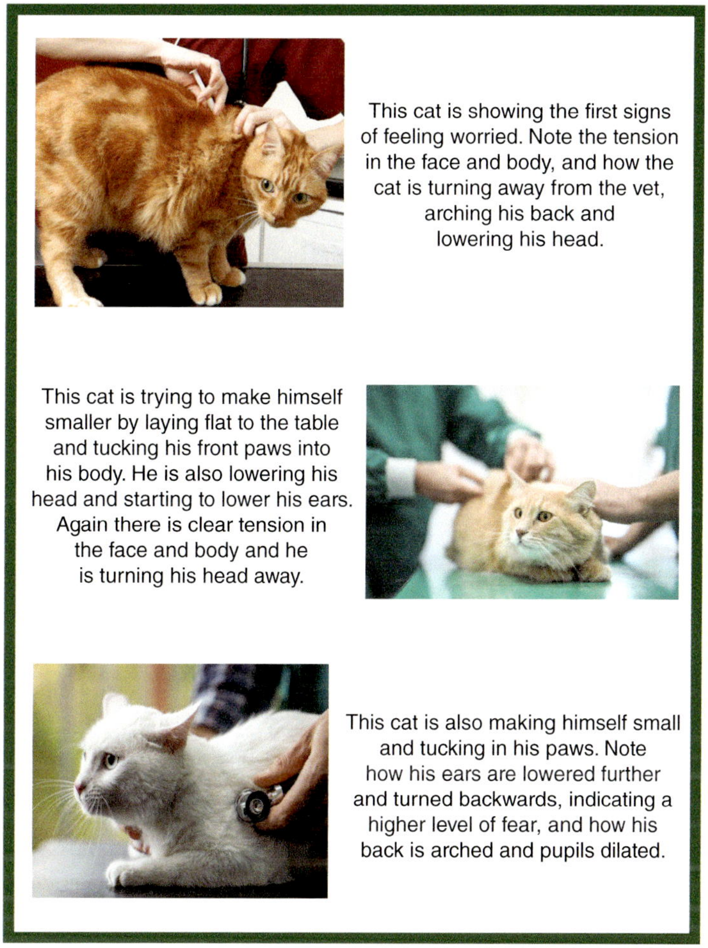

Fig 14.4. Appeasement signals and signs of fear seen in the cat. (Images from iStock.)

yet have learnt what humans consider acceptable behaviour. Natural needs and drives can also pose problem behaviour, especially in hospitalized animals. For example, housing predator and prey species or potential mates close to each other may result in attempts to escape, frustration or vocalization. Animals separated from their owners may vocalize, refuse to eat or show signs of stress; and those deprived of normal levels of exercise or stimulation or outlets for natural behaviours such as digging or clawing may become frustrated, destructive or direct their drives in other ways. The key to managing these behaviours is understanding the reasons for them and then adopting alternative approaches that address or manage the underlying cause (see Chapters 2–5; also see 'Strategies for Managing Challenging Behaviour', below).

Strategies for Managing Challenging Behaviour

It is unrealistic to suggest that we never need to use physical restraint with challenging patients. However, a range of techniques widely used by Clinical Animal Behaviourists can be adapted to the practice setting to reduce fear and manage unwanted behaviour. Physical restraint only needs be used as a last resort when all other non-coercive options have been exhausted, and

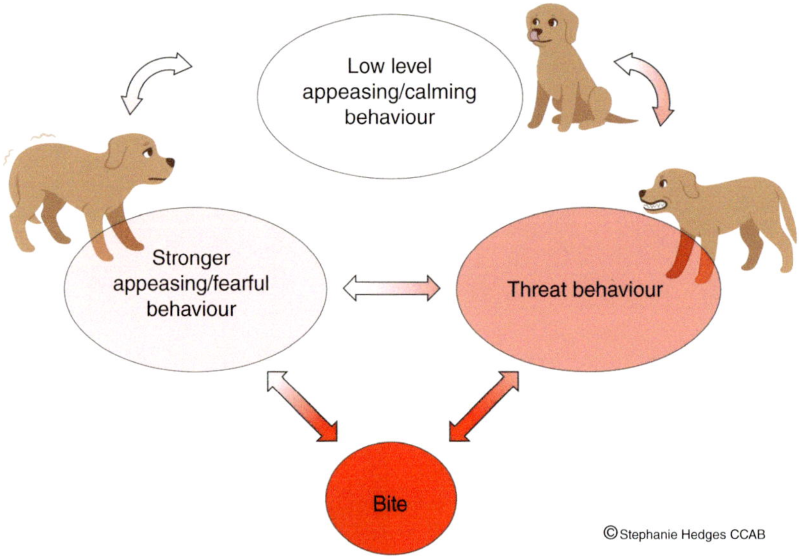

Fig 14.5. Escalation of aggression in dogs. Dogs initially use low-level appeasing signals to ask to be left alone. If these fail, they may escalate to more obvious appeasing, or may use threat. The precise nature of the signals used and speed with which they escalate will depend on many factors. They usually only escalate to causing injury if all other strategies fail, or if the dog has learnt in the past that they will. (Figure: author's own.)

then (in most cases) purely to allow for the administration of medication to facilitate examination and treatment where this cannot safely be given without it. The techniques that can be used when adopting a 'minimum restraint' approach are summarized in Box 14.2.

Take a brief behavioural summary

Although taking a full behavioural history can be time consuming, it is worth spending a few moments at the start of a consultation, or even at the time of booking, to discuss any behavioural concerns that may be relevant to examination or treatment. This demonstrates to the client that the practice is considering the needs of their pet and taking steps to make the visit as positive as possible. It also allows any potential triggers for fear or unwanted behaviours to be identified and pre-emptively addressed. Any relevant behavioural concerns reported or observed should then be recorded, to save time during subsequent visits. Useful behavioural information relevant to the practice setting is summarized in Box 14.3. Keep in mind the owner's ability to observe and report their pet's behaviour when interpreting this.

Managing the environment

The veterinary practice is full of things pets may find worrying. Being aware of and managing these reduces patient fear and stress; it also reduces the risk that the animal may no longer be able to cope and so may behave defensively (see Fig. 14.2). Known triggers can be identified by taking a brief behavioural history (see Box 14.3). However, owners may not always be aware of their pet's triggers. It is therefore also important to continually monitor body language so that any so far unidentified triggers can also be noted and addressed. Common triggers and suggestions for managing them are discussed in Table 14.1.

Box 14.2. A 'minimum restraint' approach to managing the challenging patient.

- Take a brief behavioural summary.
- Identify and manage possible triggers in the environment.
- Read and respect the animal's communication signals.
- Be aware of how the animal will interpret human behaviour and body language.
- Avoid non-essential handling and manipulation.
- Consider appropriate use of PPE to facilitate safe handling without restraint.
- Consider trial therapies.
- Provide pain relief.
- Trial anxiolytic support.
- Consider sedation or general anaesthesia.
- Use physical restraint as a last resort, ideally only to administer sedation/GA where this can't be given without it.
- Implement behaviour modification for future visits.

Box 14.3. Behavioural information useful in the practice setting.

- Is the animal fearful of anything (e.g. strangers, other animals, noises, new things)?
- Has the animal shown aggression towards people in the past?
- Is the animal comfortable being picked up or handled by strangers?
- Are there any parts of the body the animal does not like being touched on?
- Is the animal happy to be held by the collar?
- How has the animal responded to being in the practice in the past?
- Is the animal happy on the examination table?
- What are the animal's favourite treats/toys?
- Does the animal know any basic cues?
- Has the animal had any training to facilitate examination or treatment?
- Is the animal used to wearing a muzzle?

Approach and handling

An animal's choice of behaviour is strongly influenced by that of those around them. Modifying the way we approach and interact with patients can therefore go a long way towards reducing fear and the animal's decision to behave defensively. The foundation of a 'minimum restraint' approach to handling in practice is understanding how each species communicates or responds in situations they find worrying. This allows the practitioner to modify their own body language to ensure it does not appear threatening to the patient and to spot early signs that the animal is distressed and respond appropriately. Avoiding any handling or manipulation that is not absolutely essential, including physical restraint, also reduces the risk of triggering a defensive response.

Species-specific techniques for a minimum restraint approach to the handling of dogs and cats are outlined in Tables 14.2 and 14.3. Links to similar techniques for other species can be found under 'Further Resources'.

Alternatives to handling

If an animal cannot be safely examined or treated without the use of restraint, there are a number of further options that can be considered before resorting to this.

Rearrange the consultation

Once an animal has become distressed it can take some time for them to calm down again. If there is a later appointment available, some dogs may be calmed sufficiently by a walk to be rebooked for later the same day. However, others may take up to 72 hours to fully recover from a distressing experience, and cats and exotics are unlikely to become calm whilst confined to a carrier. In this case, if the appointment is routine or interim treatment is an option it may be possible to rearrange for another day.

When rebooking, it is important to identify why the animal became distressed and develop a plan to avoid a repercussion. This may include changing the practitioner (perhaps where the animal is fearful of certain types of people), managing identified environmental triggers, or changing the way the animal is approached or handled. Where the animal is distressed as soon as it enters the practice, a home visit could be considered or, if this is not practical, the client may be able to administer an anxiolytic medication at home, so that the animal feels more relaxed when it arrives. Clients can also be asked to

Table 14.1. Managing environmental triggers in the veterinary setting (also see Chapter 6 for practice design). (Author's own data.)

Trigger	Management
People: all people; certain types of people (e.g. gender, appearance); in response to their behaviour	• Identify and avoid exposure to types of people the animal is fearful of • Allow time to acclimatize before approaching or handling, including with each new person entering the room/coming to help • Read and respond to signs that the animal is worried by the person or their behaviour
Other animals: same species; different species, especially natural predators	• Separate species areas in waiting rooms, prep room and hospital areas • Separate entrance and exits to prevent meeting face to face (especially in dogs) • Provide space and hiding places around reception and in the waiting room • Provide height for cats in all areas, including reception • Cover open carriers • Cover kennels, provide hide boxes and site fearful animals away from others
Smells: illness; pheromones from conspecifics; smell of fear from people; learnt associations	• Maintain good ventilation, especially between patients • Clean surfaces, wash hands and empty bins • Pheromone therapy such as Adaptil® and Feliway® • Managing own and owner's stress
Noise: sudden, loud, unfamiliar, or unidentifiable noises; distress calls in other animals	• Keep doors and windows closed • Manage environmental noise (e.g. equipment, banging doors) • Talk quietly • Move/support vocal patients
Novelty	• Allow time to acclimatize before approaching or handling • Allow the animal to explore new areas, equipment before using, etc.

undertake training that would help the animal cope better during future consultations (also see 'Prevention' and 'Desensitization and counter-conditioning', below).

Steps can also be taken to reduce the need for approach or handling at the next visit. For example, clients could be asked to obtain samples, or to take video recordings of gait or images of lesions or injuries. In some cases, medication such as analgesia could be prescribed and the animal's response to it assessed at the next visit.

Pharmaceuticals

Animal welfare legislation places equal weight on protecting an animal's physical and emotional welfare. The use of sedation or general anaesthesia to protect an animal from physical pain in practice is routine. Its use, alongside anxiolytic medication, is equally justified to protect the animal from emotional distress. Given the choice many owners would prefer this, even with the associated cost, to seeing their animal fearful or being physically restrained.

Appropriate medications are discussed in Chapter 22. Ideally these would be given whilst the animal is calm, either at home or when they first arrive at the practice. Where the need is not apparent until examination has been started, options for rearranging the consultation should be considered.

Use of PPE

Although personal protective equipment (PPE) can offer some protection if an animal behaves

Table 14.2. Strategies to minimize restraint of dogs. (Author's own data.)

Strategy	Comment
Manage human signals	• Dogs are particularly attuned to people and are known to synchronize with owner stress (Sundman *et al.*, 2019) and show heightened fear if they sense fear in people (D'Aniello *et al.*, 2018) • Take time to reduce your own stress. Take deep breaths and try to release body tension before approaching • Alleviate owner stress by reassuring them, providing distraction, or allowing them to wait in a quiet area
Allow the dog to acclimatize	• Give the dog time to explore and acclimatize to its environment • Use this time to take a behavioural summary and assess the dog's overall demeanour • Use a loose lead and/or cage muzzle if there is any risk the dog may behave defensively
Keep them comfortable	• Provide padded bedding on floors or tables for dogs who are sore or sensitive to cold • Provide non-slip matting on floors and tables, especially for elderly or struggling animals • Proved pain relief if not contraindicated. In many cases the need for this is immediately apparent and it can be given on arrival and before examination
Entertain	• Provide toys or distractions for young, bored, or excitable animals, e.g. Lickimat® • Direct mouthing in puppies on to suitable toys • Provide suitable chew items for puppies in hospital • Encourage owners to teach appropriate ways to obtain and keep their pet's attention
Avoid trapping	• Dogs that feel trapped are more likely to behave defensively • Allow the dog to choose where they would prefer to stand or lie • Avoid blocking their (perceived) exit • Wherever possible, allow the dog to approach you • Avoid using treats to lure the dog towards you, as he may then suddenly feel out of his depth. Toss any treats behind the dog, using minimal arm movement
Be aware of your own body language	• Speak quietly and avoid sudden movements • Leaning over is a canine threat signal: stand up straight or crouch down at an angle. If crouching, keep one foot flat on the floor so you can move away easily. Lean away as you stand up again to avoid inadvertently looming over the dog • Eye contact is a canine threat signal: avoid staring or holding eye contact. Look to one side, or look away intermittently if examining the eyes • Touching the back of the head or neck is a canine threat signal: avoid wherever possible
Read and respond to the dog's signals	• Monitor for any appeasing or low-level threat signals throughout all interactions (see Fig. 14.3) • If seen, stop what you are doing and withdraw. Meet appeasement by turning at a slight angle, narrowing your eyes, and relaxing your mouth. Monitor for any escalation
Minimize physical contact	• Not all dogs enjoy human handling, especially from strangers • Ideally wait for the dog to approach you • If it does not and you want to examine it, initially offer a hand for the dog to sniff. If it is happy with this, touch for a few seconds under the chin or on the front of the chest, then move your hand away slightly. If the dog is happy, it will reinitiate touch

Continued

Table 14.2. Continued.

Strategy	Comment
Minimize manipulation	• Only use manipulation when absolutely essential • Support the body and only move in ways it is naturally designed to go when positioning • If the dog resists or shows appeasing or threat signals, stop and consider alternative options • Consider whether the owner can safely assist with basic examination, for example placing the stethoscope on the dog's chest, lifting the lip and pressing the gum, or taking a temperature
Avoid lifting if possible	• Lifting forces the dog to be closer to the handler than it may feel comfortable with • Try examining or carrying out procedures with the dog positioned on the floor • Lower an electric or hydraulic lift table to allow the dog to step on, then elevate again. Observe the dog as you do so, as some may find this unsettling • Dogs may accept being lifted by their owner more readily than by a stranger • Put some treats on the table for the dog to find
Using walk-on scales	• Steady walk-on scales with a foot to prevent them rocking or clanging • Use food lures to encourage the dog to step on • Give the dog a clear exit from them, rather than positioning them in a corner
Use food lures	• Treats can be used to encourage and reward desirable behaviour • Place treats or something sticky on a table to reward the dog for accepting an injection into the scruff • Hold food above the dog's nose to facilitate examination of the ventral neck • Smear something sticky on the syringe for intranasal injection • Use something sticky on a spoon to guide the head away when examining a paw • Only hold the dog in place with the food lure for a few seconds before giving and then introducing another if needed • Make sure the dog is aware of what you are doing. Avoid using food to distract, as the dog may spook and react defensively once it realizes what is happening
Use trained commands	• Use trained commands to aid positioning, e.g. 'sit', 'down' and 'paw' • Owners can use trained commands to direct focus or reward calm behaviour, e.g. 'stay' or 'watch' • Teach the dog names for various parts of the body, and use them to let it know where you are going to touch and to expect a treat when you do so
Make the kennel a safe space	• Avoid performing interventions in the kennel, and intersperse removing the animal for interventions with positive interactions such as tossed treats and fuss (if it enjoys this) • Keep in mind that dogs feel safe when surrounded by their own smell. Try to keep the dog in the same kennel throughout its stay, and spot-clean gross soiling without disinfecting the whole kennel • Provide multiple pieces of bedding and rotate washing these • Three-quarters cover the front of the kennel for nervous dogs, so they can hide but still see out • If leaving a trailing lead on a dog, make sure it is long enough to reach outside the kennel. Never tie to anything or attach it to a slip lead or check chain
Avoid punishment	• Never shout at, physically punish or intimidate the dog, e.g. staring at or standing over • Never use 'dominance' methods such rolling the dog on to its back

Table 14.3. Strategies to minimize restraint of cats. (Author's own data.)

Strategy	Comment
Allow the cat to hide	• Worried cats like to hide in safe dark spaces • Allow the cat to stay in the carrier as long as possible and to return to it as soon as examination/treatment is complete • Allow nervous cats to hide under a blanket or towel and only expose the parts needed for examination. Ideally use the cat's own bedding so that it has their familiar smell
Avoid lifting from the carrier	• Lifting or physically removing a cat from a carrier can be worrying for the animal • Top-opening carriers: Open or remove the top half of the carrier and allow the cat to stay in the carrier if it wants to. Allow the cat to hide under its own bedding, or cover with a large towel and just expose the parts of the body that need to be examined • Front-opening carriers: Open the door and allow the cat to step out in its own time. Reward with food when it does, if not contraindicated. If it is reluctant to step out, try sliding the cat out on its own bedding, or covering with a large towel and supporting underneath as you lift it out • Remove the carrier until examination is complete, to prevent the cat scrabbling to get back into it
Allow the cat to decide where it feels most comfortable	• Once out of the carrier, allow the cat to explore and decide where it feels most comfortable sitting. Allow it to mark with facial pheromones • Use the time to take a brief behavioural history and to assess the cat's demeanour • Provide an elevated shelf in consulting rooms for cats that like height • Provide seats for the owner and staff so that the cat can choose to sit on their lap • Ensure there are suitable locks on doors to prevent them being opened unexpectedly
Keep them comfortable	• Provide padded bedding for animals who may be sore or sensitive to cold • Provide non-slip matting, especially for elderly or struggling animals • Provide pain relief if not contraindicated. In many cases the need for this is immediately apparent and it can be given on arrival and before examination
Entertain	• Provide toys or distractions for young, bored, or excitable animals, e.g. Lickimat® • Provide suitable toys for kittens in hospital
Be aware of your own body language	• Be aware that the cat may pick up on your stress or the owner's. Take deep breaths and try to release body tension before approaching. Alleviate owner stress by reassuring them, providing distraction, or allowing them to wait in a quiet area • Move slowly, and speak quietly and calmly • Avoid making eye contact with the cat • Stand to the side and aim to greet at the cat's level. Avoid looming over • Allow the cat to approach you. If it initiates contact, touch briefly on the area in front of the ears and above the eyes and then wait to see if the cat reinitiates contact. Be careful not to confuse scent marking you with a desire to be fussed
Read and respond to the cat's body language	• Monitor for any appeasing, displacement or low-level threat signals throughout all interactions (see Fig. 14.4) • If seen, stop what you are doing and withdraw. Consider alternative approaches

Continued

Table 14.3. Continued.

Strategy	Comment
Support when lifting	• Wrap in a towel and support from underneath when lifting • Some cats feel comfortable resting along your forearm with their head in the elbow when being carried
Use minimal restraint	• Only use manipulation or restraint when essential • Support the body and only move it in ways it is naturally designed to go when positioning • Monitor the cat's response and stop if it shows appeasing or threat signals • Consider whether the owner can safely assist with basic examination, for example by placing the stethoscope on the cat's chest or lifting the lip and pressing the gum • Some cats are calmed by being firmly but not tightly swaddled in a towel. However, do not persist if the cat is struggling • Avoid scruffing other than when unavoidable, and only then to sedate/ secure
Weighing	• Allow the cat to step voluntarily on to flatbed scales, lured with treats if necessary • If the cat is nervous and being admitted, weigh in the carrier then subtract the weight of the carrier once the cat has been transferred to the kennel
Use food lures	• Treats can be used to encourage and reward desirable behaviour. These may need to be smelly to motivate the cat to want them • Place treats or something sticky on a table to reward accepting an injection into the scruff • Hold food above the nose to facilitate examination of the ventral neck • Use something sticky on a spoon to guide the head away when examining a paw • Only hold in place with the food lure for a few seconds before giving and introducing another if needed • Make sure the cat is aware of what you are doing. Avoid using food to distract, as the cat may spook and react defensively once it realizes what is happening
Make the kennel a safe space	• Provide a place for the cat to hide. Ideally place its own open carrier in the kennel. Alternatively use a 'Hide, Perch and Go'™ or adapted cardboard box, or cover three-quarters of the front of the kennel with a towel so that the cat can hide but still see out • Avoid performing interventions in the kennel or in sight of other cats. Intersperse removing the cat for interventions with positive interactions such as tossed treats and fuss (if they enjoy this) • Keep in mind that cats feel safer around their own smell. Use their own bedding, and provide multiple pieces of bedding and rotate washing these • Avoid moving the cat between kennels • Spot-clean gross soiling around them, allowing them to hide in their carrier, or hide under a blanket if they are worried. Secure in their own carrier if they have to be removed • Avoid removing scent marks on kennel surfaces (seen as oily brown patches)
Putting back in a carrier	• Allow the cat to step back in voluntarily • If in a kennel, place the carrier in the kennel and allow the cat to step in
Avoid punishment	• Never shout at, physically punish, or intimidate the cat

defensively, this is not absolute – especially in the case of large dogs. Defensive behaviour also indicates that the animal is in distress and so using PPE simply to facilitate restraint carries all the risks to health and welfare discussed above. The pros and cons of use therefore need to be considered against the alternatives in each case (see Box 14.4). If PPE is to be used it is important that it is used properly and steps are taken to minimize any potential negative effects.

Muzzles

Dogs tend to tolerate cage/basket muzzles more readily than cloth ones and can be encouraged to put their nose in voluntarily by using peanut butter or cream cheese smeared inside. Ideally dogs will be taught to feel relaxed wearing a muzzle before it is needed, so that this does not add to their stress at the time (see 'Further Resources' for how to do this). Always check a muzzle's integrity before using it and check that it is fitted correctly by asking the owner to attempt to pull the muzzle over the dog's nose after all the straps are secured. Bear in mind that a dog can still cause bruising whilst wearing a muzzle, so owners should also have their dog on lead if there is a chance that the animal might lunge.

Box 14.4. Considerations when using PPE.

- Is there any suggestion the animal may attempt to cause injury, based on previous history and current behaviour?
- Will the PPE offer adequate protection if the animal does attempt to cause injury?
- Does the PPE allow continued use of a 'minimum restraint' approach in cases where the animal's possible reaction is unclear?
- Does the animal tolerate the PPE without signs of distress (see Figs 14.3 and 14.4)?
- Has the animal been habituated to the PPE (e.g. a muzzle), or can it be?
- Can the examination or intervention be delayed (see 'Rearrange the consultation', above).
- Are pharmaceuticals contraindicated?
- Can the pharmaceutical be administered without the need for restraint (e.g. by the client at home, in food)?

Cats are generally less tolerant of muzzles, especially those that cover the eyes. Although they may appear to calm the cat, this is more likely to be because of extreme fear due to not being able to assess their environment or even breathe properly than actually feeling more relaxed. Air muzzles may be better tolerated if the cat has been gradually taught to feel comfortable wearing them. However, each cat's response to any type of muzzle needs to be assessed and muzzles offer no protection against claws. Other methods, such as swaddling or 'crush' cages, may therefore be both safer and less aversive for the cat.

Gauntlets and protective gloves

Gauntlets and protective gloves or sleeves can offer protection to the hands and arms when handling smaller dogs, cats, or exotics. Leather gauntlets offer protection against bites and bruising but can restrict movement and so make handling clumsy. Kevlar reinforced gloves and sleeves typically protect against bites, but are not as effective against bruising. Both can pick up and hold the odour of stress from the handler or previous patients and so need to be cleaned regularly.

Swaddling

Swaddling small dogs, cats and exotics can help to restrict movement. This is especially useful for protecting the handler from a cat's claws. Some cats also find that being swaddled firmly, but not tightly, is calming. The blanket or towel needs to be large enough to fully wrap around the animal and overlap so that it is not easily unravelled. Covering the head can calm some animals and worry others, as they are no longer able to survey their environment. Which the animal prefers can be assessed by laying the towel loosely over the head and seeing whether they choose to keep it there or try to shake it off.

Dog catcher

Dog catchers or snare poles can improve safety where a dog's behaviour is unpredictable, and they do not need to be distressing if they are used correctly. Ensure that they are properly functioning; incorporate a locking mechanism to prevent over- (or under-) tightening and a quick-release mechanism. When adjusting the noose to fit, it should be no tighter than a standard collar. This will ensure that the dog cannot escape without triggering fear due to pain or a sensation of being asphyxiated. There is nothing to be gained by making the noose tighter than this. It must be kept in mind that if a dog is restrained on a dog catcher the animal's 'flight' option has been removed, increasing the possibility of 'fight'.

Crush or squeeze cage

The unfortunately named 'crush' or 'squeeze' cage can help restrain a very fractious animal into a small area for sedation; if used correctly it does not need to compromise welfare. Ideally cats would have learnt to feel comfortable being restrained in a carrier before a crush cage is used. In any event the cage should be lined with soft bedding and covered, and the cat should be given time to acclimatize to the cage with the movable panel fully open. The panel should then be moved gradually to encourage the cat into as small an area as is comfortable for the animal. It should never be closed to the point of causing discomfort or distress to the cat, as this not only compromises welfare and teaches the cat to be fearful of the practice and of carriers, but the stress it causes can also reduce the efficacy of the sedation.

Physical restraint

There may be occasions when traditional methods of forced restraint and manipulation are unavoidable. However, these are best considered a last resort and reserved for the administration of medication to enable further examination or treatment to keep it to the minimum. Use of PPE when restraining for sedation is discussed above.

Prevention and behaviour modification for future visits

The strategies discussed above are aimed at short-term management and facilitation of treatment. In the longer term, fear of the practice, or any of the triggers within it, is best addressed through prevention or a programme of desensitization and counter conditioning.

Prevention

Prevention is always better than cure. Managing early experiences at the practice not only stops challenging behaviour developing but also 'inoculates' against unavoidably distressing experiences in the practice in later life. All species can be taught to enjoy practice visits and a range of cues or trained behaviours that can facilitate management, examination, or treatment. Examples of prevention strategies are given in Box 14.5 and discussed in Chapter 12. How to choose where to refer clients who require additional support with this is discussed in Chapter 21.

Desensitization and counter-conditioning

'Desensitization and counter-conditioning' is the process whereby an animal is exposed to something that worries it at a level that does not trigger fear and in a way that enables the animal to make associations between the trigger and something pleasurable, such as food. The level of exposure is then very gradually increased so that the animal can eventually cope with it at a level that is normal for the trigger. For example, if an animal is worried by the noise of the clippers, it would be exposed to the clippers far enough away not to scare it whilst being fed treats. The clippers would then be brought gradually closer, usually over a number of days or weeks, until the animal is relaxed even when the clippers are next to and eventually touching it.

The process only works if the animal's triggers are identified and carefully managed, and the training plan progresses at a speed it can cope with. This requires a plan tailored to the individual animal's needs and responses. Unless the practice has appropriately trained staff it may be necessary to refer the client to a Clinical Animal Behaviourist (CAB) to develop a plan that can then be delivered in collaboration with the practice and owner. How to make a referral is discussed further in Chapter 21.

Box 14.5. Prevention.

- Learning during early development lasts a lifetime. Arrange extra 'socialization' visits to the practice during the species' most sensitive period of development and manage them to ensure that they are a completely positive experience for the animal. Avoid any examinations or interventions, manage environmental triggers, monitor body language throughout and give food, toys, or fuss, according to the animal's preference, to create positive associations with the practice.
- Encourage owners to practise mock examinations at home so that these are familiar to the animal. Cheap home equipment, such as stethoscopes or nail clippers, can add realism.
- Encourage owners to teach their pet the names of various parts of the body, e.g. ear, paw, tail, etc. These can then be used to let the animal know where the vet is going to touch and to expect this to be followed by a treat.
- Encourage owners to teach basic cues that can be used to guide preferred behaviour and facilitate handling. For example, 'sit', 'watch', 'stay' and 'settle' can be used to encourage an animal to remain still. 'Paw', 'down' and 'roll over' can facilitate examination of a foot or ventral abdomen. This can be applied in a modified form to all species (see 'Further Resources').
- Use practice visits or socialization classes to teach appropriate behaviours in practice, for example to 'sit' rather than jumping up to greet.
- Teach cats and exotics to feel comfortable in a carrier.
- Treat any behaviour of concern that will aggravate problem behaviour in the practice (e.g. fear of travel, other animals, attention-seeking behaviour).

Conclusion

The key to successful management of challenging behaviour in the practice is recognizing the motivation underpinning the behaviour, and then modifying the environment and how the animal is approached and handled accordingly. The most common cause of problem behaviour in the practice is fear. Managing environmental triggers, ensuring our own behaviour and body language do not unintentionally appear threatening, and learning to read and respect the first signs that an animal is worried can all help reduce overall fear at the time and prevent this escalating over successive visits. The need to physically manipulate the animal for examination or treatment can often be minimized by using food lures, training, or (where safe to do so) help from the owner, or avoided through the use of trial therapies. Where this is not possible, early use of pain relief, anxiolytic medication, sedation or general anaesthesia, as appropriate, is justified on both welfare and safety grounds, and can be administrated with careful non-aversive use of PPE. Coercive physical restraint then becomes the last resort for administration of sedation where this cannot be safely done without it.

Further Resources

Books

Atkinson, Trudi (2018) *Practical Feline Behaviour. Understanding Cat Behaviour and Improving Welfare*. CAB International, Wallingford, UK.

Hedges, Stephanie (2021) *Practical Canine Behaviour for Veterinary Nurses and Technicians*, 2nd edn. CAB International, Wallingford, UK.

Yin, Sophia (2009) *Low Stress Handling, Restraint and Behavior Modification of Dogs and Cats: Techniques for Developing Patients Who Love Their Visits*. CattleDog Publishing, Davis, California. (Also available at: http://drsophiayin.com/lowstress/online)

Videos and websites

- Muzzle training video. Available at: http://www.bluecross.org.uk/99144-109679/muzzle-training.html.
- Cat handling videos. Available at: https://icatcare.org/veterinary/resources/
- Training techniques for exotics. Available at: https://barbarasffat.com/
- Rabbit body language. Available at: https://www.rspca.org.uk/adviceandwelfare/pets/rabbits/behaviour/understanding
- Rabbit handling in the practice. Available at: https://veterinary-practice.com/article/minimising-stress-during-rabbit-examinations

References

Asher, L., England, G.C.W., Sommerville, R. and Harvey, N.D. (2020) Teenage dogs? Evidence for adolescent-phase conflict behaviour and an association between attachment to humans and pubertal timing in the domestic dog. *Biology Letters* 16.

D'Aniello, B., Refik Semin, G., Alterisio, A., Aria, M. and Scandurra, A. (2018) Interspecies transmission of emotional information via chemosignals: from humans to dogs (*Canis lupus familiaris*). *Animal Cognition* 21(1), 67–78.

Edwards, P.T., Hazel, S.J., Browne, M., Serpell, J.A., McArthur, M.L. and Smith, B.P. (2019) Investigating risk factors that predict a dog's fear during veterinary consultations. *PLoS ONE* 14(7).

Lloyd, J.F.K. (2017) Minimising stress for patients in the veterinary hospital: why it is important and what can be done about it. *Veterinary Science* 4(2), 22.

Mariti, C., Bowen, J., Campa, S., Grebe, G., Sighieri, C. and Gazzano, A. (2016) Guardians' perceptions of cats' welfare and behavior regarding visiting veterinary clinics. *Journal of Applied Animal Welfare Science* 19(4).

Sundman, A., Van Poucke, E., Svensson Holm, A., Faresjö, A., Theodorsson, E. *et al.* (2019) Long-term stress levels are synchronized in dogs and their owners *Scientific Reports* 9, 7391.

Turcsán, B., Wallis, L., Berczik, J., Range, F., Kubinyi, E. and Virányi, Z. (2020) Individual and group level personality change across the lifespan in dogs *Scientific Reports* 10, 17276.

15 Considering Medical Influences on Behaviour

Helen Zulch

Introduction

The interrelationship between physiology and behaviour has long been recognized, as attested by ethological texts that discuss behaviour change at important life stages. Sexually dimorphic behaviour that becomes more pronounced after puberty in unneutered animals and maternal behaviour are two clear examples. The relationship between physical health and behaviour in broad terms has also been accepted for many years, with references to 'sickness behaviour' (for example, Dantzer and Kelley, 2007). However, the extent to which physical health in pets can impact on the behaviour that they demonstrate has been given more prominence relatively recently, with the past few decades seeing a substantial increase in literature examining this area of companion animal behavioural science.

This topic is large and complex; therefore, this chapter will briefly discuss:

- The key indicators for a higher index of suspicion of a medical condition playing a role in a behaviour problem.
- Different approaches that can assist a veterinary surgeon to rule in or rule out the presence of a medical condition.
- A few key presentations for which there is peer-reviewed evidence in support of the need to more exhaustively rule out an underlying medical condition.

This chapter can never be an exhaustive exploration of the subject, not least because of new information constantly emerging in the field, so more than anything it aims to raise awareness of the fact that every pet presenting with a behaviour problem should undergo at least a basic medical assessment in order to determine, to the best of the veterinarian's ability, the likelihood that a medical condition is contributing to the behaviour problem reported. In addition, the veterinary surgeon needs to retain an open mind about the impact of physical conditions on the behaviour of each patient and be willing to reassess this over time as new evidence emerges or behaviours progress.

How Do Medical Conditions Influence Behaviour?

Medical conditions can influence behaviour in a number of ways. Direct influence is easy to see and understand; for example a clearly painful lesion is likely to lead to a defensive response by an individual when that lesion is manipulated. However, less direct relationships can be more challenging to understand and, in some cases, links can only be hypothesized with our current data. For example, the reason for the onset and generalization of noise fear in dogs with musculoskeletal pain (Lopes Fagundes *et al.*, 2018) is

 Companion Animal Behaviour Problems: Prevention and Management of Behaviour Problems in Veterinary Practice (eds. R. Casey, S. Heath and H. Zulch)
DOI: 10.1079/9781800621312.0015

hypothesized to be caused through a dog startling to a noise, which, as it flinches and tenses muscles, triggers pain so that noise itself becomes associated as the source of the pain.

Associative learning is strong in all individuals, especially when the learning is to avoid harm or anticipation of harm. This is the reason that some animals may have recovered from a medical condition by the time they are examined by a veterinarian or seen by a behaviour practitioner, but are still exhibiting the behaviour that they learnt would avoid a painful or uncomfortable experience. Examples include the dog who, months after a healed otitis, will still snap at any hand approaching his head, or a cat who avoids certain toilet characteristics even months post resolution of cystitis. For these reasons it is important to consider the medical history as well as the immediate medical presentation when trying to determine the likelihood of an influence of a past or current medical condition on behaviour.

For a very useful overview of the range of ways in which pain is thought to contribute to the presentation of behaviour problems, see Mills *et al.* (2020).

When to Consider a Medical Influence on Behaviour

Reports in the literature for medical conditions present in a behavioural referral case load vary, with the most recent paper citing a range of 28%–80% of cases seen across specialist behaviour clinics in a range of countries (Mills *et al.*, 2020). This encompasses a range of behavioural presentations and medical conditions. For this reason, it is always important to consider the risk of a medical condition playing a role when a client mentions a behaviour problem or concern in their pet. However, the following situations should raise the index of suspicion and potentially lead to more in-depth medical investigation either before or concurrent with behavioural referral.

Sudden onset of a behaviour problem in a middle-aged or older animal where there has been no known trigger

If an older animal develops a painful condition, perhaps because they have a joint abnormality which then leads to the development of osteoarthritis and associated pain, it is possible that this pain may result in a change in behaviour. For example, the dog may have been sociable in the past but may now no longer tolerate other dogs' invitations to play and may exhibit aggressive behaviour in order to discourage interactions. This sudden onset of behaviour change is a not uncommon indication of a medical condition underlying the problem behaviour. However, there is of course a caveat to this: medical conditions, including osteoarthritic changes and degenerative joint disease, can occur even in young animals (Kealy *et al.*, 2000) and some congenital conditions may be more likely to be seen in a young individual, for example 'house soiling' that is a result of an ectopic ureter or a porto-systemic shunt that leads to post-prandial behaviour change.

Waxing and waning signs or what an owner may describe as a 'Jekyll and Hyde' personality

Anecdotally, behaviour practitioners have reported over a number of years that patients with a medical condition implicated in their behavioural presentation frequently show waxing and waning behavioural signs. Barcelos *et al.* (2015) demonstrated for the first time that this is a description more often seen in pets showing aggressive behaviour associated with a locus of pain, as opposed to those showing aggression with no locus of pain present. For some of these patients we can only assume that a chronic health problem causes them to experience better and worse days, which then translates to behavioural manifestations. However, in other patients a careful review of the medical history together with diaries of behaviour change may enable the clinician to determine a clear link between a waxing and waning physical condition and changeable behaviour. For example, the author has seen a dog where, with careful analysis of behaviour and health diaries, it became clear that aggressive episodes correlated with recurrent anal gland impactions.

Unusual clusters of behavioural signs

An example might be the dog who has started to show aggressive responses to owners when he is resting on his bed, who is also less eager to go for walks and has never really liked grooming over his hindquarters. The risk for a patient such as this is that he could be treated independently for guarding his bed, dislike of handling and either the reluctance to go for walks is overlooked, or if extreme may be treated as 'fear of going out'. However, if he is carefully examined and found to be experiencing degenerative joint disease in his hips, all three apparently disparate signs can be tied together and a more appropriate holistic treatment plan put in place.

Nature of progression of the problem

Typically, in a behaviour problem that is not associated with a medical condition, we would expect a behaviour to be triggered by some event and then worsen in a more or less predictable manner as further experiences and learning impact on the pet (this is put very simplistically; more detail on the reasons for behaviour problem development can be found in the chapters in Parts I and II of this volume). However, when medical conditions underlie, not only will the waxing and waning as described above potentially be seen, but additionally unusual progression, such as a very fast generalization as described in Lopes Fagundes *et al.* (2018), may be reported. The meaning of the term 'fast generalization' is illustrated by the following example. Typically, a pet who shows a fear of thunder gradually becomes frightened of more and more things associated with a storm. However, in fast generalization, he may after only one or two experiences of the thunder become totally avoidant of any area in which the noise was heard.

Assisting an owner in putting together a timeline of the development of the behaviours of concern can greatly assist in identifying the situations described above, or alternatively clarify a progression that can, with relative certainty, be explained by experience and learning. Additionally, establishing a timeline early enables better tracking of behaviour change into the future.

What Approach Should the Primary Care Clinician Take?

The fact that suspicion of a medical condition underlying a behaviour problem is frequently raised by a behaviourist (veterinary or non-veterinary), even post referral, is not difficult to explain. A first-opinion consultation is usually time constrained and takes place in a confined environment not conducive to the pet being able to express normal behaviour. Added to this, a large proportion of pets are frightened at the veterinary practice and do not therefore exhibit normal responses and may be very challenging to examine fully or in a manner likely to elicit a natural reaction. This latter aspect can be improved by interventions that promote the pet's positive emotional state and cooperation in the practice: see Chapters 6, 7 and 14 for more information.

However, there are steps that the primary care veterinary surgeon can take to better assess the risk of an underlying health condition, especially one associated with pain. This can then allow more targeted further investigation should the index of suspicion be raised early.

It will become clear from the approach suggested that obtaining even a basic history that will enable you to move forward with your assessment may take some time. For this reason, it is likely that in order to manage these patients appropriately in practice, once an owner has raised their concern with you, you may wish to consider one of the following.

- Make use of longer appointment slots, which can have the added benefit of more time to observe the patient, assuming the patient can cope adequately in the environment. However, do be aware that for many pets, even a double or triple appointment slot in the practice environment may not enable them to relax sufficiently to exhibit behaviour that is normal for them, or react to examination and manipulation in a manner that delivers useful information.
- Arrange multiple appointments to move through the process stepwise.
- And/or utilize questionnaires that owners can complete ahead of appointments so that you have details regarding the problem (including a timeline), the pet's history and lifestyle and anything else of relevance and can prepare in advance of seeing them.

For these patients the whole-practice approach discussed in Chapters 7 and 22 where non-veterinarians within the practice can assist in information gathering can greatly assist the vet in appropriately assessing these patients.

Taking the history

There are a number of questions that can be asked in order to better rule out a medical component, although none are fail-safe. As a minimum the following areas should be considered and potentially explored in some depth.

- Known temperament of the animal. For example, has it always been a very fearful or anxious pet and therefore development of a specific fear may be easier to understand?
- General life experiences. For example, is there an early life history that makes development of some behaviour problems more likely?
- Potential external events coinciding with the onset of the problem. Identifying a key triggering event is often the clearest rule-out of a medical component, but for many owners they may have been unaware of the significance of an event when it happened.
- Sudden onset of a behaviour in a pet that has never previously shown a tendency to exhibit the behaviour of concern.
- Progression of the problem (as above).
- Waxing and waning nature of the problem (as above).
- General activity levels of the pet on a day-to-day basis.

There is a caveat here though: if a pet is extremely motivated to engage in a given activity this may, in the moment, overrule discomfort. For example, if a dog is highly motivated to chase a ball, it may do so despite fairly severe musculoskeletal disease. So, engaging in desired activities is not a rule-out for the presence of pain or discomfort.

- Changes in habits. For example:
  - Less keen to go for walks
  - Reluctant to jump on to furniture or into the car
  - Less playful, with toys or with other individuals (be that conspecifics or individuals of other species)
  - More tired than usual
  - Eating differently (posture, head posture, behaviour around mealtimes, etc.)
  - Toileting differently (sites, postures, behaviour around the event, etc.)
  - Sleeping differently – possibly on a bed on the floor rather than on the sofa as previously
  - Grooming itself differently
  - Not liking being touched or groomed in a specific area.
- Other physical signs that the owner may not have thought are significant enough to mention:
  - Occasional bouts of diarrhoea or periods when defecation is more frequent
  - Occasional episodes of vomiting
  - Appetite changes or fluctuations
  - Asking to go out to wee more frequently.

The use of adjunctive diagnostic tools

Despite having an index of suspicion for a medical condition, it may be challenging to gather clinical data because of the patient's behaviour in the consulting room. To help you to determine whether or not further diagnositics should be embarked upon, you can:

- Ask the client to obtain video footage. Videos of the pet engaging in a range of activities over a period of time, especially those that are likely to elicit pain or discomfort, so for example:
  - The patient eating if you suspect dental pain
  - The patient walking, jumping, climbing, changing positions, for example sit/down/stand and the reverse, resting (duration, posture, etc.) if you suspect a musculoskeletal condition.

Of course, we would never request that a client deliberately elicit a behaviour of concern or put themselves, others or a pet in a situation of risk to obtain video footage, but much of the above can be gathered in day-to-day life without risk. If a video of the behaviour of concern can be obtained in a safe and welfare-friendly manner, then this can also be really valuable, especially if it can demonstrate the behaviour in context with preceding events as well as events post behaviour also captured.

- Ask the client to keep a diary. This can be particularly useful for conditions that wax and wane. The diary should include details of day-to-day maintenance behaviours such as sleep, eating and voluntary activity engaged in, as well as the incidence of behaviours of concern and how these relate to environmental events or social interactions. For example, it may be possible to pick out a pattern of a 'bad day' after a day of activity. One still needs to decide whether this relates to the physical activity itself, or to exposure to external stimuli that have a psychological impact, but it is a step in the right direction.

- Consider performing your assessment in an area other than the consulting room, if the species allows. For example, if your practice has a car park, will a dog allow assessment there that it would not tolerate within the consulting room? Can you assess its movement in an alternative environment, for example can you join the owner and dog on a short walk? Would a house visit be more valuable for any pet, but especially for those more challenging to assess 'hands off' in the practice, for example birds and small prey mammals who could be assessed in their housing?

Once you have gathered the information you believe is required for initially ruling in or out a medical component to the behavioural presentation, you can then make a decision regarding the next steps. If you believe that the pet is clinically healthy, or that any medical conditions of which you are aware are currently under control, behavioural first aid can be implemented by the practice either pending or in place of referral to a behaviour practitioner, depending on the practice's expertise and policies. And a decision regarding referral can be made.

If you believe that further diagnostic investigations are indicated to determine the presence or absence of a medical condition, then a decision can be made regarding the urgency of the behavioural referral. If the behaviour is posing an immediate and serious risk to the pet or others, then timely implementation of behavioural advice is recommended; this may mean immediate referral pending the outcome of further investigations. Alternatively, if it is felt that it is more appropriate and the risk allows further medical interventions before detailed behavioural interventions, then behavioural first aid advice can be given while these are carried out. Once any medical condition uncovered has been treated, or should no such problem be found, referral can take place at this point should further behavioural advice still be needed.

It should be noted that even if there is a low index of suspicion for a medical component to the behaviour problem at the point of initial assessment, or if nothing is found on further investigations, it may be necessary for the pet to return for further diagnostics at a later date once behavioural interventions have been implemented. This is particularly true if the pet has been challenging to examine at presentation. In cases such as these, working together with the clinical animal behaviourist or animal behaviour technician can be invaluable. They can assist an owner in teaching their pet to cooperate with future medical examination or to perform behaviours that make assessment easier, such as wearing a muzzle for safe physical manipulations or blood draws, or walking on a loose lead to facilitate gait analysis.

Fig. 15.1 is a summary of the suggested approach to dealing with a patient presenting with a behavioural concern.

Medical Conditions Impacting on Behaviour

While potentially any medical condition can impact on behaviour, either through pain or discomfort, direct impact of physiological change on behaviour (for example, onset of house soiling in a diabetic or cushingoid pet due to increased urination) or the as yet incompletely understood impact of inflammatory mediators on the brain, the literature discusses a number of specific conditions where links between behavioural presentation and underlying disease have been found. This section will examine a number of these in more detail. An exhaustive description of all is not possible, but Table 15.1 at the end of this section includes additional illness links to bear in mind.

Note that it is likely that a number of the conditions mentioned below cause behaviour change due to pain, but we do not yet have the evidence for all, and for many we do not yet fully

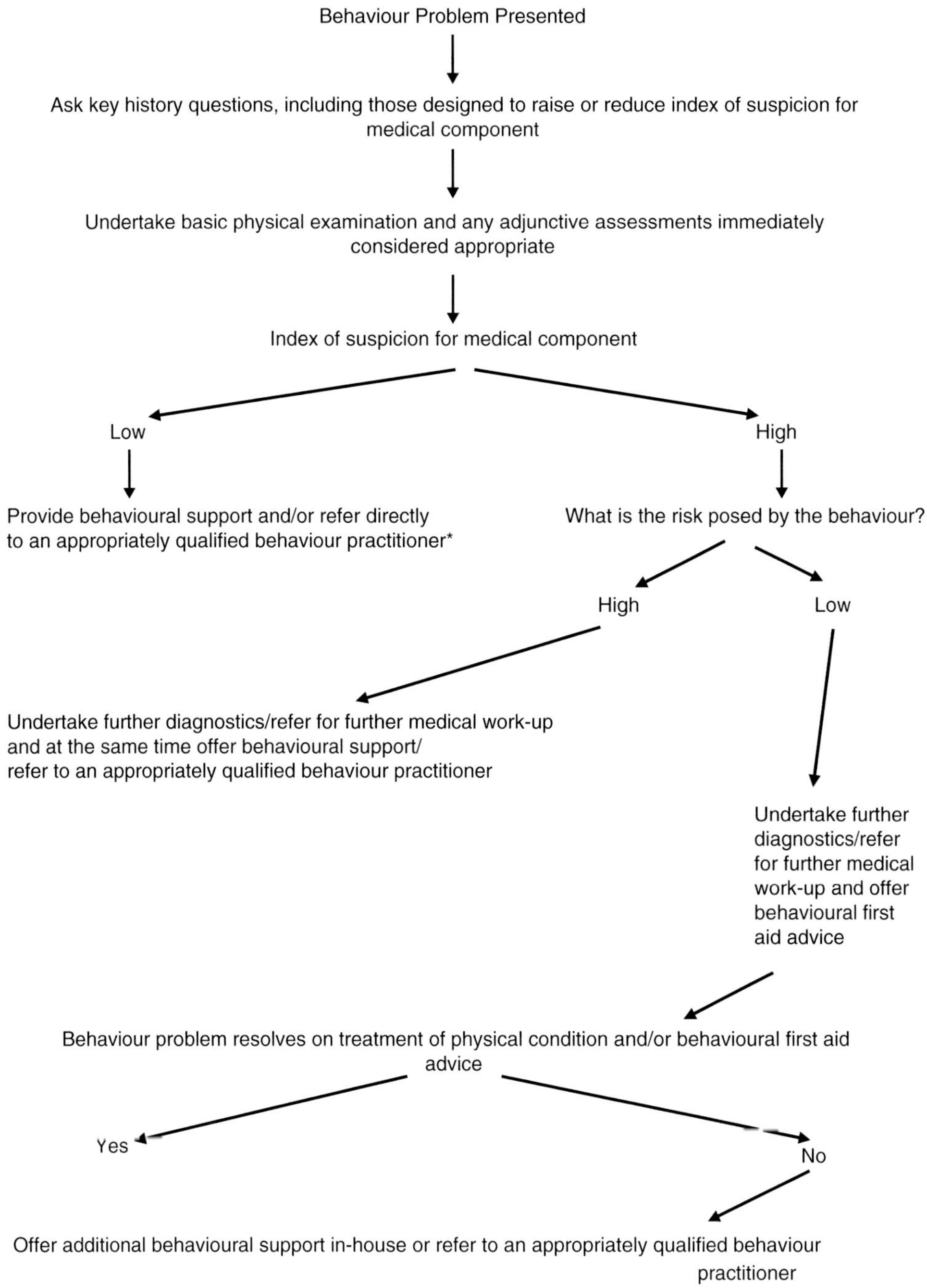

*Further diagnostic tests may still be required at a later date.

Fig. 15.1. A suggested decision-making process when faced with a behaviour problem in first-opinion practice. (Figure: author's own.)

Table 15.1. Key medical differentials for behaviour presentations. (Please note: no list can be exhaustive, but this table contains key differentials at the time of writing. Data sourced from References and Further Reading, below.)

Overarching behaviour presentation	Details/subcategories	Medical conditions to consider
Aggressive behaviour		Pain Hypothyroidism Partial seizures?
Repetitive behaviour	Oral not self-directed (including pica)	Gastrointestinal disease Food intolerance
	Self-directed including self-trauma	Locus of pain underlying area Skin disease causing overgrooming
	Locomotor	Partial seizures Pain Hepatic encephalopathy
	Hallucinatory (light chasing, fly snapping, etc.)	Ocular disturbances Painful conditions, e.g. syringomyelia Gastrointestinal disorders (fly snapping)
House soiling	Urine	**Anything giving rise to polyuria/polydipsia**. For example: Diabetes mellitus through increased need to urinate Diabetes insipidus through increased need to urinate Hypoadrenocorticism/hyperadrenocorticism through increased need to urinate Chronic renal failure through increased need to urinate **Cystitis** through increased urgency or through learnt avoidance of toilet areas when pain is experienced **Painful conditions** making access to toilet areas challenging **Reduction in sensory capacity/cognitive decline** – cannot find usual toilet **Incontinence** for any reason can present as house soiling
	Faeces	**Any condition causing increased need or urgency in defecation** In cats, conditions causing pain on defecation, learnt aversion to litter tray **Reduction in sensory capacity/cognitive decline** – cannot find usual toilet
Fearful behaviour		Sensory compromise Painful conditions
Separation-related behaviour		Painful conditions
Other	Periods of 'confusion' Night-time vocalization in cats	Hepatic encephalopathy Hyperthyroidism Cognitive dysfunction syndrome High blood pressure

understand what it is about the physiological change of the disease process that correlates with behaviour change. For this reason, this chapter will largely address behaviour change linked to systems without discussing the underlying causative link. Pain will be considered as separate from the systems discussions, but always bear in mind the potential occurrence of pain in any condition, as this may impact on treatment decisions.

Neurological conditions

It is easy to see that neurological disease, especially if located in the brain, is likely to cause changes in behaviour (see for example Uccheddu *et al.*, 2018). However, in the absence of other indicators of neurological disease, it is probably more likely that most behaviour is not related to a physical condition within the brain or central nervous system (neuropathic pain excepted – see section on pain below). Some exceptions include those discussed briefly below.

Cognitive dysfunction syndrome is probably the neurological condition impacting behaviour for which we have most evidence. This is a condition caused by physiological brain changes different to those of normal ageing in dogs and cats and having some features in common with dementia causing disease in people. Cognitive dysfunction may be seen as early as at 7 years of age in dogs, although more commonly it presents in dogs over 11 years old, and in cats it can be seen from around 11 years of age, with increased prevalence in cats over 15 years of age (Sordo and Gunn-Moore, 2021). Salvin *et al.* (2010) demonstrated that only a very small proportion of dogs surveyed as presenting with signs of canine cognitive dysfunction syndrome (CCDS) have a formal diagnosis. As there is now evidence that the rate of deterioration can be slowed through interventions such as prescription medications (for example selegiline), diet, some specific neutraceuticals and environmental enrichment, it is key that these pets are diagnosed early. The primary care clinic is the ideal place to do this. Making routine use of one of the available tools for assessment of cognitive functioning (e.g. Salvin *et al.*, 2011) in mature dogs and asking relevant questions in older cats has the potential to improve the quality of life for both pet and owner if interventions are implemented as soon as decline is measured.

Partial seizure activity has been postulated as a cause of tail chasing in bull terriers, and partial or full seizures are also talked about as a cause of other behaviours, such as episodic severe aggressive behaviour in dogs. With respect to aggressive behaviour, there is currently little evidence for this assessment and a good clinical history must be taken, ensuring that a clear and detailed description of the behaviour and timeline is obtained in order to rule out other behavioural explanations. In addition, the presence or absence of other factors that are frequently associated with seizure activity, such as when the suspected seizure activity occurs and the presence of pre-ictal and post-ictal phases, must be ascertained. A diagnosis of seizure activity underlying behaviours of concern is, as with idiopathic epilepsy, largely a diagnosis of rule out.

Anecdotally, **cerebrovascular accidents** (CVAs) have been reported to present as behaviour change in cats and therefore a CVA can be considered in cases of sudden onset of behaviour change in an older cat, especially where other risk factors for a CVA exist.

Finally, although the direction and details of the relationship are not yet fully understood, patients with **idiopathic epilepsy** may experience co-morbid anxiety and fear-related conditions (Shihab *et al.*, 2011). Patients diagnosed with epilepsy may therefore benefit from investigation as to the presence of behaviour problems so that if any come to light a holistic approach to managing them (such as enlisting the support of a behaviourist) can be instigated.

Endocrine disease

The two endocrine disorders that are most commonly linked to behaviour change are hyperthyroidism in cats and hypothyroidism in dogs. Hyperthyroidism in cats may present behaviourally with signs such as hyperphagia, hyperactivity, aggressive behaviours and vocalization, so when a cat presents with any of these, unless a clear alternative explanation is present, hyperthyroidism should be considered. Hypothyroidism in dogs has been linked to a range of behavioural presentations including various anxieties, such as fear of noises, as well as aggressive behaviour. It is unlikely that hypothyroidism is playing a role in behavioural change in the absence of physical indicators of thyroid disease, but it is not impossible and so in cases of intractable behaviour change, especially in a breed known to be predisposed to hypothyroidism, a full assessment of thyroid function should be considered. However, it is also important to note that in cases where low thyroid function is diagnosed, supplementation does not automatically resolve behaviour problems (Dodman *et al.*, 2013) and may in fact cause worsening in some cases (Ogi *et al.*, 2016). For further details on the hypothyroid/behaviour association in dogs, see Fatjó *et al.*, 2002; Beaver and Haug, 2003; Carter *et al.*, 2009; Radosta *et al.*, 2012.

Other endocrine conditions that may be linked to behaviour change are: hypoparathyroidism and hypo- and hyperadrenocorticism, which have been associated with anxiety; diabetes mellitus, which through the onset of polyuria/polydipsia (PU/PD) may be associated with house soiling and through increasing appetite may lead to stealing of and defence of food items, or places where food is stored or prepared; and diabetes insipidus, leading to house soiling through the onset of PU/PD.

Skin disease

In all species an underlying skin condition can lead to overgrooming, even leading to self-trauma (acral lick granulomas, for example), and therefore this always needs to be ruled out. Psychogenic alopecia is a syndrome often referred to in cats who overgroom. However, an investigation by Waisglass *et al.* (2006) determined that the majority of cats presenting with overgrooming (76%) had an underlying skin condition, 14% had a combination of a skin condition and a psychogenic component and in only 10% of patients was the assessment that the overgrooming was purely behavioural. For this reason, a comprehensive medical work-up should be undertaken on all cats presenting with overgrooming, either before or in conjunction with behavioural referral.

It is also important to recognize that common medications prescribed for the treatment of chronic pruritic conditions can themselves lead to behaviour change (see below).

Musculoskeletal disease

Pain originating in the musculoskeletal system (including axial skeleton and the myofascia) is anecdotally the most common cause of pain-related behaviour change in dogs and cats. A number of papers have examined the relationship between pain and aggressive behaviour (for example: Camps *et al.*, 2012; Barcelos *et al.*, 2015; Affenzeller *et al.*, 2017) and the paper by Lopes Fagundes *et al.* (2018) specifically investigated the relationship between pain and noise fears, while the author's paper (Zulch *et al.*, 2012) reported a case of self-mutilation potentially linked to a painful tail lesion. In addition, anecdotally, numerous reports of other behavioural presentations are discussed within the field; for example, separation-related destructive behaviour related to osteoarthritis of the hock in a collie, intractable frustration-related behaviours related to hip dysplasia in a wire-haired dachshund, and a range of fears and anxieties in a number of individuals. For a selection of case studies see Mills *et al.* (2020).

Unfortunately, the diagnosis and assessment of chronic painful conditions can be challenging because, as was shown by Hielm-Björkman *et al.* (2003), radiographic change does not always correlate with the pain assessment of owners and veterinary surgeons. For this reason, even for pets where suspicion of locus of pain is in a bony structure, it is not possible to state on the basis of the radiographic evidence alone that an animal is or is not likely to be in pain. This leaves aside all of the rest of the system, which is challenging to image.

In addition, some disease processes will only become evident with more detailed investigation (see Affenzeller *et al.* (2017) for two cases of discospondylitis presenting as human-directed aggression), which may be a restriction in the case of some pets due to financial or other considerations.

It is therefore essential that for any pet presenting with a physical condition that might be painful, whether this is an identified lesion or a subtle physical change such as in gait, the most complete possible assessment is undertaken in order to try to elucidate whether or not pain is in fact present and may be contributing to a behaviour change. In addition, in any patient fulfilling the criteria for high index of suspicion for a medical condition contributing to a behaviour presentation, a full musculoskeletal assessment needs to be undertaken.

Finally, if suspicion remains that an occult locus of pain may be a contributory or underlying factor in behaviour change or a behaviour problem, an analgesia trial should be considered. If an analgesia trial is to be undertaken, two factors need to be considered: (i) what is the most appropriate drug or drug combination to choose; and (ii) for how long should the trial be undertaken? The answer to the first question will depend on the locus of pain that one suspects. The answer to the second question will depend on a range of factors, particularly the drug in question. References pertaining to NSAIDs suggest that a minimum trial of 2–4 weeks (MacFarlane *et al.*, 2014) is required, while Reid *et al.* (2018) showed that although the greatest change in quality-of-life scores occurred from day 0 (no analgesia) to day 14 of analgesia, quality-of-life scores in their study peaked at 60 days after provision of carprofen and referred to evidence that when measuring peak vertical force as an outcome measure a study also showed peak improvement at 60 days. There are many reasons to ensure that an analgesia trial is of sufficient duration for improvement to be seen and this largely relates to factors around the physiology of pain. However, it must also be remembered that animals in pain will have learnt to respond in a specific manner in an attempt to protect themselves from pain. Learning that a predicted action or stimulus no longer causes pain takes time, so behaviour change may take time to become evident.

Key points to remember when undertaking an analgesia trial are as follows.

- Choose the right drug/s.
- Use them for long enough.
- Be prepared to amend the drug/s used if required.
- Start with a baseline assessment of the behaviours that you believe are likely to be influenced by analgesia (for example, a behavioural diary completed for a week or more pre-trial, unless analgesia cannot be delayed for some reason).
- Continue to record behaviours of interest over time in order to track changes or lack thereof so that the success of the analgesia trial can be assessed more objectively.

Gastrointestinal disease

Gastrointestinal disease as a cause of or contributing factor to behaviour problems has been highlighted in publications by Bécuwe-Bonnet *et al.* (2012) and Frank *et al.* (2012) where behaviours such as licking surfaces and fly biting were shown to correlate with a range of gastrointestinal diseases. In the study on licking of surfaces, the gastrointestinal problems diagnosed included eosinophilic and/or lymphoplasmacytic infiltration of the GI tract, delayed gastric emptying, irritable bowel syndrome, chronic pancreatitis, gastric foreign body and giardiasis. After treatment for the diagnosed disease process, improvement of the licking was seen in 10 of 17 dogs and resolution in 9 of 17 dogs. In the fly-biting study, seven dogs presented with a medical condition, six of which were gastrointestinal and one of which was diagnosed as chiari-like malformation. Six dogs improved on medical treatment. It is hypothesized that the repetitive behaviours occur either as a direct result of the gastrointestinal pain or discomfort, or as an attempt to relieve the pain, for example through stimulation of saliva flow.

Additionally, Poirier-Guay *et al.* (2014) described star gazing in a dog with erosive gastritis with reflux oesophagitis, the hypothesis being that the oesophageal pain led to the postural changes, which could be described in behavioural terms as an abnormal repetitive behaviour.

From the evidence in these papers, the author would recommend that any patient presenting with oral behaviours directed at the environment, including pica, as well as any other repetitive behaviour that may potentially be indicative of or offering relief from abdominal discomfort, should be investigated for an underlying gastrointestinal disease.

Finally, Suñol *et al.* (2020) described a case study where a hydrolysed, gluten-free diet resolved aggressive behaviour in a dog and where relapse was seen with dietary lapses. Although a single case study needs to be treated as only a first step in understanding the relationship between diet and behaviour, in a field with relatively little evidence base it should be noted.

Urinary tract disease

A range of studies have shown that both upper and lower urinary tract disease may correlate with house soiling in cats. Ramos *et al.* (2019) showed a higher association of urinary tract disease in cats that were spraying or showing latrine-related periuria, as well as the behavioural control housemates of the cats that were living with the cats with latrine problems. Interstitial cystitis was found to correlate with both types of periuria. As the interrelationship between these conditions is challenging to disentangle and other studies have demonstrated the role of psychological stress in interstitial cystitis, it is really important that cats either presenting with evidence of urinary tract disease without behaviour change or presenting with house soiling even in the absence of overt urinary tract disease have both physical and behavioural assessments and are treated to ensure minimal environmental and social stress and optimal positive welfare, as well as to resolve any underlying physical condition.

In any dog where an owner reports onset of house soiling, or where a puppy is presented as apparently intractable to house training, a full assessment of urinary tract health should be done as well as an assessment of any condition that might cause polyuria/polydipsia. Anything leading to an increased need to urinate or an increase in urgency of urination may mean that a dog cannot hold urine until it has access to or can reach its usual toilet area. In addition, it is possible for owners to mistake urinary incontinence for house soiling and careful questioning regarding where and when urine is found, as well as volume, should be undertaken to assess whether further diagnostics or medication trials to rule out incontinence are required.

Other painful conditions

As mentioned throughout this section, it is likely that pain is playing a role in many of the behavioural manifestations that we see associated with medical conditions. However, on top of that it is important to recognize that there are other conditions that may cause pain and where this may be challenging to diagnose. A full treatise on pain in pets is beyond the remit of this chapter, but it is important always to consider the potential for any behaviour to be caused by or influenced by pain. In patients where the suspicion of pain remains despite not being able to reach a conclusive diagnosis of pain in the usual loci, or caused by more common disease processes, both neuropathic pain and visceral pain could be considered as follows.

- Neuropathic pain has been shown to correlate with a range of behavioural presentations (including stranger-directed and non-social fear as well as separation-related behaviour) in Cavalier King Charles Spaniels with chiari-like malformation and syringomyelia (Rutherford *et al.*, 2012). Neuropathic pain may be suspected more readily when a lesion that may cause it can be identified, but it may also occur when it is impossible to identify a lesion and therefore response to a targeted analgesia trial may be needed to rule it out.
- Visceral pain can again be difficult to diagnose unless a clear source such as pancreatitis or peritonitis is present, but potentially may occur in other patients too. The author has previously treated a young dog presenting with fear and anxiety in a range of situations, sometimes presenting as aggressive behaviour and which, although showing improvement, never demonstrated the resolution expected despite a range of interventions. A year or so after first presentation, polycystic kidney disease was diagnosed when physical health severely deteriorated and the dog was euthanized. It is an open question whether or not the dog was experiencing chronic pain (described in people with polycystic kidneys), but visceral pain is something that should not be forgotten.
- In addition to pain per se, the influence of inflammatory mediators on a range of behaviours and potentially on emotional and psychological states (in humans, both depression and anger have been studied) means that it is important for veterinary surgeons to adequately address inflammatory states in pets.

Medication and behaviour change

It is always important that a pet that presents with behaviour change and which is on medication is assessed for the likelihood that the behaviour change may be related to the medication. There are many medications that are known to influence behaviour. Some to be aware of, as described by Siracusa (2016), include phenylpropanolamine, diphenhydramine and oclacitinib. Even drugs used to help treat behaviour problems can have side effects of altered behaviour, albeit largely temporary, such as reduced appetite, increased sleepiness, the risk of hyperexcitability and disinhibition.

One of the groups of medication that has been investigated in more detail is the corticosteroids. Notari *et al.* (2015, 2016) investigated a range of behaviours correlating with the use of corticosteroids in dogs and reported that owners of dogs on corticosteroids reported an increase in some anxiety behaviours as well as some aggressive behaviours, although they did report some potential confounding factors in one of the studies. They also undertook a laboratory-based study with pet dogs that showed a significantly poorer ability of dogs on corticosteroids to cope with a potentially negative and startling stimulus (the recording of a dog growl played while they were eating). It is of course not suggested that the use of corticosteroids should be avoided, as even in patients with complex behavioural needs the benefits of the corticosteroid to treat a medical condition may outweigh the risk of increase in other behaviours of concern. However, given the frequency with which this class of drugs is used in veterinary medicine it is very important that clinicians are aware of the behavioural risk and discuss this as well as mitigation with owners prior to dispensing.

The bottom line is that many medications can potentially be implicated in behaviour change and this needs to be considered if the timing of onset of behaviour change coincides with the prescription of a new medication.

A Few Additional Thoughts

There are many aspects of the interrelationship between physical health and behaviour that are outside the scope of this chapter. However, a few additional points that are important to bear in mind are listed below.

The ageing pet and behaviour change

Many owners expect to see behaviour change in their ageing pet and therefore are unlikely to raise with their vet changes that they notice such as getting more tired on walks, reduction in appetite, being less playful or being wakeful at night. However, it is always possible that these signs are in fact indicative of a medical condition, such as degenerative joint disease, the pain of which may be manageable through the use of analgesics and adjunctives, or another condition such as endocrine dysfunction, which may also be manageable, if not curable. In addition, as mentioned above, the earlier that cognitive dysfunction syndrome can be ruled out, the better will be the prognosis for slowing decline and improving quality of life over the rest of the animal's lifespan. For these reasons initiating discussions with owners of older pets regarding behaviour change can help the practice to support these pets and their owners more effectively.

Additionally, through providing age-related support for older pets, owners can be helped to see how small adaptations such as the introduction of non-slip mats on slippery floors, thick bedding to lie on, lower-sided litter trays, etc., can improve their pet's senior years. Ageing does not always have to mean a reduction in quality of life for the pet and proactive interventions by the practice can help ensure good quality of life for longer.

The role of learning

Parts I and II of this book have extensively explained the role of learning in the behaviour of pets. The reason for mentioning it here is to remind the practitioner that even when a medical condition has been identified as driving a behaviour change, pets can very quickly learn new ways of coping with physical conditions and this learning can drive behaviour even once the medical condition has been resolved. For this reason, it is advisable always to offer behavioural first aid at the time of the first consultation to minimize the practising of the unwanted behaviour and thus the establishment of learnt patterns and habit. Additionally, in many patients, post resolution of the medical problem, support from a clinical animal behaviourist may be needed to bring about new learning and thus behaviour change.

The converse relationship

Another key area that this chapter has not explored is the increased risk of certain medical conditions when behaviour problems and negative emotional states (potentially correlating with psychological stressors being experienced) exist. Some of these are easy to see: the pet that shows aggressive behaviour towards conspecifics may receive injuries in a fight, and of course, the impact of long-term distress on immune function has long been acknowledged. However, as the field of behaviour grows, other less obvious relationships are coming to light, such as the development of idiopathic cystitis in cats, which is believed to have a component of underlying psychological stress (Defauw *et al.*, 2011). A recent study (Harvey *et al.*, 2019) showed that pruritis severity in canine atopic dermatitis correlated with prevalence of a range of other behavioural manifestations which the authors hypothesized could be indicative of underlying psychological stress. This is just one more reason that primary care veterinary surgeons can benefit the holistic health of their patients by considering the interrelationship between health and behaviour.

Further Research

Finally, there are a number of aspects of the relationship between physical health and behaviour that are not yet understood, especially as much of the research in this field is still correlational. We can only hypothesize in some cases why a specific condition leads to a specific behavioural manifestation. Further research will hopefully clarify some of these relationships.

One key area of research for the future is the impact of personality on expression of pain and possibly other medical signs. For example, Ijichi *et al.* (2014) demonstrated correlation between personality factors and pain expression in horses. Work in dogs by Reaney *et al.* (2017) also showed correlation between personality traits and reported recent medical conditions. Although it is impossible to assess the causal relationship between the two, these are interesting areas to investigate as in future they may be able to assist veterinary surgeons in better understanding why two patients with the same condition demonstrate different levels of behaviour change. It may well be that temperament or personality factors have complex relationships with physical aspects such as pain and thus some patients will always need more medical support, for example analgesia, than others.

Conclusion

As a substantial proportion of behaviour conditions have an underlying or contributing medical component, it is essential that the primary care veterinary surgeon, as well as any referral clinicians working with the pet, look to rule these out either before referral to a clinical animal behaviourist or while working through the behaviour problem together with the behaviourist. Making a medical diagnosis can be additionally challenging in these patients, due to behaviours that they may be showing, and therefore creative approaches to assessment as well as close working with the behaviourist are essential. In order to best serve the pet's physical as well as emotional and psychological health, good relationships and communication between the veterinary team, the pet owner and the behaviour practitioner are essential. Time spent in fostering this is likely to pay dividends to all, especially the pet.

References and Further Reading

Affenzeller, N., McPeake, K.J., McClement, J. and Zulch, H. (2017) Human-directed aggressive behaviour as the main presenting sign in dogs subsequently diagnosed with diskospondylitis. *Veterinary Record Case Reports* 5(4), e000501.

Barcelos, A.-M., Mills, D.S. and Zulch, H. (2015) Clinical indicators of occult musculoskeletal pain in aggressive dogs. *Veterinary Record* 176(18), 465.

Beaver, B.V. and Haug, L.I. (2003) Canine behaviors associated with hypothyroidism. *Journal of the American Animal Hospital Association* 39(5), 431–434.

Bécuwe-Bonnet, V., Bélanger, M.C., Frank, D., Parent, J. and Hélie, P. (2012) Gastrointestinal disorders in dogs with excessive licking of surfaces. *Journal of Veterinary Behavior: Clinical Applications and Research* 7(4), 194-204.

Camps, T., Amat, M., Mariotti, V.M., Le Brech, S. and Manteca, X. (2012) Pain-related aggression in dogs: 12 clinical cases. *Journal of Veterinary Behavior: Clinical Applications and Research* 7(2), 99–102.

Carter, G.R., Scott-Moncrieff, J.C., Luescher, A.U. and Moore, G. (2009) Serum total thyroxine and thyroid stimulating hormone concentrations in dogs with behavior problems. *Journal of Veterinary Behavior: Clinical Applications and Research* 4(6), 230–236.

Dantzer, R. and Kelley, K.W. (2007) Twenty years of research on cytokine-induced sickness behavior. *Brain, Behavior, and Immunity* 21(2), 153–160.

Defauw, P.A., Van de Maele, I., Duchateau, L., Polis, I.E., Saunders, J.H. and Daminet, S. (2011) Risk factors and clinical presentation of cats with feline idiopathic cystitis. *Journal of Feline Medicine and Surgery* 13(12), 967–975.

Dodman, N.H., Aronson, L., Cottam, N. and Dodds, J.W. (2013) The effect of thyroid replacement in dogs with suboptimal thyroid function on owner-directed aggression: a randomized, double-blind, placebo-controlled clinical trial. *Journal of Veterinary Behavior: Clinical Applications and Research* 8(4), 225–230.

Dreschel, N.A. (2010) The effects of fear and anxiety on health and lifespan of pet dogs. *Applied Animal Behaviour Science* 125(3), 157–162.

Fatjó, J., Stub, C. and Manteca, X. (2002) Four cases of aggression and hypothyroidism in dogs. *Veterinary Record* 151(18), 547–548.

Frank, D., Bélanger, M.C., Bécuwe-Bonnet, V. and Parent, J. (2012) Prospective medical evaluation of 7 dogs presented with fly biting. *Canadian Veterinary Journal* 53(12), 1279–1284.

Frank, D.F., Erb, H.N. and Houpt, K.A. (1999) Urine spraying in cats: presence of concurrent disease and effects of a pheromone treatment. *Applied Animal Behaviour Science* 61(3), 263–272.

Harvey, N.D., Craigon, P.J., Shaw, S.C., Blott, S.C. and England, G.C. (2019) Behavioural differences in dogs with atopic dermatitis suggest stress could be a significant problem associated with chronic pruritus. *Animals* 9(10), 813.

Hielm-Björkman, A.K., Kuusela, E., Liman, A., Markkola, A., Saarto, E. *et al.* (2003) Evaluation of methods for assessment of pain associated with chronic osteoarthritis in dogs. *Journal of the American Veterinary Medical Association* 222(11), 1552–1558.

Ijichi, C., Collins, L.M. and Elwood, R.W. (2014) Pain expression is linked to personality in horses. *Applied Animal Behaviour Science* 152, 38–43.

Kealy, R.D., Lawler, D.F., Ballam, J.M., Lust, G., Biery, D.N. *et al.* (2000) Evaluation of the effect of limited food consumption on radiographic evidence of osteoarthritis in dogs. *Journal of the American Veterinary Medical Association* 217(11), 1678–1680.

Lopes Fagundes, A.L., Hewison, L., McPeake, K.J., Zulch, H. and Mills, D.S. (2018) Noise sensitivities in dogs: an exploration of signs in dogs with and without musculoskeletal pain using qualitative content analysis. *Frontiers in Veterinary Science* 5, 17.

MacFarlane, P.D., Tute, A.S. and Alderson, B. (2014) Therapeutic options for the treatment of chronic pain in dogs. *Journal of Small Animal Practice* 55(3), 127–134.

Mills, D.S., Demontigny-Bédard, I., Gruen, M., Klinck, M.P., McPeake, K.J. *et al.* (2020) Pain and problem behavior in cats and dogs. *Animals* 10(2), 318.

Mills, D., Karagiannis, C. and Zulch, H. (2014) Stress – its effects on health and behavior: a guide for practitioners. *Veterinary Clinics of North America: Small Animal Practice* 44(3), 525–541.

Notari, L., Burman, O. and Mills, D. (2015) Behavioural changes in dogs treated with corticosteroids. *Physiology & Behavior* 151, 609–616.

Notari, L., Burman, O. and Mills, D.S. (2016) Is there a link between treatments with exogenous corticosteroids and dog behaviour problems? *Veterinary Record* 179(18), 462–462.

Ogi, A. *et al.* (2016) Possible behavioural effects of the normalisation of thyroid hormone levels in hypothyroid dogs: a preliminary investigation on 27 dogs. Poster Presentation: *ECAWBM/ESVCE Annual Congress*, Cascais, Portugal.

Olm, D.D. and Houpt, K.A. (1988) Feline house-soiling problems. *Applied Animal Behaviour Science* 20(3), 335–345.

Poirier-Guay, M.P., Bélanger, M.C. and Frank, D. (2014) Star gazing in a dog: atypical manifestation of upper gastrointestinal disease. *The Canadian Veterinary Journal* 55(11), 1079.
Radosta, L.A., Shofer, F.S. and Reisner, I.R. (2012) Comparison of thyroid analytes in dogs aggressive to familiar people and in non-aggressive dogs. *The Veterinary Journal* 192(3), 472–475.
Ramos, D., Reche-Junior, A., Mills, D.S., Fragoso, P.L., Daniel, A.G. *et al.* (2019) A closer look at the health of cats showing urinary house-soiling (periuria): a case-control study. *Journal of Feline Medicine and Surgery* 21(8), 772–779.
Reaney, S.J., Zulch, H., Mills, D., Gardner, S. and Collins, L. (2017) Emotional affect and the occurrence of owner reported health problems in the domestic dog. *Applied Animal Behaviour Science* 196, 76–83.
Reid, J., Wright, A., Gober, M., Nolan, A.M., Noble, C. and Scott, E.M. (2018) Measuring chronic pain in osteoarthritic dogs treated long-term with carprofen, through its impact on health-related quality of life (HRQL). *Veterinary and Comparative Orthopaedics and Traumatology* 31(S 01), A0015.
Rutherford, L., Wessmann, A., Rusbridge, C., McGonnell, I.M., Abeyesinghe, S. *et al.* (2012) Questionnaire-based behaviour analysis of Cavalier King Charles spaniels with neuropathic pain due to Chiari-like malformation and syringomyelia. *The Veterinary Journal* 194(3), 294–298.
Salvin, H.E., McGreevy, P.D., Sachdev, P.S. and Valenzuela, M.J. (2010) Under diagnosis of canine cognitive dysfunction: a cross-sectional survey of older companion dogs. *The Veterinary Journal* 184(3), 277–281.
Salvin, H.E., McGreevy, P.D., Sachdev, P.S. and Valenzuela, M.J. (2011) The canine cognitive dysfunction rating scale (CCDR): a data-driven and ecologically relevant assessment tool. *The Veterinary Journal* 188(3), 331–336.
Shihab, N., Bowen, J. and Volk, H.A. (2011) Behavioral changes in dogs associated with the development of idiopathic epilepsy. *Epilepsy & Behavior* 21(2), 160–167.
Siracusa, C. (2016) Treatments affecting dog behaviour: something to be aware of. *Veterinary Record* 179(18), 460–461.
Sordo, L. and Gunn-Moore, D.A. (2021) Cognitive dysfunction in cats: update on neuropathological and behavioural changes plus clinical management. *Veterinary Record* 188(1), 30–41.
Suñol, A., Perez-Accino, J., Kelley, M., Rossi, G. and Schmitz, S.S. (2020) Successful dietary treatment of aggression and behavioral changes in a dog. *Journal of Veterinary Behavior* 37, 56–60.
Uccheddu, S., Gallucci, A., Briguglio, P., Cossu, G., Pais, M. and Pirrone, F. (2018) Tail chasing in a dog with brain atrophy: a case report. *Journal of Veterinary Behavior* 25, 52–55.
Waisglass, S.E., Landsberg, G.M., Yager, J.A. and Hall, J.A. (2006) Underlying medical conditions in cats with presumptive psychogenic alopecia. *Journal of the American Veterinary Medical Association* 228(11), 1705–1709.
Watson, F., Rusbridge, C., Packer, R.M., Casey, R.A., Heath, S. and Volk, H.A. (2018) A review of treatment options for behavioural manifestations of clinical anxiety as a comorbidity in dogs with idiopathic epilepsy. *The Veterinary Journal* 238, 1–9.
Zulch, H.E., Mills, D.S., Lambert, R. and Kirberger, R.M. (2012) The use of tramadol in a Labrador Retriever presenting with self mutilation of the tail. *Journal of Veterinary Behavior: Clinical Applications and Research* 7(4) 252–258.

16 Making the Decision Whether to Treat, Rehome or Euthanize

Julie Bedford

Living with a pet with problematic behaviour can be stressful for both pet and owner. Before help is requested, the owner will often have struggled with the problem for many months. The early warning signs may have been unnoticed, misinterpreted or considered something the pet will 'grow out of' and it is only following a recent escalation of the behaviour (such as a bite), or a change in circumstances which prevents the problem being avoided or managed, that the owner finally seeks advice. Sadly, by the time help is sought, the relationship between the owner and pet may have broken down and the owner may be desperately looking for a 'quick fix', a guarantee of a successful outcome or simply want to get rid of the pet as well as the problem.

Faced with a desperate client, a rapid assessment may have to be made to decide on the best course of action for all involved. Many problem behaviours are driven by negative emotional states such as fear, anxiety or frustration and this has to be taken into account when considering how the need to exhibit the behaviour reflects the quality of the pet's life and in turn how the pet's behaviour impacts on the owner's quality of life. This consideration may not be straightforward. For example, the risk of injury from an aggressive pet may not impact as badly on the owner's quality of life as a pet who causes damage to the home. In fact, many owners can successfully prevent aggression (by avoidance or muzzling) but find problems that occur when they are absent, such as destructive behaviour, excessive noise or house soiling, are far more stressful (Fig. 16.1). The type, frequency and intensity of the problem, the owner's view of it and how safely or effectively the animal can be managed will all play a part in the decision whether to treat, rehome or euthanize.

If the welfare of the pet, people and other animals is not considered to be at immediate risk, and the owner is willing to work to try to resolve the problem, then referral to a veterinary surgeon specializing in behavioural medicine or to an appropriately qualified clinical animal behaviourist should be offered. Reassuring the client that most problems can be reduced and managed, and that even cases that are seemingly hopeless can have successful outcomes, can encourage the owner to accept a referral and work with their pet. No two cases are the same and there are many factors that influence the outcome, most noticeably the distinction between behavioural responses that are the result of an emotional health issue in the pet and those that result from misunderstanding between humans and the species they live with. The client may also be reassured that problem behaviour, although expressed in situations or at an intensity that they may find unacceptable, may be normal and may be successfully and safely channelled and managed.

If the owner is unwilling, or unable, to commit the time and meet the costs of professional

DOI: 10.1079/9781800621312.0016

Fig. 16.1. Problems can occur when owners are absent, such as destructive behaviour, excessive noise or house soiling. (Image: author's own.)

intervention, then rehoming may be considered. In some of these cases, it should be pointed out that, unfortunately, the problem that has caused them to want to relinquish their pet may also be the reason that a new home cannot be found, in which case euthanasia may be the only option. Ideally, by explaining the options available, and by having access to veterinary colleagues working in behavioural medicine or to competent, appropriately qualified clinical animal behaviourists, the veterinary surgeon will be able to reassure the client that things can be improved and encourage them to accept a referral for specialist behavioural help.

Referral of Behaviour Cases

General practitioners are familiar with the procedure of referral to veterinary colleagues and the process is no different for behavioural medicine cases than for any other veterinary discipline. If the client is referred to a non-veterinary behaviourist, a professional referral procedure is also needed. The behaviourist will need to be informed of any health problems that have been identified and any treatments the pet is receiving. Prior to referral the animal must be clinically examined by a veterinary surgeon to check for any underlying medical conditions or medications that may be creating or contributing to the problem behaviour, as this can influence the behavioural diagnosis and behaviour modification plan. In return the referring veterinary surgeon can expect to receive a post-consultation summary of the behavioural history, report and treatment plan. Good communication between the referring veterinary surgeon and the behaviourist is important and this is especially the case when referring to a non-veterinary behaviourist, as the duty of care of the animal remains with the veterinary surgeon. (Information about arranging referral and deciding who to refer to and when can be found in Chapter 21.)

Is It Possible to Manage and Resolve the Problem Behaviour?

Although most situations can be improved, it is important to be realistic and give the owner achievable expectations. When behaviours have been long standing, this can have a negative impact on the final outcome. The animal will have repeatedly performed the unwanted behaviour and learning will be an important factor in the persistence of the problem. In addition, the owner will have lived with the behaviour for some time and may have failed in attempts to manage the pet, which is likely to lead to frustration and leave them feeling powerless. The behaviourist's first job may be to rebuild the owner–pet relationship. This can usually be achieved by giving the owner a thorough understanding of the emotional state of their pet and how that influences behaviour. This will be followed by some simple tasks and goals to start the process of modifying the pet's behaviour and building the owner's confidence in their ability to positively influence and change it.

Influencing factors, including the cause, frequency, target and severity of the behaviour, will be considered in all cases. This is particularly important in cases involving aggressive behaviour. If there are clearly identifiable triggers and the owner can learn to recognize warning signs prior to the aggression (bite), then it will be more straightforward to intervene. The individual situation, confidence and compliance of the owner will play a key part in safely managing an aggressive pet. For some, the simplest solution will be to avoid the triggers. For example, if the pet is aggressive to visitors, they may simply choose to put the pet in a separate room with something enjoyable to occupy it. Such an approach may not always be realistic and in very busy households, particularly where children are frequent or unannounced visitors, the pet could become increasingly isolated, leading to unacceptable compromise to the animal's welfare. Excessive isolation may also increase the pet's distress and lead to increasingly unmanageable behaviour. For these pets rehoming may be the best option.

Rehoming may also be the better option for the cases where the owner cannot change the environment or routine in ways that would be necessary to resolve the problem or manage it in the short term while a treatment plan is implemented. Examples where this may be the case might include a cat living in a multi-cat household that is spraying or a dog that is constantly barking when the owner is out at work. Constantly cleaning soiled areas or dealing with neighbour complaints about noise can be very stressful for the owner and these behaviours are a clear indicator that the pet is under considerable stress. It is not realistic for the owner to rehome all the other cats in the household or give up work, and if short-term solutions cannot be found while a long-term behaviour modification plan is implemented, it may be concluded that, in the long term, these pets may be happier if rehomed to a more suitable situation.

Another situation in which rehoming may be a more realistic option is in cases where the behaviour places considerable restrictions on the pet and owner. For instance, a dog with a very high predatory drive that lives adjacent to farmland may have to be prevented from having free-running exercise, due to the presence of sheep. While this may be solvable for some owners, who have access to sufficient places where free running is possible, for others the restriction of exercise may lead to unacceptable compromise of the dog's welfare.

Benefits of Referral

The major benefit of referral is that it provides the opportunity for a consultation in which there is sufficient time to explore and explain the emotional aspect of the pet's behaviour. By learning what motivates their pet, the owner can focus less on trying to stop the unwanted behaviour (which in the past may often have resulted in the pet being punished) and more on increasing the pet's positive emotional state so that it no longer needs to display the behaviour. This approach gives the owner a better understanding of how the problem developed and how to manage it effectively. Understanding also facilitates greater tolerance and empathy and motivates the owner to work actively with the behaviourist to devise an appropriate management routine and make the necessary changes. In turn this can help to strengthen the owner's attachment to the pet, which has often been adversely affected by the behaviour. The benefit of an improved relationship is that, even if the problem is not entirely resolved, it can be safely managed in a manner that does not compromise the pet's welfare or the safety of others and that is accepted by the owner. This not only ensures that the veterinary practice retains a client but also the client has none of the guilt associated with relinquishing their pet for rehoming or euthanasia.

Many behaviourists are happy for a veterinary surgeon or nurse from the referring practice to accompany them on their home visits, provided that the client gives their permission. This can give an excellent opportunity for veterinary staff to gain an insight into the work of the behaviourists and increase their behavioural knowledge.

In some cases if, despite everyone's best efforts, the problem cannot be kept at an acceptable level and the welfare of the pet or client, in both short and long term, is compromised, then at least the client has the knowledge that they have tried to resolve the problem and they can be supported through the process of considering the alternative options of rehoming or euthanasia.

Rehoming

Rehoming may seem to be the best alternative to retaining the pet in its current home, but it has to be acknowledged that this is not always the case. There is a significant shortage of suitable homes and sadly many animals have to be euthanized every day in rescue centres. Rehoming should only be considered if it is likely that a change of ownership will result in an improvement in the long-term welfare of the animal, that the problem can be successfully and safely managed and that there is a realistic expectation that a suitable owner can be found. If these criteria cannot be met, there is a risk that animals can find themselves in long-term kennel accommodation, which is detrimental to their welfare and does not represent a suitable solution.

Rehoming organizations are often overwhelmed with the number of animals requiring assistance and have more animals than they can accommodate. Accepting animals with behaviour problems will place additional strain on already stretched resources (Fig. 16.2). To ensure that as many animals as possible are helped, the organization will often restrict how many 'problem' animals it has on site. This is because the 'problem' pets take longer to rehome and, once rehomed, they are more likely to be returned. The result can be that facilities quickly become full with 'difficult to rehome' pets, leaving no space for other animals in need of help and very little choice for families wishing to adopt a rescue pet.

It also has to be acknowledged that animals with behaviour problems are often the ones who cope least well with a change in their environment. This may limit the success of the rehoming process and even result in a deterioration in the individual animal's welfare. Regardless of how well done the rehoming is, the animal has to cope with the loss of everything familiar, such as family, other animal companions, home and routine. Their problem behaviour is often caused by a lack of appropriate coping strategies, which may be the result of various factors, including poor breeding, lack of or inappropriate socialization and training or poor management. These animals may have become overly anxious or react in ways unacceptable in a pet or may simply have never learned suitable behaviours. If placed in a rehoming centre the pet will have to adjust to a new diet, interaction with strangers, a busy environment that often prevents sufficient rest, change in quality and quantity of exercise opportunities or access to outdoors and a lack of privacy, as they are surrounded by many other animals and people (both staff and public looking for a new pet). Additional discomforts such as cold kennels (when compared with a

Fig. 16.2. Accepting animals with behaviour problems will place additional strain on already stretched resources in rehoming organizations. (Image: author's own.)

centrally heated home) and excessive noise levels will all have the potential to act as stressors. Kennels often exceed noise safety guidelines for people and therefore the exposure to these noise levels is likely to be unpleasant for any dogs or cats housed on the site. The overall result is that these animals are being taken into a stressful environment, which may increase or create feelings of anxiety and insecurity (Fig. 16.3). The change of environment and routine may also increase the animal's fear or frustration, thereby creating or intensifying behaviour problems. Even when the pet is rehomed into a suitable environment the potentially negative effects of change cannot be ignored, as they will need to adjust to a new set of routines and build up a bond with a new family.

This said, there are many cases where behavioural issues will be readily resolved by moving the animals into a different environment. For example, a nervous cat who is constantly followed around the house by children who are keen to pet and cuddle it may present a very significant risk if it responds by scratching and biting. If this cat is rehomed to an all-adult household that offers a quieter environment with privacy and less social stress, it is very likely that the cat will no longer need to behave in this defensive manner and will be able to live successfully in a domestic environment. Similarly, a dog who is constantly chasing, barking and nipping when it is living in a high-arousal household with small children may be rehomed to a situation where these misplaced hunting skills can be channelled into suitable activities and games.

Fig. 16.3. Animals being taken into a stressful and noisy environment may lead to an increase in or create feelings of anxiety and insecurity. (Image: author's own.)

The fact that rehoming has the potential to be stressful increases the need to investigate behavioural issues thoroughly before going down that route. Where possible the current owner, who will hopefully have more emotional investment in the pet, will be better placed to resolve or manage the issues, provided that they receive appropriate professional help. There are exceptions and in cases where there has been an inappropriate selection of pet, or where there have been unforeseen and unavoidable changes in the owner's circumstances, it is possible that the welfare of the animal will be better served in a new environment. Provided that the professional assessment suggests that an individual animal will be able to cope successfully with the changes experienced during rehoming, then short-term distress of being moved from one home to another, probably via a busy rehoming centre or foster home, may be justified by the long-term improvement in their welfare and a reduction in the behavioural problem to a safe and acceptable level.

Owners do need to give their pet the best possible chance of being found a successful new home by disclosing full details of the animal's behaviour, but sadly this is not always the case. Accurate information about the animal's behaviour, upbringing, experience, training and lifestyle will help the staff to formulate a suitable plan to rehabilitate and rehome. Covering up or withholding information is often done in the mistaken belief that the animal will more easily be rehomed. The reality is that lack of information places the animal at more risk of an unsuitable match to a new family and a resulting increase in the likelihood of being passed on again. Even if the pet does remain in the new home, a mismatch may result in the new owner not being able to take appropriate control or avoidance measures and the pet being placed in a situation where the problem behaviour is displayed again. This time rehoming may not be an option and the pet may be euthanized.

If a rehoming organization accepts the pet, the owner should expect to donate towards the cost of the pet's care. Caring for and rehoming animals is expensive and costs such as food, veterinary treatment, staff, heating, building maintenance, administration and advertising to potential new owners all have to be covered. Without donations from the public the rehoming organizations would be unable to offer the service. Additionally, reputable charities will neuter, vaccinate, microchip and worm animals prior to rehoming, which significantly increases the costs involved. If the present owner is willing to ensure that vaccination, worming and microchipping have all been done before their pet is handed over to the charity, it will reduce costs and can also shorten the length of time the animal is in the centre.

Rehoming Options

If rehoming is considered the best option, there are various ways in which this can be achieved.

Internet sites

This is the most high-risk rehoming option but for many people it is the most accessible. Sadly, there are many reports of this form of rehoming creating more problems than it solves. For desperate owners who are trying to offer their pet a new start in a better home, this option may seem like the easiest but in a number of cases these animals have been collected by people who have false but convincing stories and credentials. The pet may then be used for breeding or illegal activities. If the owner is concerned about the future welfare of their pet and is rehoming in the hope that their pet will have a better future, internet rehoming is not a suitable option.

Breeders and breed rescue

When owners acquire their pet from a pedigree breeder they will often sign a contract in which it states that the breeder will take the animal back if the owner finds, for whatever reason, that they can no longer keep it. Ethical breeders will take this approach but if a breeder refuses to take the pet back the next port of call may be the breed rescue society.

Breed rescues are usually run by volunteers who take the animal into their own home or put potential new owners in touch with the current owner. There are many benefits of using this route. These organizations are familiar with particular traits that are strong in the breed and often have a network of contacts to assist in the rehoming process. They are also likely to have a list of people looking for that specific breed who will be well informed, experienced and able to accept and provide for any breed traits that might present a problem to less experienced or less well-informed owners. For dogs the Kennel Club produces a list of breed rescues and for cats the Governing Council of the Cat Fancy publishes a cat welfare rescue directory.

Private rehoming

In some cases, a trusted friend or family member is able and willing to take the pet on or the client may, via their network of friends, source and place their pet directly into a new home. By being personally involved, the current owner can check the new home and maintain contact with the pet. Potential disadvantages of this approach are that finding a home may take some time and if the rehoming breaks down the original owner may have to take the animal back. In addition, the original owner will have to accept that, once the pet is sold or given away, they have no control over what the new owner does with the pet. If the owner does not want to accept these risks, then private homing is not a sensible option.

Rehoming charities

Most rehoming charities will take the animal into kennels or catteries or occasionally rehome via a foster home. A few will be able to rehome the animal direct from one home to the other and this can be advantageous, because the animal stays in the current, familiar home whilst a new home is found. This reduces the number of changes that the pet is exposed to and also removes the need for it to live in a kennel or cattery environment. Even if the pet is brought into the charity's own facilities, most organizations have long waiting lists and owners must be willing to keep the pet until space is available in the kennels or cattery or a home can be found.

Procedures and policies vary between rehoming charities and owners should research carefully what standard of care the organization can provide. Although a decision has been made to rehome, the welfare of the animal is still the owner's responsibility. The accommodation where the animal is kept should be safe, clean, comfortable and warm with suitable environmental enrichment and space to exercise and retreat for privacy. The staff should be skilled, take an interest in each individual animal and be able to handle them kindly and effectively. Equally important is how the staff behave towards owners when they enquire about giving up their pet. Friendly, professional, non-judgemental staff who have the ability to put people at ease, gain their trust and gather all of the important information are more likely to interact well with potential new owners and be successful at finding suitable new homes for the animals in their care.

It can be helpful to visit the website of the rehoming organization to get some indication of how proactive the organization is in seeking suitable new homes for their animals. Trying to find new homes for pets is challenging and does not happen without active input. Accurate but positive descriptions and pictures of animals in their care will help to attract potential new owners. The rehoming charity will have a variety of pets available and, if the owner of a 'problem pet' chooses to rehome their pet via this route, potential new owners will have the choice of many other, non-problematic animals. Without any emotional attachment and investment in the animal, potential new owners may, sensibly, choose a more straightforward companion. It can be beneficial for the charity website to outline any challenges the pet presents but also include information about any work that is currently being done with the pet and an indication of the support and advice that will be provided to new owners.

While a good standard of accommodation, staff care and rehoming is desirable, the owner should also enquire about the success of rehoming and what steps the charity takes to ensure the best possible long-term welfare of the pet. For example, do the staff take care to match the pet to the new owner and their situation and do they visit the new home to ensure that the environment is suitable for the animal? Many organizations carry out home visits, as these offer an opportunity to build relationships with the new owners and to offer personalized advice. It is important for new owners to be comfortable speaking to the centre staff so that they will readily ask for their help if the animal presents a problem. Failing to seek advice early in the new relationship increases the risks of the rehoming failing.

Many charities train staff to recognize and deal with behaviour, welfare and training issues and will be happy to explain their qualifications and what interventions, assessments, training, management and monitoring is undertaken whilst the pet is in their care. The organization's website is likely to include information about the approach to pets with behaviour problems and the quality of generic advice given there can be a useful reflection of how up to date their approach is and indicate the general quality of advice that will be given to new owners and the level of support that will be offered. The first month after taking on a rescue pet is a critical time. As the relationship between owner and pet is being formed, any new owner will need support but this is even more important if they are dealing with challenging behaviour. Ideally the charity will be able to help new owners to get to know, train and settle the animal into their home. This can be done via websites that contain helpful information on caring for pets, but it is preferable for new owners to be offered more personal support by telephone, email or face-to-face appointments.

If, despite the charity's best efforts, the animal cannot be rehomed or is returned to them, there are a number of possible outcomes. In some cases, charities will keep the animal indefinitely in a kennel or cattery environment, others will return the pet to the original owner and others will organize for it to be humanely euthanized. If the owner has strong feelings about the long-term outcome of the rehoming process and in particular about the acceptability, or not, of euthanasia, they will need to be clear about the charity's policies and choose one that best fits their views.

Euthanasia

In cases where the client is unable to work with or rehome their pet, especially when the animal is a risk to people or animals, or significant emotional health disturbance is seriously compromising the animal's welfare, then euthanasia may be the pragmatic option (Fig. 16.4).

Fig. 16.4. In some cases, when the animal is a risk to people or other animals, or significant emotional health disturbance is seriously compromising the animal's welfare, then euthanasia may be the pragmatic option. (Image: author's own.)

However, making the decision to euthanize a pet is very difficult and it is important to ensure that the owner has thoroughly considered the implications of that decision. They will often be feeling overwhelmed and uncertain and be seeking professional advice at this time. If it is possible for the veterinary practice, or a local kennel or cattery, to board the pet for a week or two, this can give the owner time to experience life without the pet. For some clients this will make the decision to euthanize much easier, as they realize how much their quality of life improves. For others the time without their pet makes them realize that, despite the difficulties, they would rather live with the pet, and seek help to resolve or manage the problem, than live without them. Following such a decision to keep their pet, owners may be more motivated to improve the situation. For instance, the owner of a cat in a multi-cat household may decide that it is possible to redistribute resources within the home so that all of the cats feel more secure and the indoor marking issue can be resolved. Similarly, the owner of the dog who cannot cope with being isolated all day may find a solution in the form of day care for their pet, either professionally or from a friend or neighbour who is retired or works from home.

When considering the option of euthanasia, it is important to consider the welfare of the pet but also the welfare of the humans in the family. Euthanasia of very old or physically sick animals can be of some comfort to owners, as they are able to spare the animal physical pain, but euthanizing a young or physically fit animal for behaviour problems, however severe, may present ethical concerns and lead to additional feelings of guilt and intensified grief. When the behavioural signs are linked to an underlying emotional health issue it can be easier for the client to come to terms with their decision, but there may be cases where the behavioural issue relates to normal behaviour that is not acceptable to the owners and rehoming has not been an option. In these situations, the client may need additional support to come to terms with the situation. Some practices are able to offer bereavement support in-house but where this is not possible there are a number of sources of help, including voluntary organizations such as the Pet Bereavement Support Service. They train volunteers to support individuals prior to and after the death of a pet. When attempting to support clients at this difficult time it can be helpful to emphasize the fact that their pet is no longer struggling and has been spared the possible distress of being moved from home to home. By taking responsibility for their pet's euthanasia, owners have certainty about the outcome and may be able to take some comfort from knowing that their pet was with familiar people right to the end.

Summary

In recent years there has been an increase in awareness of pet behavioural problems within the general public and an increase in the scientific study and understanding of animals' emotional health and learning. Veterinary behavioural medicine and behaviour therapy are far more widely available and there have been great improvements in the understanding and management of problem behaviour. Behavioural intervention is often effective and can improve the welfare of the animal and the relationship the owner has with their pet. Sadly, it is not all positive and increased media attention to the issue of dog behaviour in particular has also perpetuated some myths that are potentially dangerous and detrimental to animal welfare. The fact is that owners need to be given appropriate guidance when seeking help with their pet's behaviour and veterinary practices are in the ideal position to offer advice or refer their clients to someone who is appropriately qualified to do so. Even with appropriate professional advice the prognosis for behaviour problems may be uncertain, particularly as changes to a pet's behaviour can only be made by changes in the owner's behaviour and attitudes. These changes can be difficult to implement and often have to be maintained for the lifetime of the animal. If the owner is unwilling, or unable, to commit time and meet the costs of professional intervention, it may mean that rehoming has to be considered.

Unfortunately, rehoming is not always a suitable option and passing on a 'problem' pet may result in compromised welfare and increased distress for the animal. Rehoming charities are overwhelmed by the demand for their services; they often have very long waiting lists and may only take in animals that are more straightforward to rehome. Taking in animals with behavioural problems, who tend to take longer to rehome and are more likely to be returned, will 'block' kennel or cattery space, preventing other, less challenging animals from having the opportunity to be taken in and rehomed. If no one comes forward to offer a new home, many organizations will have a policy to euthanize, rather than keep the animal in a kennel or cattery environment for a long time or even for the rest of the animal's life.

In cases where the animal presents a clear and immediate risk to people or other animals and the owner is unable to prevent or meet any future liability issues relating to damage to people, other animals or property, euthanasia may be the most appropriate option. This is also the case if the pet is suffering from an emotional health issue which severely compromises its welfare or there is very little possibility of the pet finding a suitable home within a reasonable time period. Rather than pass the animal on to an uncertain future and risk further compromising its emotional health and welfare, the decision to euthanize may need to be taken.

Making the decision whether to treat, rehome or euthanize a pet with a behavioural problem is not easy. Veterinary practices may have the expertise in-house to help clients through the decision-making process, but in many cases it is necessary to involve external professional support. Euthanasia is a final and drastic option and one that many owners find too distressing to consider. With expert help the owner can be reassured that the welfare of the animal, both short- and long-term, will be central to any professional advice that is given to them. Time will be needed to investigate the case fully but with reassurance and support from a veterinary behaviourist or appropriately qualified clinical animal behaviourist it may be possible to offer a way forward. Restoring the pet's emotional health will be key to reducing the frequency and intensity of the unwanted behaviour and good communication skills will be necessary to help the owner achieve a deeper understanding and a stronger relationship with their pet. The insight gained will also help to ensure that owners will have the confidence and ability to choose and manage future pets more effectively and continue to enjoy all the benefits that pet ownership can bring.

Further Resources

Animal Behaviour and Training Council
www.abtcouncil.org.uk

Animal behaviourists
Association for the Study of Animal Behaviour
http://www.asab.org/ccab/
RCVS Advanced Partitioner Register
https://findavet.rcvs.org.uk/find-a-vet-surgeon/by-advanced-practitioner/
RCVS Specialist Register
https://findavet.rcvs.org.uk/find-a-vet-surgeon/by-specialist/

ECAWBM(BM) Specialist Register
http://www.ecawbm.com/diplomates-2/

Pet Bereavement Support Service (provided by Blue Cross)
http://www.bluecross.org.uk/2083/pet-bereavement-support-service.html

Rehoming animals in the United Kingdom
All pets (including pure-bred animals):
The Association of Dogs and Cats Homes
http://www.adch.org.uk/adch-members-list.htm
Cats (pure-bred):
The Governing Council of the Cat Fancy
http://www.gccfcats.org/welfare.html
Dogs (pure-bred):
The Kennel Club
http://www.thekennelclub.org.uk/breedrescue

17 First Aid Advice for Common Behavioural Signs: Dogs

Caroline Warnes

Introduction

Dog owners experiencing problems with their pets' behaviour will often report these initially to their veterinary surgeon or other members of practice staff. Whether the behaviour problem is assessed and treated 'in-house' or referred to a suitably qualified behaviourist outside the practice will depend on the nature of the problem itself and the availability of members of practice staff suitably qualified and experienced to deal with it. In all cases the dog will also need a veterinary examination to identify or rule out possible medical causes for the problem behaviour before the behaviour consultation. There may be a delay before the dog can be properly assessed by a behaviourist, and first aid advice will be needed to help the owner manage the behaviour problem in the meantime.

First aid advice for dog behaviour problems must serve two main functions, as follows.

1. Ensuring the safety of people, the dog and other animals and reducing the risk of legal repercussions

Some problem behaviours in dogs are potentially dangerous to people, other dogs or other animals, especially those involving aggressive or predatory/chase behaviour. The most important function of first aid behavioural advice is to ensure the safety of all people and animals that the dog is likely to come into contact with until they are seen and assessed by a suitably qualified behaviourist. The safety of the dog itself must also be considered. For example, dogs that are likely to panic when they hear loud noises must be prevented from injuring themselves as a result of escaping from the house or bolting on walks.

It is also important to be aware that if someone is injured as a direct consequence of advice given by a member of practice staff, they may bear some legal liability for the injury. This is also the case in relation to advice given by a non-veterinary behaviourist if the practice has made a referral to them. The referring veterinary surgeon will retain duty of care for the case. This highlights the importance of being fully aware of the qualifications and experience of anyone that the practice is considering referring behavioural cases to. If they refer to a veterinary behaviourist, the duty of care will be transferred to the referral-level veterinary surgeon.

All members of practice staff who are giving behavioural advice to owners must be aware of the potential legal implications associated with certain behaviour problems in dogs. For example, dogs that show aggressive behaviour to people or assistance dogs, and those that demonstrate any other behaviour that could be regarded as being 'dangerously out of control', which can include non-aggressive behaviours

DOI: 10.1079/9781800621312.0017

such as accidentally running into someone and knocking them over or jumping up and scratching them, can have implications under the Dangerous Dogs Act 1991. Owners of dogs presenting with behaviour problems must also be made aware of their responsibilities with regard to keeping their dogs under control both in public places and in their own homes. Any advice given to the owners must minimize rather than increase the risk of legal repercussions, which can include a criminal conviction if an owner is found guilty under the Dangerous Dogs Act 1991 (available at http://www.legislation.gov.uk/ukpga/1991/65/contents, accessed 1 May 2022).

2. Preventing the behaviour getting worse as a result of learning

Advice that protects a dog in the short term from the situations in which it performs an unwanted behaviour can be an important factor in increasing the success of behavioural therapy in the longer term. For example, if a dog is reactive to visitors, it is beneficial to ensure that it is shut securely in a separate room when people arrive. Owners must make a carefully considered decision whether or not to let the dog out while the visitor is present. It may be safest to leave the dog shut away until the visitor leaves, ideally with items to keep it occupied, such as activity feeders or things to chew.

General Principles of First Aid Advice for Dog Behaviour Problems

Avoiding triggering stimuli if at all possible

Many behaviour problems are either stimulus- or context-specific, and if these can be identified and avoided the problem behaviour may not occur. Owners may need advice regarding how to do this. Sometimes quite minimal changes can have a profound effect with relatively little effort on the part of the owner. Other changes may be more significant and require a substantial alteration in owner routine or in the way that they interact with their dog. In these cases, careful attention to the manner in which advice is delivered to the client will be essential in gaining compliance. Examples of the ways in which owners can protect dogs from triggering stimuli and situations include the following.

1. If a dog is showing aggressive behaviour only to people who approach it when it is eating, the response of allowing the dog to eat undisturbed by leaving the room and closing the door until it has finished, or even feeding it outside in the garden on its own, can significantly reduce the risk of the dog showing aggression.

2. If a dog is fearful of or highly reactive to things encountered outside the house, such as unfamiliar people, other dogs, traffic or loud noises, owners should be advised to walk in quiet locations and at quiet times, to minimize exposure to these triggers. If it is not possible to minimize exposure in this way it might be best to advise the owners not to take the dog out for walks at all, and to provide exercise and enrichment in the form of games, activity feeding and training at home, until the dog can be seen and assessed by a suitably qualified behaviourist. If it is not possible to reliably identify and/or avoid all triggering stimuli, additional safety and management strategies will also be needed.

Safety/management strategies

If triggering stimuli cannot be reliably avoided, it is important to put in place safety and management strategies to minimize the dog's opportunity to display the problem behaviour, to increase the owner's ability to control the dog and, in the case of dogs showing aggressive behaviour, to prevent them being able to bite or cause injury. Obviously, the particular strategies required will vary according to both the problem being shown and the particular situations in which it is occurring, as discussed in more detail below.

Wearing a muzzle

Dogs that display aggressive behaviours toward people or other dogs, and particularly those with a history of biting, should be muzzled if they are likely to be in close contact with the targets of their aggression. A muzzle can also deter unfamiliar people from approaching or attempting to interact with a dog who might be scared of them, and owners are often less tense in potential problem situations when they know their dog cannot bite, which in turn can help the dog to feel more relaxed.

If a dog needs to wear a muzzle for anything longer than a few minutes, the Baskerville™ plastic basket muzzle is preferable to a nylon mesh or 'tube' muzzle, because the basket design allows the dog to open its mouth to pant and drink and thereby reduces the risk of it becoming overheated or dehydrated. It also encases, rather than restricts, the face and the dog is therefore able to give facial expressions and vocalizations that provide information about its emotional state, while being prevented from causing a bite injury. This is important to avoid escalating frustration, caused by the dog's inability to communicate, and worsening of the behaviour (Fig. 17.1).

Dogs must be carefully accustomed to wearing a muzzle and members of practice staff should be practised in both fitting muzzles accurately and muzzle-training dogs, as well as in coaching owners how to do this if practice staff are unable to do so directly, due to the behaviour that the dog is displaying. See Appendix 17.1 for a guide to choosing and fitting a muzzle, a step-by-step muzzle-training guide for owners, and links to videos that demonstrate this process particularly well.

No muzzle is infallible and muzzled dogs can still cause injuries through muzzle-punching, causing bruising, knocking people over or scratching with their claws. Therefore, care must always be taken when handling muzzled dogs, to reduce the risk of injury to people or other dogs and also to minimize the risk of escalating fear, anxiety and frustration for the dog during interactions.

Head collars and harnesses

There are various head collars available and these are marketed for the purpose of increasing physical control and particularly for reducing the dog's ability to pull. Ideally it is better to teach a dog to walk correctly from early puppyhood, but it must be recognized that adult dogs with behavioural reactivity will often be difficult for their owners to manage physically.

A head collar can make it easier to turn a dog's head away from someone or another dog, which can reduce the risk of them lunging towards other people or dogs. For maximum safety and control a head collar can be used in addition to a Baskerville™ muzzle: the muzzle is fitted over the top of the head collar and the lead strap is fed between the plastic bars on the underside of the muzzle. If necessary, one of the plastic bars underneath the muzzle can be removed to facilitate this (Fig. 17.2).

Fig. 17.1. Dog wearing Baskerville™ muzzle. (Photo: author's own.)

Fig. 17.2. Dog wearing Baskerville muzzle and head collar. (Photo: author's own.)

Some dogs will tolerate wearing a head collar as long as it is fitted and introduced correctly (see Appendix 17.2. Accustoming Dogs to Wearing a Head Collar). It is imperative that head collars are fitted well, as some dogs can get out of them and the head collars can also break. It is strongly advised that if a head collar is used, there must be a secondary attachment to the dog's collar for safety, such as a proprietary quick-release linking strap or a metal carabiner attached from the lead ring of the head collar to the D-ring on the dog's normal flat collar. A head collar must never be used to pull a dog's head around quickly, as this may cause or exacerbate neck problems. For the same reason they should never be used attached to an extending lead or long-line.

Some dogs show a significant level of resistance to wearing a head collar. This may be due to discomfort, or in some cases because the head collar impedes the dog's vision. If the dog pulls away from something it is frightened of, such as another dog, the head collar may move and block the dog's view of the other dog. This will lead to an increase in negative emotional arousal through the fear–anxiety system.

Harnesses are another piece of equipment that can be used to increase physical control. There are many varieties of harness on the market but key considerations should be comfort and fit, so that the harness does not rub. Another important factor is ease of fitting, as some dogs can get very frustrated if it takes too long for the owner to put it on. If a dog needs to be manhandled to put the harness on, there is also a risk of development of fear and anxiety. In these situations, frustration may occur when the dog cannot use an avoidance behavioural response to deal with those negative emotions. Harnesses where the lead clips on the front can be useful for animals that pull as they are unable to put their full weight into the harness.

Both head collars and harnesses are commonly used for dogs that have developed a problem of pulling on the lead, but the use of such equipment is not an alternative for establishing the emotional motivation for the pulling behaviour and offering long-term solutions.

Increasing safety in the home

House-lines

The house-line is a very light lead usually about 2.5 m long, with no handle, that can be clipped to the dog's collar and left trailing. This allows an owner to guide their dog off furniture, take them into another room or move them away from people without having to grab the dog's collar, which can cause dogs to become fearful or defensive. If there is any risk that a dog might bite the owner when they take hold of the line, the owner should be advised to stand on the line close to the end and then pick up the end furthest away from the dog. A house-line must never be left on a dog when unsupervised, in case it gets caught up under furniture, doors, etc. Soaking the line by fully immersing it in something unpleasant-tasting such as TCP or Bitter Apple before use can reduce the chances of it being chewed.

Stair-gates and dog-gates

Adding these gates across doorways in the home can restrict a dog's access to certain parts of the house or to certain people or other animals. They can be especially useful if a dog is showing aggressive behaviour to visitors, children, other dogs or animals in the house or in situations where the owner is experiencing the problem of indoor toileting (but do note that in this case the barrier will most likely only restrict the problem to one area, not eliminate it, especially if the problem results from the dog not coping when left alone). Stair-gates or dog-gates across doorways or passageways can also be used to create 'air-locks' to keep a dog safely away from exit points (see below). However, some dogs can jump over stair-gates, and occasionally even the taller dog-gates, although this is uncommon. In all cases the gate must be securely fixed to ensure that it will not collapse if the dog jumps against it.

Crate training or indoor pen

Dogs can be safely confined to a smaller area in the home using a crate or indoor kennel or a playpen. This can be useful in some aggression cases and also for dogs that are fearful of things such as loud noises, because covering a crate in blankets or duvets can help reduce exposure to external sounds. Some dogs appear to feel more secure in a small enclosed area and a crate can be an extremely effective 'safe' area both in the home and in other places. However, it is important to remember that some dogs can find being enclosed in a confined space very stressful. It is essential that all dogs are introduced to being confined in a crate or pen very gradually so that they learn to accept this calmly. Dogs that have had a bad experience with being confined may never learn to tolerate this in future. See Appendix 17.4 for general principles of crate training.

Increasing security at entry/exit points

This is particularly important for owners of dogs showing aggressive behaviour to people coming into or passing the house, or any dog who might bolt through an open door and escape.

Fencing

Dogs should not be allowed in the garden unattended or off-lead unless the fencing is secure and high enough that the dog cannot climb or jump over it. Solid fencing 2 m high will contain most dogs but some can still clear this. If this is likely, additional strategies will be needed, for example fitting extra fencing on inward-angled brackets along the top of the existing fence. It is also important to ensure that dogs cannot escape through gaps in the fencing or by digging underneath. All gates into and out of the garden must be kept securely locked unless there is a secure 'air-lock' system in place.

'Air-locks'

One way of increasing security around entry and exit points is by making sure the dog is behind a second closed or locked door, or other secure barrier, before an external door or gate is opened. This is essential for dogs that show territorial or escape behaviours as above, but can also be very useful if there are children in the house who might accidentally leave external doors or gates open. Stair-gates or dog-gates across internal doorways or passageways can act as secure barriers as long as the dog cannot jump them. If an internal door is used this must always be closed securely before the outer door is opened, and it is important to ensure that the dog cannot open the door itself: if it can, the door may need to be locked or bolted from the outside. Air-locks can also be created around gates in the garden, for example using an opened-up playpen or other secure barrier. However, these barriers must be totally secure and if there is any doubt it is far safer to keep gates locked or to ensure the dog is never allowed in the garden off-lead or unattended.

Other safety/management strategies

Boarding

For behaviour problems that are only occurring in the home, for example aggressive behaviours directed towards family members or to another dog in the household, temporary boarding can be a useful way of removing the dog from the problem situation and ensuring the safety of the target of the aggression until a behaviour consultation can be arranged. This also gives both the owners and the dog time to calm down as they are removed from the stress of the situation. This may enable the owners to see the problem more objectively and allow them to make appropriate decisions regarding their willingness and ability to keep the dog and work on the behaviour problem. Obviously if a dog has shown aggressive behaviours to people or other dogs this must be fully disclosed and discussed with the kennel owner and staff before the dog is boarded: some kennels are better prepared to manage dogs that show aggressive behaviours than others.

Day-care or dog sitter

Depending on the nature of the behavioural concern, it may be possible for dogs to go to a day-care centre or be cared for by a dog sitter. Examples may include dogs with separation-related problems that are not specifically related to over-attachment to one particular person. Such dogs may be fine at a day-care facility surrounded by other dogs and people at times when the owners have to go out. Other options might include leaving the dog with another family member, friend or neighbour, or having someone come and sit with the dog in their own home while the owners are out. If the owner wishes to use a doggy day-care or dog sitter, it is very important that an emotional assessment of the dog has been made. If the dog has low self-confidence and resulting anxiety issues, there may be a detrimental effect from leaving them in one of these facilities. It is also important for owners to research local facilities well before using them, as some will be able to offer very carefully managed care that is more suitable for dogs with existing behavioural challenges.

As outlined in Chapter 16, rehoming or even euthanasia may also be an appropriate safety or management strategy in certain behaviour cases.

Avoiding confrontation or physical punishment

Many owners have been encouraged to believe that confrontation and physical punishment are appropriate ways of dealing with undesirable behaviours and that dogs showing aggressive behaviours should always be made to back down. When owners come to a veterinary practice for professional advice, it is essential that they are advised never to confront a dog that is behaving in a manner that is causing the owner concern, particularly if it is showing aggressive behaviour. Any form of confrontation, such as attempting to take items away from a dog that is guarding them, pulling a dog off the sofa by their collar if they refuse to get off when asked, or any form of physical punishment, such as shouting or smacking the dog, is likely to cause the dog's behaviour to escalate and will increase the risk of the owner being bitten. Owners need to understand that it is far safer to move away from a dog that is growling, for example because it is guarding something or because it feels its personal space is being invaded, as this should reduce the likelihood of the dog escalating the level of aggressive behaviour to snapping or biting. If necessary, distraction (such as ringing the doorbell, making interesting food-related noises in the kitchen, picking up and playing with the dog's favourite toy) or exchange (offering something of higher value from the dog's point of view) can be used to make the dog lose interest in an item it has stolen or encourage it to move away from a particular area. Once the dog is busy in another room the door can be closed and the problem item removed.

Inappropriate use of physical punishment often leads to a worsening of the problem behaviour, or the emergence of other, potentially more severe behaviour problems. Dogs often show unwanted behaviours because they are anxious, frightened or frustrated and the use of physical punishment will exacerbate all of these emotional responses. Dogs will often respond to physical punishment by showing defensive aggression and are likely to become wary of their owners in the future.

Keeping a behaviour diary

If the owner can keep a note of all incidences of the problem behaviour until the time of the behaviour consultation, including when and where they happen, any associated events, people or other animals and an indication of other things going on at the time, this will give the behaviourist a good idea of the frequency and severity of the problem. It can also help them to identify particular triggers (things that cause the behaviour to occur) or exacerbating factors (things that make the behaviour more likely to occur).

Video recording and live video monitoring

For some behaviour problems, video recording or live video monitoring can be helpful. For dogs with separation-related behaviour problems, live video monitoring can allow the owner and behaviourist to see exactly how the dog behaves when it is left alone. It is possible to monitor behaviour remotely using a smart-phone linked to a web-cam in the home, via apps such as FaceTime or Skype, or using specialist remote monitoring cameras. Video recording or remote monitoring can also be useful in dogs showing other types of problem behaviours. For example, they can be used to see if dogs that show repetitive behaviours when their owners are present also perform them when they are absent. However, dogs must never deliberately be put into situations so that a behaviour can be recorded if there is any chance that this may put themselves, or any person or other animal at risk of either physical or psychological injury.

Summary

- Avoid triggering stimuli if at all possible.
- If triggering stimuli cannot be avoided, ensure some form of safety/management strategy is in place to prevent the dog showing the problem behaviour, or causing injury if it does.
- Avoid confronting the dog or using physical punishment.
- Keep a behaviour diary.
- Take video recordings if safe and appropriate.

Safety: First Aid Advice for Specific Dog Behaviour Problems

Aggressive behaviours

- Any dog displaying aggressive behaviours, including lunging, barking, growling, snapping or biting, directed at people or other dogs should have a thorough veterinary examination before the behaviour consultation, particularly if the problem appears to have started suddenly or there is no obvious learning-related reason for their behaviour. Any medical condition that causes a dog to feel unwell or uncomfortable can increase the likelihood of them showing aggressive behaviours as a result of increasing irritability and reducing their tolerance of close interaction with people or other dogs.
- The owner must be made aware of their legal liability to ensure the safety of other people and dogs, and to keep their dog under control at all times in public places and in their own home.

Aggressive Behaviour Towards Familiar People (Owners, Family Members)

General advice

- Owners should be helped to identify, and if possible avoid, all situations that are likely to trigger their dog to show aggressive behaviours.
- If problem situations cannot be avoided, the dog should be muzzled or preferably securely separated from the person(s) likely to be targeted, by a barrier such as a stair-gate or dog-gate, or a securely closed door.
- Owners must avoid any form of confrontation or the use of physical punishment.
- Owners should be shown how to recognize signs that their dog is starting to look worried or feel threatened during an interaction, as this should increase the chances of ending the interaction before the dog reaches the point of growling, snapping or biting. Explaining the 'Ladder of Aggression' (Shepherd, 2009) and giving the owners a copy to take home may make this easier.
- If unable to ensure the safety of family members or if the owners are not prepared to keep the dog in the home, short-term boarding may be an option until the dog can be assessed by a behaviourist and the prognosis for behavioural improvement and ability to keep family members safe in the longer term can be carefully determined.

Aggressive behaviour directed towards children in the family

This is potentially very serious, because it can be more difficult to avoid problems than when adults are targeted, and the risk of injury is far greater.

- Owners must be advised never to leave young children and the dog together unsupervised even for a few seconds: they must be safely separated before the owner answers the phone, opens the door to visitors, etc. This rule may need to apply to older children too, especially if the child cannot be relied upon to behave appropriately around the dog at all times.
- Owners should create a 'safe' place for the dog away from children using physical barriers, such as indoor crates and baby-gates. The dog's bed, favourite toys, items to chew, etc. should be located here. Owners must ensure that the dog can always access this safe place easily and that children are never allowed to follow the dog into the safe place or disturb it when it is there.
- If the dog attempts to move away from a child, the child must be prevented from following.
- If there is any risk of the dog biting or snapping, it may be necessary to separate the child and dog completely using a secure barrier, such as a crate, stair-gate, dog-gate or a closed door. If the dog is not behind the barrier, it should be muzzled and on a lead when children are present.
- If it is not possible to ensure the safety of the children in the home, either because the owner cannot manage their behaviour or because the dog is actively approaching a child to show aggression rather than trying to move or stay away, temporary boarding may ensure safety until a qualified behaviourist can assess the dog, the situation and the risks.

Aggressive behaviour associated with food, bones, chews or stolen items

- Owners should avoid giving the dog high-value items like bones or chews, particularly those that last a long time and that the dog is most likely to guard.
- People should not approach the dog when it is eating. The dog may need to be fed behind a closed door, dog- or stair-gate, in a closed crate or even out in the garden, especially if there are young children in the house.
- Once the dog has finished eating, the owners can open the door or barrier and either wait for the dog to leave the area voluntarily or encourage it to leave the area and engage it in another activity in a different part of the house. Once the dog is busy in another room, they can close the barrier again and remove the food bowl.
- Owners should avoid leaving items within the dog's reach that are either dangerous for the dog or valuable to the owner. Anything else the dog picks up can be ignored, unless it is one of its own toys, in which case the dog should be given lots of attention and praise (at a distance that the dog will enjoy without feeling the need to guard the toy). It may be necessary to restrict where the dog is allowed access to in the home, especially if there are young children in the family.
- Owners must never confront a dog that has stolen something. Instead, if they need to take an item away from the dog because it is valuable or potentially dangerous, they should be advised to use distraction (for example ringing the doorbell, making interesting food-related noises in the kitchen, picking up and playing with the dog's favourite toy) or exchange (offering something of higher value from the dog's point of view) to make the dog lose interest in the stolen item or to encourage it to move away from a particular area. Once the dog is busy in a different room the door can be closed and the owner can then pick up the item safely.

Aggressive behaviour when asked to move from furniture (e.g. sofa or bed)

- Owners should offer the dog an alternative, comfortable resting place, for example a bed on the floor, and make a point of praising and rewarding the dog when it settles there.

- They should use distraction rather than confrontation to encourage the dog to get off the sofa or bed.
- A house-line attached to the dog's collar will allow the owners to guide the dog off the sofa gently without having to grab its collar, which can cause dogs to become defensive. While using the house-line the owners can introduce a cue word, such as 'move', which can then be used in the future
- Owners may need to restrict the dog's access to certain problem areas of the home, for example by keeping bedroom or sitting-room doors closed or putting stair-gates across doorways and at the foot of the stairs.

Aggressive behaviour when groomed or handled

- If possible, owners should avoid the specific situations that lead to aggression until the dog has been assessed by a qualified behaviourist.
- Many problems are context-specific: if handling is unavoidable the dog may tolerate it better if it occurs in a different place or at a different time to normal. Some dogs are more tolerant of being handled by one family member than others and, if so, this person should handle the dog in all potentially problematic situations until the behaviour consultation.
- Using a muzzle will mean that the owners can handle the dog safely in problem situations, as long as the muzzle is introduced carefully and the dog is happy to wear it. However, this should only be advised before the behaviour consultation if handling is unavoidable, as this approach should ideally be combined with careful desensitization and counter-conditioning to the handling process that the dog currently dislikes.

Aggression Shown to Unfamiliar People

Aggression to visitors to the home, people passing the home, delivery people

This generally occurs because the dog is fearful of some or all unfamiliar people or has become frustrated by repetitive arrivals at the property, as occurs with regular delivery people, or repetitive passing of the home by people walking to the shops, to a school or to their own homes.

- The dog must not be allowed to meet visitors at the front door: it should be shut away securely in another room before the front door is opened and visitors are let into the house. Ideally this should apply to known people as well as strangers, so the dog no longer meets any visitors at the front door.
- If the dog is very reactive to visitors (likely to bark, lunge, growl, snap or bite once visitors are inside the house), owners should be advised to keep the dog shut away securely in a separate room while visitors are in the house, or to avoid having visitors altogether until after the behaviour consultation. This is particularly important if young children are likely to be visiting.
- If the dog is usually fine once people are in the house, it may be possible to allow the dog out of the room once the visitors are settled, as long as the visitors are comfortable with this. However, the dog should not be encouraged to approach visitors unless it appears relaxed and happy to do so. Fearful dogs may be happier to remain at a distance from the visitors. If there is any likelihood that the dog might show aggressive behaviour, it would be safer to keep it shut away as above. If this is not possible for any reason, the dog should be on lead or muzzled while visitors are in the house.
- It is extremely important to prevent the dog escaping and encountering people on the doorstep or outside the house.
- All external doors, gates from garden, etc. should be kept shut and locked whenever the dog is loose in the house or garden.
- 'Air-locks' should be created at all exits and entrances to reduce the risk of the dog running out of an open door or encountering someone on the doorstep.
- Ensure all garden fences are high and secure; if they are not the dog should not be allowed in the garden unsupervised.
- The dog should not be allowed in the front garden or outside the front door unless it is on a lead.

- If the dog attacks the post or is reactive to the postman, it would be sensible to close off or tape up the letterbox and put up an outside postbox away from the front door.
- The dog should not be allowed access to windows or glass doors from where it can see people arriving at or passing the property.

Aggressive behaviour shown to unfamiliar people outside the home

- The owner should aim to keep the dog far enough away from unfamiliar people so that it does not become emotionally aroused or reactive, and must not allow unfamiliar people, including children, to approach or interact with the dog.
- The dog should be kept on a lead at all times outside the home and should be muzzled if it is not always possible to avoid close encounters with unfamiliar people or children. In addition to preventing bites, a muzzle tends to encourage people to keep away from the dog, which can make it easier to avoid close encounters.
- If a dog has shown aggression to children, it is sensible to avoid walking it in areas where children are likely to be playing unsupervised, or close to schools at busy times.
- If the dog is reacting to unfamiliar people frequently on walks and the owners cannot avoid this by walking in quieter places or at quieter times, it would be better to stop walking the dog altogether until the behaviour consultation. The owners can concentrate on providing exercise and mental stimulation at home via training, games, fun agility, activity feeding, etc.

Aggressive Behaviour to Other Dogs

Aggressive behaviour to other dog(s) within the household

- Owners must try to identify, and if possible avoid, all triggers for aggression or fighting.
- If triggers cannot be avoided, the dogs must either be safely separated or muzzled whenever the triggers are likely to occur, until the behaviour consultation. For dogs showing intermittent mild aggression (growls, snarls, snaps but no biting), having both dogs muzzled and on leads when they are together will ensure that they cannot injure each other if they do show aggression, and that they can be more easily separated. Dogs showing more severe aggressive behaviours including biting, or any aggressive behaviour that escalates to fighting, should be separated behind stair-gates or dog-gates, or behind a solid barrier such as a securely closed door if they are likely to intimidate each other or fight through the gate. The dogs may need to be separated at all times if they have reached the point of fighting on sight, or only at problem times if the fighting is predictable and intermittent.
- Owners may need advice on how to separate fighting dogs to reduce the risk of getting bitten, although unfortunately this can be difficult and potentially dangerous. It is far better to avoid fights than to deal with them when they occur.
- Throwing a heavy coat or blanket over one of the dogs, or putting a board between them, may allow the owner to grab one dog and move it away from the other, but there is still a risk of them getting bitten. Holding on to a dog's hind leg to pull it away rather than grabbing the collar will avoid their hands being close to the dog's head and therefore reduce the danger of being bitten, but dogs can still turn around and bite, so this is not completely safe. If there is a history of conflict between the dogs it is advisable for them to wear collars and trailing house-lines when they are together in the house so that the owner can take hold of the lines to regain control.
- Distraction is often recommended and examples include the use of loud noises, such as crashing saucepan lids, an air horn, pet corrector spray, carbon dioxide fire extinguisher, etc., or throwing water over the dogs or turning a hose on them. However, there is a real risk that these interventions will increase arousal and the intensity of aggression. If they can be used to startle the dogs without detrimentally increasing arousal, it is possible that the dogs will stop

fighting long enough to allow the owners to move them away from each other quickly but this is less likely with very startling sounds or sounds that are directed toward the dogs in a manner that might be misinterpreted as threatening. Ringing the doorbell might distract the dogs momentarily, as long as this has not previously triggered them to show aggression toward each other.

- If the risk of fighting is very great or both dogs are stressed when they are separated, removing one dog from the household completely, for example by arranging temporary boarding, will avoid the risk of further fights and reduce stress for both the dogs and the owners until a behaviour consultation can be arranged.
- If the dogs are entire, neutering should not be considered until the dogs have been seen and assessed by a suitably qualified behaviourist (for more information see section on Neutering and Problem Behaviour in Dogs, p.240).
- Fighting between two bitches can be particularly serious, with a significant risk of injury and occasionally even death (Overall, 2013). Owners must be made aware of the risks and the fact that the long-term prognosis for resolution is not always good, because in some cases fighting will continue despite careful control of resources and other triggers. Consideration of the involvement of the hormonal state of the bitches is important but a decision regarding neutering should not be made until a behavioural consultation can be arranged. In cases where the behaviour coincides with times when a high prolactin level is a potential factor (i.e. if one or both bitches could have a pseudopregnancy) then appropriate medication should be considered.

Aggressive behaviour towards other dogs encountered on walks

- The dog should be kept on a lead on walks and encouraged to keep away from other dogs until the behaviour consultation. If it is not possible to avoid close contact with other dogs it would be advisable to ensure that the dog is muzzled on walks, especially if it has previously bitten another dog.
- Using a front-attachment harness or head collar (see Appendices 17.2 and 17.3) can improve the owner's control and their ability to prevent the dog lunging towards other dogs, and also make it easier to turn them away from other dogs if necessary.
- The owner should be advised to walk the dog at quiet times or in places where they are less likely to encounter other dogs. However, they must be aware that they might meet other people with dog-aggressive dogs at these times or in these places.
- If it is impossible to avoid encountering other dogs in situations that cause the dog to become emotionally aroused and reactive, it would be better for the owners to stop walking their dog altogether, and provide exercise and enrichment at home including games, training and activity feeding, until the behaviour consultation.

Other Behaviour Problems

Predatory behaviour towards other pets within the household (e.g. cats, rabbits, hamsters)

- The dog must be securely separated from other animals when not supervised, ideally behind a closed door or a stair-gate or dog-gate.
- Animals in cages should be kept on raised surfaces well above the dog's eye level, and should be provided with secure hiding places within the cage so that they can hide from the dog.
- Owners must ensure that cats always have plenty of escape routes (e.g. stair-gates across doorways) and plenty of raised surfaces to jump up on so that they can easily get out of the dog's way. Cats also need 'safe' areas inaccessible to the dog and should be able to access all important resources, including food, water, resting places, toilet areas and outdoors, without having to enter the dog's area.
- The dog should be offered high-value treats, chews, activity feeders, etc. when it is around other animals and owners should notice and reward all appropriate calm, relaxed behaviour

- Keeping a lead or house-line on the dog when it is around small animals will make it easier to move the dog away if it shows too much interest in the animal or attempts to chase. If there is any risk of the dog biting, it should also wear a muzzle.

Predatory/inappropriate chase-related behaviour problems on walks

- Until the behaviour consultation the dog should be kept on a lead, or possibly an extending lead or long-line (see Appendix 17.3), whenever it is likely to encounter targets for chase behaviour. This may well be all of the time when it is outside its house and garden.
- Ideally the dog should be walked in areas where it is unlikely to encounter targets for chase/predatory behaviour, as all attempts to chase, even on a lead, can reinforce this behaviour.

House soiling

- Medical problems are very common causes of indoor urination and defecation and should be ruled out before assuming that a toileting problem is purely behavioural in origin. However, it is not uncommon for toileting problems to persist as learned behaviour problems after a medical problem has been treated.
- Owners should establish an appropriate toileting location, ideally outside in the garden, and aim to take the dog to this location every time it needs to go to the toilet, and reward the dog as soon as it has toileted there.
- The dog must never be punished for toileting in the wrong place, especially after the event. Even if the owners do catch the dog in the act of toileting indoors, it is better to ignore this rather than risk scaring the dog. Dogs that have been startled or told off for toileting indoors can become scared of their owners and this can lead to further problems, including toileting in secluded places indoors or reluctance to toilet in the owners' presence even outdoors.
- All toileting 'accidents' should be thoroughly cleaned to remove all traces of scent, which may otherwise encourage dogs to continue to toilet in this location. They may use a proprietary enzymatic cleaner or a 10% solution of biological washing powder or liquid dissolved in hot water, then rinsed and allowed to dry before wiping over or spraying with surgical spirit to remove any last traces of scent. Owners will need to test the detergent as well as the surgical spirit on a discrete area of carpet/soft-furnishings to ensure colour-fastness before using them on the soiled area. Normal household disinfectants should not be used, because they do not completely remove all traces of toileting scent and they sometimes appear to encourage further toileting in that area.
- If the dog is being left alone for a long period, a suitable toileting area should be provided indoors (e.g. a puppy pad) to prevent urine seeping into the floor below.
- If the dog is toileting in secluded areas of the house, it may be necessary to restrict where they have access to in the house: this can prevent them accessing preferred toilet areas and also make it easier for owners to observe the dog carefully, notice when they show signs of needing to go to the toilet and gently encourage them to the 'correct' location in time.
- Confinement in a small area, such as indoor kennel or playpen, when owners are busy can also help housetraining problems, as most dogs are naturally inhibited about soiling their own bed. However, this must be done carefully: the dog must be taken outside to the correct location frequently, at the times when they are likely to need to go to the toilet, and must not be left confined for longer than they can last without needing to toilet. If this is likely to happen, a separate toileting area must be provided within the kennel or playpen. Dogs must also have plenty of opportunities to play and exercise outside the confined area, and should be given activity feeders, toys and chew items to keep them occupied when they are confined.

Separation-related distress: vocalization, house soiling and destructiveness when the owner is absent

- No matter how desperate they are, owners must be strongly advised against using a 'quick fix' approach to stopping their dog barking or howling, such as a citronella spray, or ultrasonic or electronic collars. These devices seldom work and are more likely to make a problem worse, especially if the dog is vocalizing due to anxiety.

- Ideally provision should be made to avoid leaving the dog on their own at home during the period leading up to the behaviour referral. This can sometimes be achieved through the use of family members, neighbours or friends as informal dog-sitters, professional dog-sitters or dog day-care facilities, as long as the dog is happy to be with other people and dogs. Some owners may be able to take the dog to work with them on a temporary basis. Some dogs will settle happily when left in the car, for example when the owners go shopping, although care needs to be taken to ensure that the car does not overheat in hot weather.

Additional strategies that can help dogs cope better when left alone include the following.

- An Adaptil™ diffuser plugged in and left running, ideally as close as possible to the dog's bed or other preferred location for settling when left alone.
- An item of the owner's clothing that has been worn and not washed, again left in the dog's bed or its preferred location for settling when left alone (this can be stored in the owner's dirty laundry basket between uses to ensure it retains the owner's scent).
- Activity feeders such as stuffed kongs, tasty chew items, treats hidden around the room: if the dog is not too anxious to eat altogether, eating and chewing can help to reduce anxiety as well as relieve boredom.
- Items to tear up and destroy: for dogs that enjoy tearing things up or need an outlet for frustration, cardboard boxes stuffed with crumpled paper and treats and then taped up can provide safe and cheap outlets for this behaviour. For very strong chewers, boxes can be taped inside other boxes to make a 'pass the parcel' type of toy with a few treats inside each layer.
- Large cardboard activity boxes containing a number of different toys, treats, chew items and activity feeders can keep dogs occupied for relatively long periods of time. Many dogs will also enjoy tearing up the large cardboard box, so it is important to make sure it does not contain any staples etc. This is particularly useful for dogs that experience boredom when left alone for long periods, but it can also help dogs that are anxious, by keeping them occupied with constructive activities.
- Background noise such as a radio or music CDs can help mask sounds outside the house, which is useful for dogs that tend to bark at these sounds. Some dogs settle better if they can hear people talking (e.g. a talk radio station or a recorded audiobook) while others may prefer calm background music such as classical music: this can be checked using a video camera or remote monitoring. It is usually best to use background noise that is familiar to the dog, such as the radio station that their owner listens to when they are at home.

If it is not possible to avoid leaving the dog alone before the behaviour consultation, medication with a short-acting drug with anxiolytic properties may help the dog cope better with being left and reduce the likelihood of the problem getting worse. See Chapter 22 for more information about the availability and use of short-acting behaviourally active drugs in dogs. Before doing this the dog must be given a test dose of the chosen drug at a time when the owner is going to be at home, to check for any adverse reactions such as the dog becoming hyper-excitable or uncoordinated, which would necessitate trying a different drug. It may also be necessary to experiment to find the most appropriate dose for the individual dog. Once this has been determined the appropriate dose should be given about an hour before the owner is due to leave the house.

Fear of noises (e.g. fireworks or thunderstorms)

- Owners may need advice regarding how to avoid exposing their dogs to scary noises unnecessarily. For example, dogs scared of bangs should not be walked after dusk or dark over the firework period, or in areas where there are likely to be bird-scarers or people shooting. If it is impossible to avoid scary noises on walks it might be better for the owner not to walk the dog at all, and to provide exercise and enrichment at home until the behaviour consultation.

- Dogs should be kept on a lead or line outside the house if there is any risk of them panicking and bolting if scared by a sudden loud bang.
- For dogs that may panic and try to escape from the house, exit doors should be kept locked during scary events, or 'air-locks' created so that the dog is behind a secure barrier before exit doors are opened.
- All dogs should be provided with a safe, sound-proofed place at home where they can retreat to escape from scary noises and which they can access 24 hours/day. The dog may already have a safe place (for example under or behind furniture or in a shower room or cupboard), but if their chosen location is not safe or convenient an alternative can be created. Crates covered in thick blankets or duvets can make good 'safe places' as long as the dog is happy to be in a confined space when scared. However, the owner must not attempt to shut the dog into the crate, as this can lead to panic.
- A safe place can be made to feel even safer by:
  - Providing a comfortable bed, and possibly also extra bedding or towels for the dog to dig in or hide beneath.
  - Providing an item of the owner's clothing that has been worn and not washed (this can be stored in the owner's dirty laundry basket between uses to retain the owner's scent).
  - Providing activity feeders, items to chew and destruction boxes: for dogs that are only mildly anxious, chewing and tearing things can keep them occupied and can also reduce anxiety.
  - There should be a bowl of water in or close to the 'safe' place, as dogs that are frightened may pant a lot and can get very thirsty.
  - Adaptil™ products may also help. A diffuser can be plugged in as close as possible to the safe place, or Adaptil™ spray used on the bed. Because this has an alcohol carrier that must evaporate before the dog has access to it, it is best to spray the Adaptil™ on to a clean cloth and wait about 20 minutes before putting the sprayed cloth into the dog's safe place, on or beside the bedding. The spray lasts for about 2 hours and if needed for a longer period another piece of cloth can be sprayed in the same way.
- Owners can further reduce sounds from outside by shutting all windows, pulling curtains and playing music at a reasonable volume.
- Owners may need advice regarding how to respond to their dog when it is fearful.
  - If the dog is only mildly anxious, playing a favourite game or doing some fun, reward-based training may help keep it occupied and take its mind off the noises outside the house.
  - If a dog is too fearful to play or take treats it should be taken to its safe place and encouraged to settle there. As long as the dog can settle, the owner should try to remain fairly relaxed and behave normally rather than making a big fuss of the dog. The owner should remain close by, especially if the dog feels more secure with its owner present. After a scary event dogs should not be made to leave a 'safe' place until they are ready to do so voluntarily.
  - If a dog is very worried and does not settle in its safe place the owner can try to reassure them through talking, stroking and/or cuddling. This will not increase fear in an already fearful dog, and for some dogs can make all the difference between coping and panicking: it is inhumane to suggest that an owner ignores a dog that is extremely distressed. However, in the longer term these dogs will require additional help, possibly including medication, to help them learn to cope better in scary situations, particularly if they need to face them alone at any point, and teach them to be less fearful.
- For dogs that are extremely fearful of sounds, medication may be necessary to help them cope better. Licensed drug options for short-term medication where exposure to scary sounds is occurring infrequently, and particularly for predictable situations such as over the fireworks period, include dexmedetomidine oromucosal gel (Sileo), and imepitoin (Pexion). Longer-term fear- or

anxiety-reducing medication may be necessary for dogs whose fears have generalized to a wide variety of sounds, or where exposure to scary noises is occurring frequently or unpredictably. See Chapter 22 for more information about the use of behaviourally active drugs.

Repetitive behaviours (e.g. spinning, tail-chasing, acral lick dermatitis)

- Dogs showing repetitive behaviours should always receive a thorough veterinary examination before referral to a behaviourist, because these may occur in response to neurological conditions or medical conditions causing pain or discomfort, as well as to stressful external events.
- Owners must be advised to identify and, as far as possible, avoid all triggers for the repetitive behaviour and to reduce stress generally as far as possible.
- Keeping the dog occupied with a variety of activity feeders and chews may help to reduce the frequency of the repetitive behaviour and may also provide an alternative, more acceptable outlet for repetitive behaviours involving licking and biting.
- Play and opportunities to show behaviours such as searching or tracking can be beneficial in keeping the dog occupied in more constructive activities, as long as this does not further increase stress.
- However, this should be a temporary measure and not seen as an alternative to a thorough behavioural investigation. A muzzle should never be left on an unsupervised dog.

Neutering and Behaviour Problems in Dogs

Although neutering can sometimes be very helpful as part of the treatment strategy for specific behaviour problems as outlined below, it should not be used as a 'first aid' strategy until the animal has been seen and assessed by a qualified behaviourist.

Most behaviour problems are not solely influenced by sex hormones and many are not hormonally motivated at all. In these cases, the behaviour is unlikely to improve after neutering. This includes most control-related problems, problems relating to boredom and under-stimulation, predatory behaviour and fear- and anxiety-related problems including defensive aggression (Box 17.1). Neutering an animal before arranging a behaviour consultation will delay the animal's access to behaviour therapy and also risks the behaviour becoming worse. Owners may also be disappointed if they are led to believe that neutering will improve a behaviour problem and then it does not.

Problem behaviours most strongly influenced by sex hormones in dogs and bitches

The behaviours most strongly influenced by sex hormones are the sexually dimorphic behaviours that occur more often in males than in females.

Box 17.1. Behaviour problems that are directly influenced by sex hormones do not always improve reliably after neutering.

Male dogs	Mounting: other dogs, people, inanimate objects Roaming in search of bitches Urine marking Competitive aggression to other male dogs, either within or outside the household
Bitches	**Behaviour associated with seasons:** • Irritability • Urine marking • Roaming in search of male dogs • Competitive aggression to other bitches, especially in the same home **Behaviour associated with pseudopregnancy:** • Sometimes quite marked aggressive behaviours to people or other dogs, usually in the context of resource guarding • Nesting/digging • Increased reactivity to stimuli, including noises, territorial triggers.

However, neutering does not always reliably reduce these problem behaviours, for the following reasons.

- Many of these behaviours can have other non-hormonal causes. For example, mounting can be a response to excitement or emotional arousal, a play behaviour or even a learned attention-seeking behaviour in male dogs.
- The tendency to show sexually dimorphic behaviours is determined by changes occurring in the brain before birth as well as by circulating sex hormones, although circulating sex hormones can lower the thresholds at which these behaviours are expressed.
- Learning can maintain behaviours after neutering.

Animals showing sexually dimorphic problem behaviours should therefore be assessed by a suitably qualified behaviourist before they are neutered. Neutering may be part of the treatment plan in some cases, as it can reduce the intensity of the problem behaviour or raise the thresholds at which it is shown, but the decision should only be made after a thorough investigation of the emotional motivation for the unwanted behaviour. Neutering will be more effective if combined with a comprehensive behaviour modification plan to address any other underlying reasons for the behaviour and any learned components. Finally, the timing of neutering in relation to the application of behavioural modification needs to be considered and thus should be decided in consultation with a suitably qualified behaviour practitioner.

Neutering can make some behaviour problems worse

This is the most important reason why neutering should not be used as a 'first aid' strategy until the animal has been carefully assessed by a qualified behaviourist.

Increasing fearfulness in already-fearful dogs

In other species, circulating testosterone has been shown to be correlated with self-confidence and castration to increase fearfulness (Vanheede and Bouissou, 1996; Frye and Seliga, 2001; van Honk *et al.*, 2005). Apparent increases in fearfulness after castration have been reported in male dogs, particularly in dogs that were already nervous before they were castrated. In addition, dogs that are fearful of strangers, of being handled or of being in unfamiliar places are likely to find being hospitalized and handled for a spay or castration operation highly traumatic, and their fearfulness in these situations may well be increased.

If a fearful dog must be neutered for either behavioural or health reasons, it would be sensible for a behaviourist to work on reducing their fearfulness and increasing their confidence generally before arranging the operation, and particularly to reduce their fear of strangers, of being handled and of the surgery environment, using desensitization and counter-conditioning. Careful management, premedication and effective use of analgesia both during surgery and post-operatively should also reduce the risk of a dog forming adverse associations with the surgery, staff and being handled.

Post-spay pseudopregnancy

Spaying a bitch during metoestrus when progesterone is raised can precipitate a pseudopregnancy that may persist long-term if not recognized and treated. The same applies to bitches spayed during an existing pseudopregnancy. Some of these bitches will show physical signs such as enlarged mammary glands and milk production but in others the only obvious sign may be a sudden change in their behaviour, such as a sudden onset of aggression to family members or other dogs, after being spayed. Owners should be made aware that they need to report to their vet any sudden change in their bitch's behaviour after spaying. Post-spay pseudopregnancies will normally respond to Cabergoline (Galastop), and, as for normal pseudopregnancies, treatment must continue until the bitch's behaviour has completely returned to normal, which often takes at least 2 weeks.

Aggressive behaviour and neutering

The effect of neutering on dogs that are showing aggressive behaviours is not always easy to predict.

Males

Male dogs are more likely than females to show aggressive behaviour in a number of situations. However, castration does not always reliably reduce aggression, because this tendency is influenced by many other factors apart from circulating testosterone, including hereditary and developmental factors and learning. Circulating testosterone has the greatest influence on aggression to other male dogs, so castration is most likely to be beneficial in dogs showing the following.

- Aggressive behaviour shown specifically to other male dogs away from home. Neutering alone may not completely resolve the aggression problem, particularly if a dog has learned to find other male dogs threatening. Behaviour modification will also be needed. Castration is less likely to improve behaviour in dogs showing fear-related or defensive aggression to both dogs and bitches and in these cases it is advisable to delay castration until the behaviour has been fully investigated and a decision about behavioural treatment has been made.
- Aggression to another male dog in the household. Problems are most likely between dogs that are very similar in size, temperament and confidence and also place similar value on particular important resources. Castration of one or both dogs has been suggested as an approach to reduce aggression, especially when combined with a comprehensive behaviour modification plan to address problems relating to the owner's interactions with the dogs, provision of resources and learning. The decision as to which dog(s) to neuter, or indeed whether to neuter either, should be made very carefully, based on a careful assessment by a suitably qualified behaviourist of their temperament, confidence and the underlying motivation(s) for the aggression in specific situations.

In both situations, getting a behaviourist involved before neutering is undertaken should improve the prognosis for successful resolution of the problem behaviour.

Aggression in other situations is less directly influenced by circulating testosterone. However, testosterone can reduce frustration-tolerance and increase self-confidence, impulsivity and risk-taking behaviours, all of which can increase the likelihood of a dog getting into a threatening situation and then responding by showing aggression (Eisenegger *et al.*, 2011; Terburg and van Honk, 2013). Castration can therefore potentially reduce the likelihood of a dog showing aggression, or at least raise the threshold of arousal at which aggression is shown (Overall, 2013). This is most likely to be successful if done alongside a comprehensive behaviour modification plan. Circulating testosterone has the least influence on fear-related or defensive aggression and as castration can potentially increase fearfulness the decision regarding whether to neuter a dog showing fear-related aggression must be very carefully considered.

Bitches

Aggression occurring outside a season or pseudopregnancy is unlikely to be influenced by sex hormones and therefore very unlikely to improve after spaying. The only situations where aggression is likely to be influenced by sex hormones, and therefore potentially improved by spaying, are when aggression occurs only around the time of a season or during a pseudopregnancy.

- Aggression between bitches living together. In entire bitches, aggression is most likely to occur around the time that one or both comes into season or during a pseudopregnancy. If aggression only occurs at these times and the bitches get on well at all other times, spaying both bitches will prevent further hormone fluctuations and may also prevent further episodes of aggression. If aggression appears to be unrelated to seasons or pseudopregnancy, spaying is very unlikely to improve it.

Unfortunately, even if aggression initially appears to be hormonally induced it will sometimes continue after spaying, possibly because bitches have learned to feel threatened by each other. Behaviour modification is essential in these cases, but the prognosis for resolution is not always good and in some cases one bitch will need to be rehomed.

- Aggression to owners associated with pseudopregnancy. Bitches showing dramatic behaviour changes including aggression associated with pseudopregnancy should be spayed to prevent further episodes in future cycles. However, it is important to ensure that the pseudopregnancy has completely resolved and the bitch's behaviour has completely returned to normal before she is spayed. Treatment with Cabergoline (Galastop) is usually effective but this may need to be given for 2 weeks or longer to resolve the behaviour signs completely.

References and Further Reading

Defra (n.d.) Controlling your dog in public. [Outline of laws relating to dog ownership with emphasis on dogs regarded as dangerous. Includes links to important laws including Dangerous Dogs Act 1991, The Dangerous Dogs (Amendment) Act 1997, Dogs Act 1871, Animals Act 1971, The Dogs (Protection of Livestock) Act 1953 and Dog Control Orders.] Available at: http://www.defra.gov.uk/wildlife-pets/pets/dangerous/ (accessed 1 May 2022).

Eisenegger, C., Haushofer, J. and Fehr, E. (2011) The role of testosterone in social interaction. *Trends in Cognitive Sciences* 15, 263–271.

Frye, C.A. and Seliga, A.M. (2001) Testosterone increases analgesia, anxiolysis, and cognitive performance of male rats. *Cognitive Affective and Behavioural Neuroscience* 1, 371–381.

Overall, K. (2013) *Manual of Clinical Behavioural Medicine for Dogs and Cats*. Elsevier.

Ryan, S., Zulch, H. and Baumber, P. (2014) *No Walks? No Worries! Maintaining Wellbeing for Dogs on Restricted Exercise*. Hubble and Hattie, Dorchester, UK.

Shepherd, K. (2009) Behavioural medicine as an integral part of veterinary practice. In: Horwitz, D.F. and Mills, D.S. (eds) *BSAVA Manual of Canine and Feline Behavioural Medicine*, 2nd edn. BSAVA, Gloucester, UK, pp. 10–23.

Sherman, C.K., Reisner, I.R., Taliaferro, L.A. and Houpt, K.A. (1996) Characteristics, treatment, and outcome of 99 cases of aggression between dogs. *Applied Animal Behaviour Science* 47, 91–108.

Terburg, D. and van Honk, J. (2013) Approach–avoidance versus dominance–submissiveness: a multilevel neural framework on how testosterone promotes social status. *Emotion Review* 5, 296–302.

Vanheede, M. and Bouissou, M.F. (1996) Effects of castration on fear reactions of male sheep. *Applied Animal Behaviour Science* 47, 211–224.

van Honk, J., Peper, J.S. and Schutter, D.J.L.G. (2005) Testosterone reduces unconscious fear but not consciously experienced anxiety: implications for the disorders of fear and anxiety. *Biological Psychiatry* 58, 218–225.

Appendices to First Aid Advice for Common Behavioural Signs: Dogs

Appendix 17.1. General Principles of Muzzle Training

Dogs that show aggression may need to wear a muzzle in certain situations in order to avoid injury to people, dogs or other animals. It is important that dogs learn to be comfortable wearing a muzzle so that they do not only associate wearing it with potential problem situations when they may be stressed or upset. Muzzle training involves teaching the dog to wear a muzzle happily at home *before* it needs to be used in potential problem situations such as the vet, groomer, etc.

Owners will need help in choosing an appropriate type and size of muzzle and then accustoming their dog to wearing it. Ideally every surgery should have at least one member of staff who is experienced and competent in fitting and accustoming dogs to wearing muzzles. It is also useful to have a printed guide for owners to follow at home.

This is a general guide to muzzle training for owners, indicating the main stages required to accustom a dog to wearing a muzzle. There are also some links to videos that demonstrate the process of muzzle training particularly well.

However, all dogs are different and for dogs that are particularly worried, or those that have had a previous bad experience associated with wearing a muzzle, it will be necessary to progress very slowly indeed, making absolutely sure the dog is completely comfortable with each stage before progressing to the next. Owners should be advised to seek further help if they have any problems with the muzzle training process.

Muzzle Training Guide for Owners

Choosing and fitting a muzzle

The most suitable muzzles are the plastic basket-type muzzles, such as the Baskerville™ muzzle made by the Company of Animals (http://www.companyofanimals.co.uk/products/muzzles) and available from a wide range of outlets. These allow dogs to open their mouth to pant and even drink, unlike the soft nylon tube muzzles, and it is also possible to feed treats through the basket.

- The Company of Animals sizing chart is a good initial guide to the likely correct size for the dog (http://www.companyofanimals.co.uk/product/baskerville-muzzle)
- However, some dogs will need a size larger or smaller than that suggested for their breed.
- The fit of the muzzle can be checked by cupping it in your hand and putting tasty treats into the bottom to encourage the dog to put their nose into it. It is not necessary to do the strap up at this point, but you can hold the straps behind your dog's head to assess how it will fit when they are done up.
- A correctly fitting muzzle should:
  - sit snugly around your dog's nose without rubbing or digging in;
  - have a basket deep enough to allow your dog to open its mouth and pant;
  - the top of the muzzle should not be too close to or rub its eyes;
  - there should be about 1 cm gap between the end of your dog's nose and the end of the muzzle when the strap is done up;
  - when properly fitted the strap should be fastened fairly tightly just behind the ears, allowing just one finger underneath, or for extra security it can be looped through your dog's normal flat, buckled collar before fastening, to prevent slipping it off over its head (this is particularly useful for dogs with narrow heads); and
  - when the basket is gently pulled forwards down your dog's nose, the 'give' in the strap should only allow the basket to move halfway down the nose. If it can be moved any lower than this, it is not fitted tightly enough
- For safety, once the muzzle has been correctly fitted and adjusted, the fabric strap running through the snap buckle should be stitched in place to prevent it loosening if your dog tries to remove the muzzle.

Initial introduction

- Hold the muzzle in your hand and praise and reward your dog if it shows any interest in the muzzle, such as sniffing it.
- Cup your hand over the basket part of the muzzle and put some pieces of your dog's favourite food inside. Or, even better, smear the bottom of the inside of the muzzle with something tasty and sticky like peanut butter, liver paste or cream cheese. Hold the muzzle near your dog's nose so it can smell the food: the dog should put its nose into the muzzle to reach the food. It is important to be patient and to wait for your dog to put its nose into the muzzle rather than pushing the muzzle on to its nose. Once it has taken the treat or had a lick at the paste, remove the muzzle again.
- Repeat this until your dog is happy to put its nose into the muzzle as soon as you present it, and will keep its nose in the muzzle for several seconds. You can encourage your dog to keep its nose in the muzzle for gradually increasing periods by squeezing extra cheese or paste into the end, or feeding a succession of treats through the basket.
- Say the word 'muzzle' in a happy voice whenever your dog puts its nose into the muzzle, both during this training and also in the longer term. This will help your dog to anticipate what is going to happen whenever you want it to wear a muzzle and the dog should also learn a pleasant association with the word 'muzzle'.
- Repeat the exercises above, but while your dog is licking at the paste gently hold the

straps of the muzzle behind its head to simulate fastening the buckle up, before removing it.

- As a separate step, your dog should be accustomed to the sound of the snap-fastener by pairing this sound with treats. This should be done initially at a distance from the dog and then gradually closer to it, so that it is completely comfortable with hearing this sound before you attempt to do the strap up behind its ears. This is particularly important for sound-sensitive dogs.
- With your dog's nose in the muzzle, accustom it to the strap being done up behind its head. Initially fasten the strap momentarily and then release and remove the muzzle before your dog stops licking. It is a good idea to have a lead attached to its normal collar so you can stop the dog running off before you can remove the muzzle.

Gradual acclimatization

- Once your dog is comfortable with having the muzzle put on and fastened, gradually increase the length of time that your dog wears the muzzle for. To reduce the chances of it trying to remove the muzzle, keep your dog occupied by doing some reward-based training and rewarding it with titbits posted through the bars of the muzzle or small squeezes of liver paste, etc. If your dog does try to remove the muzzle, try to distract it, and as soon as the dog stops trying to remove the muzzle, praise and reward it immediately with another squeeze of paste or a treat.
- Always try to remove the muzzle when your dog appears relaxed, and follow these muzzle training sessions with something your dog really enjoys, such as a walk, meal or a favourite game. The aim is for your dog to learn that wearing the muzzle is a pleasant experience and that it predicts other pleasant experiences.
- The muzzle should be put on at different times of the day and in different places so that your dog learns to accept wearing it whenever you produce the muzzle.
- These muzzle training sessions should occur frequently at home, and continue in the long term. This should ensure that the pleasant association with the muzzle remains even if the muzzle is occasionally used in less pleasant situations, such as veterinary examinations.
- If you have any problems at any stage with accustoming your dog to wearing a muzzle, please contact the practice for further help.

Links to videos that demonstrate the process of muzzle training particularly well:

- Linda Ryan (FABC): https://www.youtube.com/watch?v=l7oyh40pUh8
- Claire Stallard (Blue Cross): https://www.youtube.com/watch?v=6BjPpXer8IE
- Dog Charming: https://www.youtube.com/watch?v=_9wK-QE5v9w
- Chirag Patel: Turning muzzle training into a game: https://www.youtube.com/watch?v=1FABgZTFvHo
- There is more information on choosing, introducing and using muzzles at: https://muzzleupproject.com/

Appendix 17.2. Accustoming Dogs to Wearing a Head Collar

Head Collar: Introduction Guide for Owners

Choosing and fitting a head collar

Individual dogs tolerate different makes of head collar better than others, and some fit certain sizes and shapes of head better than others, so it may be necessary to experiment to see which make suits your dog best. Each make comes with a size guide but this is not always accurate so it is important to check the size by trying it on before buying. The biggest problem is usually the noseband riding up into the dog's eyes, most common in dogs with a short nose or those without a pronounced 'stop' below the eyes. Correct fitting can help, as per the instructions provided with each make of head collar, and ensuring the strap behind the ears is fitted quite snugly is particularly important, as this usually means the noseband can be looser without the dog being able to remove the head collar altogether. A design with more rigid cheek pieces, such as the Dogmatic™ (http://www.dogmatic.org.uk), will fit some dogs better as the noseband is less likely to ride up into the dog's eyes. However, there are some dogs that will not be able to wear a head collar safely or comfortably, either because of the shape of their faces or because they cannot learn to tolerate wearing it.

When a head collar is fitted correctly:

- The nose-piece should sit comfortably below the eyes, and not ride up into the eyes, even if your dog attempts to pull.
- The strap behind the ears should be done up quite snugly, allowing just two fingers underneath it.
- If you gently pull the noseband forwards down your dog's nose it should move to just above the nose pad (fleshy part of the nose) but no further. If it can be moved any lower, or comes off the end of your dog's nose, the head collar is either too large or not fitted correctly.

Because dogs can sometimes remove a head collar, for safety it should be used with either a proprietary quick-release linking strap or metal carabiner attached from the lead ring of the head collar to the D-ring on your dog's normal flat collar. This means that your lead will still be attached to your dog via the collar if the head collar breaks or comes off. Another option is a double-ended lead, with one end attached to the head collar and the other to your dog's normal collar.

Initial introduction

- Start by holding the head collar in your hand when you are stroking or playing with your dog but do not try to put it on. Praise and reward your dog for taking any interest in it, for example sniffing at it.
- When your dog is happy to stay near you when you are holding the head collar, start undoing and doing up the snap-fastener, at first when your dog is further away from you, then, when it is taking no notice, closer to you, so it gets used to the snapping noise (this step is not necessary if the head collar has a buckle fastening).
- When your dog appears comfortable with you holding the head collar in your hand and with the sound of the fastener, hold the head collar by the top of the noseband and use a treat in the other hand to encourage your dog to put its nose through the noseband. Reward your dog with the treat for doing this.
- Repeat this a few times, then hold the head collar up and wait for your dog to put its nose into the noseband before rewarding it.
- Once your dog is happy to put its nose in the noseband, you can start doing up the buckle or fastener around the back of its head behind its ears. Reward your dog with a couple of food treats, then undo the buckle or fastener and remove the head collar.
- As your dog becomes comfortable with having the strap fastened behind its ears, gradually increase how long this is done up for before taking it off. It is very important that you only remove the head collar when your dog is quiet and relaxed, so if your dog struggles to remove it, try and distract it and wait until it is calm before taking it off. It may help to take your dog's mind off the head collar by doing some simple reward-based training exercises. It is also a good idea to have a lead attached to the normal collar so you can stop your dog running off.
- Check that the head collar is adjusted properly according to the instruction booklet that came with it.

Gradual acclimatization

- Start to put the head collar on your dog at random times during the day, especially just before something it likes such as being fed, going for a walk or having a game. At first, remove the head collar before feeding, playing or walking. Then start to leave it on when you play with your dog or go for a walk, but at first, do not attach the lead to it.
- Finally, attach the lead to the head collar and get your dog used to the feeling of you turning its head when you want to change direction or to stop it pulling. It is very important to turn your dog's head gently, as yanking hard can damage its neck. This may feel very strange to your dog at first, and it might again try to rub its head or remove the head collar with its paws. As before, try

not to remove it until your dog is quiet and relaxed, unless it is becoming increasingly distressed, in which case you will need to remove the head collar, wait until your dog is calm and then go back to an earlier step in the process. Try doing some reward-based training or take your dog for a walk to take its mind off the head collar.

Most dogs learn to accept wearing a head collar quite readily as long as it is fitted correctly and introduced gradually, and especially once it becomes associated with going out for walks. However, if you have any problems at any stage with accustoming your dog to wearing a head collar, please contact the practice for further help.

Appendix 17.3. Additional Safety Strategies for Increasing Control Away from Home

Front-attachment harnesses

Harnesses with a front lead attachment point can increase owners' control through reducing pulling and increasing the owner's ability to turn the dog away from someone or another dog, for example. They tend to be better tolerated than head collars by most dogs, although owners of large, strong dogs may find that a head collar gives more control. For maximum control a head collar and front-attachment harness can be used together, along with a double-ended lead. Like head collars, harnesses need to be carefully fitted and adjusted. There are a number of different makes available and, just as with head collars, some makes fit some dogs better than others, so it may be necessary for owners to experiment to find the make that fits an individual dog best.

Long-line

A long-line is a long lead that allows a dog some freedom to run or roam without the risk of it running off. Widely available at 5 m or 10 m long although longer ones can be bought or made. For safety a long-line should be attached to a standard harness and not a collar or head collar, due to the risk of injury if the dog runs and suddenly hits the end of the line with a jerk. Owners should be advised to wear leather gloves to protect their hands, and boots to protect their ankles against rope-burns. Long-lines can be useful for dogs with unreliable recalls, for dogs that might bolt if they are frightened, for example by a sudden loud noise, and for some chase-related problems. Advice on how to use a long-line safely should be given to all owners where they are recommended and owners of large or strong dogs should not be advised to use them without specialist advice from a behaviourist or trainer experienced in their use. If a long-line is being suggested as part of first-aid advice for a behaviour problem the owner should be advised only to use it in safe, non-distracting environments, as there is a risk that the dog might pull the owner over if triggered to chase, for example. It is also important that a dog is accustomed to being on a long-line and has been taught some basic commands, including 'stop', before being allowed to explore to the full length of the line.

Extending lead

This can be used in similar situations to a long-line. Again, it is safest attached to a harness rather than a collar, and should never be attached to a head collar. There are both advantages and disadvantages to using an extending lead instead of a long-line.

Advantages:

- The line retracts so dog and owner are less likely to get tangled in it.
- The line is less likely to get dirty.

Disadvantages:

- The cord ones can cause quite nasty friction burns if they get caught around the owner's hand or the dog's legs. The all-tape ones are slightly better but tend to be heavier and bulkier, and can still potentially cause friction burns if they get caught around legs or arms.

- Owners must always ensure that the extending lead is suitable for the size of their dog, and also strong enough to withstand the dog lunging.
- The brake can be knocked off accidentally, so owners must be advised to use an ordinary lead when walking alongside roads, etc., and only use the extending lead in safe, open spaces.
- The handle can be heavy and dogs can be injured or frightened into bolting if the handle is dropped or pulled out of the owner's hand and either hits the dog or drags behind it.
- Because there is always a degree of tension on the line when the brake is off, it may be more difficult to teach dogs not to pull on the lead in the longer term, compared with using a long-line.

If practices are advising owners to use an extending lead as part of first-aid advice for a behaviour problem, owners must be made aware of the potential disadvantages and ensure that they are used safely.

Appendix 17.4. General Principles of Crate Training

Indoor kennels or crates can provide dogs with a safe place in the home where they can settle without being disturbed. Portable crates can also be taken to other places, providing a safe, familiar place for dogs to settle wherever the dog goes. They can also be useful for temporary confinement after orthopaedic surgery, for example. However, dogs must be introduced to using a crate very carefully to ensure that they are comfortable spending time in it. Dogs must never be forced to enter a crate if they are reluctant and must not be confined in them for long periods of time. Crates must never be used for dogs that become distressed when left alone or when confined in a small space. Below are general guidelines only and these may need to be adjusted for individual dogs and owners, depending on the dog's general tolerance of being confined and any previous bad experiences associated with confinement.

Crate Training Guide for Owners

Choosing and preparing a crate

- A crate should be large enough for your dog to comfortably sit, stand up, turn around, lie down and stretch out inside it.
- It should contain a comfortable bed, ideally one that your dog already uses.
- Putting the crate in an area where your dog usually settles and sleeps will increase the chances that it will go into it to settle and to sleep at night, but the door must be left open.
- It is a good idea to cover the top and three sides of the crate with a sheet or blanket to make it more enclosed and private. If more sound-proofing is required, for example for a dog scared of noises, heavy blankets or duvets can be used.

Initial introduction to the crate

- Put chews and treats inside the crate at random times throughout the day. This should encourage your dog to enter the crate voluntarily. Notice and praise your dog if it does enter the crate voluntarily at any time. At this point **the door must always remain open** when your dog is inside the crate. If there is a risk of the door accidentally closing on the dog, tie it open.
- Once your dog is going into the crate happily by itself you can can start taking your dog to the crate and encouraging it to go inside before offering a treat or chew. You can introduce a cue word or phrase such as 'in your bed' or 'in your crate' as your dog enters the crate.
- You can progress to asking your dog to go into the crate using your cue, offering a treat or chew, then getting it to wait for a moment before giving it a cue to come out again. If your dog already knows a 'release' cue such as 'free' or 'off you go', this can be used to release it from the crate. The time between entering and leaving the crate can be gradually increased.

Accustoming your dog to being confined in the crate

- Once your dog is voluntarily settling in the open crate for longer periods (e.g. at night) and can stay happily in the crate for at least a few minutes on cue, it should be possible to start accustoming it to being shut in the crate.
- Use your cue to ask your dog to go into the crate, offer a few treats or a tasty chew and close the door briefly while your dog is eating these. Open the door just as your dog finishes eating and give its release cue.
- Over time you should be able to increase very gradually the time that your dog waits in the closed crate before you open the door and give your release cue. However, make sure that your dog does not become unsettled at any stage. You can also start to move away from the crate while your dog is eating and then return to open the door.
- As long as your dog is comfortable with this you can start to shut your dog in the crate for very short periods during the day, particularly at times when your dog is tired and likely to settle, for example after a walk. These periods can be increased gradually but it is important to release your dog from the crate before it starts to get restless. Giving your dog a stuffed kong or something else tasty to chew should help it to settle here for longer periods.
- Your dog needs to learn that the crate is a safe place where it will not be disturbed. It is important not to reach into the crate to touch or try to grab your dog: if you want your dog to leave the crate you will need to call it to you or use your release cue instead. If your dog does not want to leave the crate for any reason, it is important not to force it to do so. Children must not be allowed to approach or disturb your dog when it is in its crate and the same should apply to visitors, especially if your dog is worried about unfamiliar people.
- If you have any problems at any stage with accustoming your dog to the crate, please contact the practice for further help.

18 First Aid Advice for Common Behavioural Signs: Cats

Trudi Atkinson

What Is Behavioural First Aid and Why Is It Necessary?

Providing advice aimed at long-term resolution or management of behaviours that are of concern or undesirable for cat owners must be based on a comprehensive understanding of the behaviour and the underlying cause (or causes). This will require in-depth investigation, plus counselling of the owner to achieve effective human behaviour change. Not only can this be very time consuming, it can also require a high level of knowledge and experience (see Chapter 21).

When owners who are experiencing a behaviour problem with their cat first approach the veterinary practice, it is very unlikely to be at a time when these requirements are immediately available and arrangements for a later in-house behavioural consultation or referral to an external behaviourist is often necessary. This can result in a significant delay between the client's request for help and getting the full support that they need.

However, the owner might expect immediate help, or even resolution of the problem, and when this expectation is not met it can have a detrimental effect on the relationship between the veterinary professional and the client. In addition, if the cat's behaviour is having a detrimental effect on the owners' lives, the time until assistance is offered could further damage the pet–owner relationship. A delay might also result in a worsening of the problem due to repeated 'practise' of the undesirable behaviour by the cat, or because of actions or reactions performed by the owner that increase the animal's need to express the behaviour.

Offering effective short-term advice can therefore be highly beneficial for the pet, the client and the veterinary practice. However, it is essential that the advice offered at this time is both relevant and correct.

Members of veterinary practice staff working in general small animal practice are likely to be more frequently approached by dog owners regarding behavioural issues, but this does not mean that they should be any less well prepared when approached for help and advice by owners of other companion animals. It is also very important to be aware that each species has its own behavioural requirements and behavioural repertoire. Behavioural advice and methods that are suitable for one species may not only be ineffective when applied to another species, but could even be potentially damaging.

Principles of behavioural first aid advice

To avoid false expectations, it must always be made clear to owners that the intention of behavioural first aid advice is not to achieve a complete and

DOI: 10.1079/9781800621312.0018

long-term resolution of the problem behaviour, but that the aims should be as follows.

- **To offer short-term relief.** Simple advice may be given to physically prevent certain behaviours from occurring, or to re-direct an unwanted behaviour. Such advice may include: closing doors to prevent a cat from repeatedly soiling in an area that is undesirable for the owner and providing litter trays in more acceptable areas; separating cats that fight severely or frequently; or refraining from handling a cat that might bite.
- **To stop or prevent worsening of the problem.** Human actions or reactions to a problem behaviour, for example attempts at punishment, may maintain or exacerbate the behaviour. Owner or handler education as to what actions might make the problem worse, and what to do instead, is therefore an important part of behavioural first aid.
- **To help promote improvement.** Behavioural first aid may only achieve temporary or incomplete respite, therefore one of the most important aims is to set the owner on the right path towards greater improvement or resolution of the problem. This should include:
- Advising the owner regarding the importance and necessity of further and more in-depth investigation.
- Further veterinary investigation to rule out or treat underlying or contributing disease (see Chapter 15).
- The need for a detailed and comprehensive behavioural assessment.
    - Arranging further appointments or referral (see Chapter 21).

It is essential that any first aid advice offered is from a reputable source, is up to date and is based on sound scientific principles. The veterinary professional must also be aware of the limitations to the advice that can be offered at this early stage before the possible causes of the behaviour have been sufficiently investigated. Advice that seriously conflicts with that given at a later stage can confuse the client and cause doubt as to what or whom they should believe. Not only might this hinder the treatment process; it can also reduce the client's trust in the veterinary professional or in the behaviourist they may later be referred to.

House-soiling

House-soiling refers to the deposition of urine or faeces within the home in a place that is undesirable for the owner, regardless of the underlying cause or motivation. Even at this early stage, gaining some basic idea of the most likely motivation or possible underlying cause is useful, to ensure that the most appropriate behavioural first aid advice can be offered.

Normal elimination, marking or underlying disease?

Normal micturition or defecation may occur in a location or on a surface that is undesirable for the owner. The cat will usually adopt a squatting position and deposit a 'normal' quantity of urine or faeces on a horizontal surface (Fig. 18.1a). Cats can feel highly vulnerable while eliminating, so they are more likely to do so in a sheltered location or somewhere they feel safe. After eliminating, most cats will attempt to cover their waste by pawing and scraping at the surrounding area.

Marking describes the various ways that olfactory and visual signals are used in intraspecific communication. Cats have various ways of marking, including the use of urine or faeces. When urine marking ('spraying'), a cat will usually adopt a standing position and 'spray' urine backwards on to a vertical surface (Fig. 18.1b). A raised and quivering tail plus treading of the back legs are also commonly associated with spraying. The amount deposited can depend on how much urine the cat has in its bladder at the time, but it is often less than is passed during normal micturition. However, it might be enough to run down a wall and create a puddle on the floor underneath. When this occurs and the owner has not observed the cat 'spraying', it can easily be mistaken as evidence of normal elimination in an inappropriate location.

Cats might also use faeces as a means of communication, a behaviour commonly known as 'middening'. This is rare and indoor defecation in undesirable locations is more likely to be due to other reasons (Atkinson, 2018). When urine or faeces are used as a form of marking, they are usually deposited in a prominent area and there is no attempt by the cat to cover it.

Fig. 18.1. One way to differentiate between (a) normal micturition and (b) urine marking (spraying) is the position that the cat adopts. (Photos courtesy of Celia Haddon.)

These are not mutually exclusive assessments and cats can present with both marking and undesirable elimination behaviour at the same time.

House-soiling can occur as a symptom of an underlying physical condition. Preliminary veterinary examination should therefore be standard practice prior to any behavioural consultation, but if the cat is demonstrating any of the following, this can highlight and triage the need for more urgent or specific veterinary investigation:

- Consistently changing locations. This may indicate that the cat is experiencing pain or discomfort when eliminating or attempting to eliminate.
- Passing variable amounts. This can be a sign that the cat is experiencing urgency or difficulty when eliminating.
- Changes in urine or faecal colour or consistency.
- A sudden breakdown in normally good house-training that cannot be easily attributed to any other cause.

If any of the above apply, the following conditions should be considered and appropriate investigation implemented (see also Chapter 15).

- Upper or lower urinary tract disease.
- Any disease associated with a change in bowel habits or an increase in the volume or frequency of urination.
- Any condition causing pain when attempting to eliminate or when attempting to access a preferred location. For example:
    - Musculoskeletal conditions
    - Lower urinary tract disease
    - Impacted anal glands.
- Age-related changes.
- Cognitive dysfunction.

First aid advice for house-soiling

General

- **Do not attempt punishment**. Attempts at punishment are likely to be ineffective and are more likely to be counterproductive. Attempts at punishment after the act are likely to be ineffective because the cat will be unable to make an association between the owner's attempted 'correction' and the act of eliminating. Even if 'caught in the act' the cat will still be unlikely to make an association with a normal and necessary behaviour, and far more likely to associate the unpleasant or frightening event with the presence of the owner. This can cause the cat to become more secretive in its elimination habits and in turn may deter the cat from using a litter tray if people are nearby.

  Indoor marking behaviour is commonly related to stress. Attempts at punishment can severely increase stress experienced by the cat and so will be more likely to increase rather than decrease marking behaviour.
- **Confine the cat to a restricted area** where the soiling is easier to manage. If this

is not possible, protect the area with an impermeable and easily cleaned covering. However, seriously restricting the amount of space that a cat has access to can be a significant stressor. Therefore, it should be emphasized that restriction to a limited area should only be used as a short-term measure.

- **Effectively clean the area.** It is important to rid the area of the scent of urine or faeces. Cats can be attracted by the scent of residual urine or faeces and are more likely to continue to eliminate in locations where they have previously urinated or defecated. The residual scent of urine deposited via spraying can also encourage the cat to over-mark in the same location or close by.

1. Remove faeces or mop up urine.
2. Test a small area first and then wash with a 10%–20% solution of a biological detergent or use a proprietary enzymatic 'odour elimination' product. Include a margin of at least 2–3 inches (5–8 cm) outside where the urine or faeces was deposited. Avoid the use of bleach or ammonia-based products, which may increase the attraction for the cat to reuse the area for elimination. Also avoid phenolic disinfectants, which are toxic to cats.
3. Rinse if necessary. Pat dry.
4. Wipe or spray over lightly with surgical spirit (again test a small area first).
5. Leave to dry for at least 30 minutes.

It is also important to advise the client that it might not be possible to remove completely the scent of urine that has soaked into soft furnishings, carpets, rugs or other porous floorings (e.g. wood or concrete). In such cases the cat must be denied access to the area, or the surface should be sufficiently well protected until it can be completely cleaned, or the flooring or item removed and replaced. New items must also be well protected until the house-soiling issue has been effectively resolved or managed.

If a carpet needs to be replaced, the underlay should also be removed, the floor underneath cleaned thoroughly, as advised above at least 2–3 times, and then left uncovered for at least a week to allow it to air. When replacing flooring or furniture it is advisable to choose replacements that have a different texture to that which the cat eliminated on previously. This is because cats can develop a preference for toileting on materials of a particular texture. It is also advisable to replace with a flooring or covering that is impermeable and can be easily cleaned. Porous surfaces, for example grouting between tiles, wooden skirting boards, etc., should also be treated with an effective sealant.

If the problem appears to be elimination in an undesirable location unrelated to disease

A common cause of house-soiling is insufficient or incorrect litter tray provision. Therefore, if the house-soiling appears to be unrelated to disease and is not predominately marking behaviour, the next stage should involve questioning and advising the client regarding what, where and how litter trays are provided. The following are the **Seven Litter Tray Rules**.

1. Is the cat provided with a litter tray?
If a cat is eliminating in the home, it *must* be provided with *at least one* indoor litter tray.

If the cat normally eliminates outside and has recently started to urinate or defecate in the house, investigation will be required to discover and address the reason for this behavioural change. One reason can be disease or injury making outdoor access difficult for the cat or making it feel more vulnerable. Conflict with neighbouring cats or other unpleasant experiences can also deter a cat from outdoor elimination. Physical examination, as well as obtaining a full behavioural history, is therefore an important part of this investigation. Until the reason for the behaviour is identified and appropriately treated, it is essential that the cat is provided with access to a suitable indoor elimination area.

2. Are a sufficient number of litter trays provided?
The general guideline is 'one per cat plus one extra' (Carney *et al.*, 2014). It is important to explain to the owner the reasons for this advice:

- Some cats prefer to urinate in one location and defecate in another. Therefore, not only should a sufficient number of litter trays be provided; they must also be positioned sufficiently well apart or with a physical barrier between them.
- In a multi-cat household, some cats can be reluctant to share a litter tray.

- If an insufficient number of litter trays are provided, they will soon become soiled, which can deter the cat(s) from using them.

3. Where are the litter trays located?

Cats will often dislike urinating or defecating near to where they eat, drink or sleep. Also, because they are very vulnerable to attack from predators or rivals when eliminating they will also choose an area where they feel safe.

Litter trays should therefore be located well away from:

- food and water;
- resting areas;
- entrances and exits; and
- anywhere that the cat is likely to be disturbed, for example in corridors or 'high-traffic' areas; near doors, cupboards, etc.; or close to areas used regularly by other animals or children.

It is also important to ensure that litter trays are easy to find, easy to get to and easy to use. This is especially important for:

- new cats or kittens who are not yet familiar with the layout of the house;
- a pregnant female, who during the latter stages of pregnancy will be carrying extra weight, making access less easy, and who might need to use the litter tray more frequently and possibly more urgently; and
- elderly or unwell cats, who might occasionally have urgent need to use the litter tray or who might have difficulty or experience discomfort when climbing stairs or when attempting to get in or out of a high-sided tray. Age-related cognitive decline can also affect the cat's ability to find a litter tray, even if it has been in the same location for a long time.

4. How big are the litter trays?

Most cats prefer a larger litter tray (Guy *et al.*, 2014), which will allow them to adopt the most comfortable position to eliminate. Many cat litter trays are far too small (Fig. 18.2), especially for larger cats such as Maine Coons. It is also not uncommon for owners to keep the same litter trays that they initially purchased when the cat was not yet fully grown. The general guideline regarding litter tray size is that it should be at least 1.5 times the length of the cat, excluding its tail (Carney *et al.* 2014).

Fig. 18.2. It is not uncommon for a litter tray to be too small for a cat to use comfortably. The ideal size is at least 1.5 times the length of the cat, excluding its tail. (Photo courtesy of Celia Haddon.)

5. Are covered or uncovered litter trays provided?

This can be an individual preference by the cat, so the owner may need to experiment to discover what the cat prefers. However, there are potential problems with the use of covered trays that owners should be made aware of.

- Smells can become trapped within a covered tray, which can deter the cat from using it.
- It is less easy to see when a covered tray needs cleaning.
- In a multi-cat household, a cat exiting a covered tray can be more at risk of being ambushed, or trapped inside by another cat.

A covered litter tray can provide a cat with an increased sense of security, but this can also be provided by using a high-sided plastic storage container rather than a completely covered tray (Fig. 18.3).

6a. What litter substrate is being offered?

Cats can develop preferences for particular litter substrates and house-soiling problems may occur if the litter substrate is changed. It is not the make or brand that is important, but the scent and texture. A litter substrate preference may develop when a kitten is very young and persevere throughout the cat's life. Or it might

Fig. 18.3. A plastic storage box can be easily converted for use as a litter tray. The high sides of a deep storage box can provide a cat with an increased sense of security without the problems that often accompany the use of a completely covered litter tray. (Photo courtesy of Heather Alvey.)

develop later in life, especially if an aversion to a previously preferred substrate has developed.

If the cat is new to the household, or if for any other reason the owner does not know the cat's preferred substrate, it is best to try a granular, 'clumping' litter as this is the type most commonly preferred by pet cats (Villeneuve-Beugnet and Beugnet, 2018).

6b. How deep is the litter substrate?

As a part of normal elimination most cats require enough litter in the tray to enable them to dig a hole into which they will deposit urine or faeces and then cover it. If there is insufficient depth of litter to enable them to do this they may then choose to eliminate elsewhere. Also, a small amount of litter substrate will very soon become soiled, making it unattractive or even impossible for the cat to reuse the tray until it is cleaned and refreshed.

The optimum depth of litter within a tray depends on the size of the cat and can also vary with individual preference, but an approximate depth of 4 cm (1.5 inches) is suitable for most cats.

6c. How clean is the litter substrate?

Cats can vary as to how fastidious they are regarding the cleanliness of their litter trays. Some may continue using a litter tray that is already heavily soiled. Others will refuse to reuse a litter tray that has only been used once. For such cats, aiming to keep the litter substrate as clean as possible is therefore essential.

Ideally, faeces and urine clumps should be removed as soon as they appear. If this is not possible the trays should be 'scooped' at least twice a day, and fresh litter added. Litter trays should be completely emptied, cleaned with plain soap or detergent and warm water, rinsed well and the litter refreshed approximately once every 5–10 days.

7. Are any of the following being used?

These are things that can deter a cat from using a litter tray.

- **Scented litter or air fresheners located close to the litter tray.** Cats have a highly sensitive sense of smell. Scents that we may regard as pleasant can be overpowering and unpleasant to a cat.
- **Plastic litter tray liners.** When digging into litter or scraping to cover waste a cat may catch its claws on a liner, which can be unpleasant, even frightening, for the cat. If the cat has any form of degenerative joint disease, it might also experience pain or serious discomfort.
- **Self-cleaning litter trays and trays with an internal grid filter.** Although these can appear to be a good idea, there are potential problems associated with the use of these litter trays:
  - The mechanism or grid usually deposits the waste in a compartment beneath or close to the litter tray area used by the cat. This may be sufficient to remove the odour of soiling for us, but the odour might still be strong enough to deter the cat.
  - Although the mechanism of automatic self-clean litter trays should not activate while the cat is using the tray, it could activate when the cat is nearby. If this should happen it can be frightening and aversive for the cat.
  - If a litter tray has an internal grid filter, this can present the same problem as a plastic liner in that the cat may catch its claws in the holes of the grid, which may deter the cat from reusing the tray.

If the house-soiling appears to be due to urine marking (spraying)

Is the cat neutered? Urine marking is hormonally influenced in entire cats, both male and female. Most sexually mature males will urine mark fairly frequently especially when aware of a female in oestrus. Entire females may also urine mark when in oestrus (Bradshaw *et al.*, 2012). Neutering can prevent or reduce the incidence of urine marking by 90–95% (Hart and Cooper, 1984).

Indoor urine marking by neutered cats appears to be predominately 'stress related' (Pryor *et al.*, 2001). In many cases a full behaviour consultation will be necessary to correctly identify and address specific sources of stress. However, as a first aid approach it can be helpful to address common stressors, as follows.

- Consider and investigate any possible underlying pain or discomfort. Ill health can be a major stressor and therefore a possible underlying cause of marking behaviour. Physical examination and investigation to rule this out or to treat underlying disease, pain or discomfort is therefore important.
- Reduce any real or perceived threats from neighbouring cats.
- If the cat has outdoor access, install an exclusive, preferably microchip-activated, cat flap. If other cats are able to enter the cat's core territory, this can be a common stressor and underlying cause of urine marking.
- Seeing neighbouring cats through windows, and more especially through glass doors, can also be a major stressor resulting in urine marking near or within sight of the door or window. Covering the lower one-third to one-half of glass doors or windows through which other cats may be seen can effectively increase the cat's sense of security and thereby help to reduce urine marking. A temporary stick-on/peel-off window frosting can be used for this purpose so that the incoming daylight is not reduced for the human members of the household
- Increase space, especially elevated space and hiding places for the cat.
- It is important that the cat feels able to get away from, or get up higher than, any real or perceived threats.
- Ensure that the cat always has access to a safe place to retreat to away from possible stressors such as loud or disturbing noises, visitors, other animals, or children.
- Increase resources.
    - Real or perceived inadequacy of resources can be a major cause of competition and conflict between cats within a multi cat household, which can be a common reason for urine marking.
    - Resources to be increased should include:
        - Food and water locations
        - Litter trays
        - Warm, safe resting areas.
- Increase enrichment.
- Provide cardboard boxes for the cat to play and/or hide in (Fig. 18.4c).
- Engage the cat in daily interactive play, for example using wand toys (Fig. 18.4a).
- Provide puzzle feeders and/or opportunities for food foraging (Fig. 18.4b).
- Respect the cat's need for control and predictability in its relationship with human family members.
    - Allow the cat to sleep undisturbed.
    - Do not start or continue to stroke or restrain the cat if it shows any of following:
        - Moving away or attempting to move away.
        - Swiping at a person or pushing them away with a paw.
        - Hissing or growling.
- Never shout at or attempt to physically punish the cat for 'bad behaviour'. Clients should be advised how this might exacerbate the unwanted behaviour as well as maintaining or increasing urine marking. Encourage them to discuss any other unwanted behaviours with you so that these issues can also be addressed appropriately.
- Avoid major disruptions to the cat's normal routine. If routines have been changed recently, is it possible for the owners to revert back to the cat's previous routine, even if temporarily until other measures can be implemented?

The use of pheromonatherapy for house-soiling

For marking behaviour: use of a synthetic feline pheromone (e.g. *Feliway*™ *Optimum* Ceva) may help to reduce marking alongside behaviour therapy. But this is less likely to be effective if used in isolation without also correctly identifying and addressing the underlying causes of the behaviour.

For inappropriate elimination: installing a plug-in diffuser close to a litter tray can sometimes help to encourage litter tray use by making the location feel more secure for the cat. However, it

Fig 18.4. Increased enrichment can help to promote positive emotions and motivations which can assist in managing some problem behaviours. (a) Image: iStock; (b) image: Celia Haddon; (c) image: iStock.

is also essential that all other possible reasons why the cat is not using the litter tray have been identified and correctly addressed.

Furniture Scratching

Scratching can be a problem for owners when it causes damage to surfaces or items valued by the owner. It is important to understand and to be considerate of the owner's concerns. It is also important that owners are made aware that scratching is a normal feline behaviour that serves essential functions for the cat's health and welfare.

- Claw sharpening/conditioning: scratching is one way that the worn outer sheaths of the claws on the forefeet are removed, exposing new sharper claws underneath. The outer sheaths from both the front and back feet are also removed during self-grooming. However, a cat may have more need to ensure that the claws on the front feet are kept well maintained because of their greater use and significance in hunting, climbing and defence.
- Marking behaviour: scratching plays an important role in visual and olfactory signalling by leaving visually evident scratches and scent marks from glands between the toes and on the pads.
- As a part of stretching: especially just after waking from sleep, stretching usually involves extension of the claws which is often followed by scratching behaviour.

- Performed at times of increased arousal: scratching is sometimes observed during bouts of play and increased excitable arousal. Owners often report that their cats scratch when they return home or when owners greet the cat first thing in the morning. It is also not uncommon for cats with outdoor access to scratch when they return home. A sense of insecurity can also be a cause of increased scratching within the home environment (Heath, 2007).

First aid advice for furniture scratching

- Advise against attempted punishment. Attempts at direct punishment, for example shouting, squirting with water, etc., can be counter-productive and such actions are also very likely to be damaging to the cat's welfare and to the pet–owner relationship.
- Provide an alternative and attractive scratching location. Cats will scratch on both horizontal and vertical surfaces. Some cats have a definite preference, while others will scratch equally on both. Scratch pads can be provided for cats that prefer horizontal scratching and scratch posts for those that prefer to scratch on a vertical surface. For most cats, provision of both is advisable.

Scratch posts and scratch pads: what to look for and what to avoid

- A scratch post must be tall enough and a scratch pad long and large enough to allow the cat to stretch as it scratches. A minimum height of 60 cm is advised for scratch posts designed for adult cats, although large or long-bodied cats are likely to require a much taller post of at least 90 cm.
- Avoid posts with toys or platforms on top that can hinder stretching and scratching.
- A scratch post must be sturdy and stable enough so that it does not move when the cat leans into it to scratch. If a scratch pad is attached to either a vertical or horizontal surface it also needs to be attached sufficiently firmly to prevent it from moving when the cat uses it.
- The surface of a scratch post or pad must be made of a material that allows the cat to leave visible scratch marks and provide enough resistance to remove the dead outer claw sheaths. However, older cats and those suffering from degenerative joint disease may find it uncomfortable to scratch on a surface that is too rough or resistant. Sisal rope can be a suitable and popular material for most young, healthy cats, whereas older cats are often more likely to prefer a softer material such as carpet. Corrugated cardboard can also be a popular choice for horizontal scratch pads.

Preparing a scratch post or pad

A cat will be more likely to use a scratch pad or post if it contains some evidence, preferably visual and olfactory, of having been scratched previously.

- Visible scratches that mimic the marks left by claws can be applied by using the tip of a screw or something similar.
- Gently rubbing the underneath of the cat's front feet with a clean, dry cloth and then rubbing the cloth over the new post or pad is one way that the cat's own scent can be transferred to the post or pad. However, not all cats will accept having their feet handled this way, so another effective means of transferring scent is simply to rub the new post or pad over the areas where the cat has been scratching most recently.

Locating the scratch pad or post

A scratch post or pad should be positioned in the same area or as close as possible to where the cat is currently scratching, even if this area is not convenient to the owner. Once the cat is using the post or pad reliably, the post or pad may then be very slowly moved to a more suitable location. However, if the location is the most important factor for the cat, it is advisable to keep the post or pad as near as possible to the area where the cat is currently scratching.

How many scratch posts or pads to provide

Cats often scratch in more than one location. Posts and pads should therefore be provided in each place where the cat regularly scratches.

Protecting the furniture and applying effective deterrents

Deterrents should only be put in place once a suitable alternative scratching area has been provided. Preventing or deterring the cat from scratching the furniture without doing so will only move the behaviour to another location that could also be undesirable for the owner.

The aim must be to protect the furniture and deter the cat by making the area less attractive for the cat to scratch, not to cause pain, discomfort or fear, all of which could be counter-productive by increasing the cat's distress and arousal, which can have the effect of increasing scratching behaviour.

- Drape a loose throw or similar over the furniture. It can be very difficult for a cat to scratch on something that is hanging loosely.
- Plastic film can act as a deterrent if wrapped around a chair leg or wherever the cat is scratching.
- If the cat's primary reason for scratching is to help 'condition' the claws, clipping the nails may help to reduce furniture scratching and the damage caused.
- If a cat cannot be deterred from scratching an item of furniture, shields that hang over or stick to the sides of furniture or corner pieces that can be attached to the arms of chairs are available that allow the cat to continue scratching whilst also providing protection from damage (Fig. 18.5).

Fig. 18.5. If a cat cannot be deterred from scratching an item of furniture, a shield can be placed over the item that allows the cat to continue to scratch, whilst also preventing damage. (Photos courtesy of Katie Spinks.)

When to seek further help

A full behavioural consultation and possible referral will be required if the presenting problem is not controlled by the above advice or if the scratching behaviour is excessive, indicating possible underlying stressors that need to be identified and correctly addressed.

Cat Conflict

Cats that live together will not automatically regard each other as being members of the same social group. Fighting between cats that live in the same household, or other signs that they are not getting along, such as chasing, avoidance, or indoor urine marking, are not uncommon. This is largely due to the ancestry and normal social behaviour of the domestic cat as a self-reliant predator (see Chapter 3 for more detail).

Other individual factors that can influence how well cats are likely to get along with other cats can include the following.

- The age of both when introduced. The younger the better (Bradshaw and Hall, 1999).
- Early experience (prior to 9 weeks of age) with other cats. Positive early interactions with adult cats other than the mother cat can increase the likelihood that the kittens will be more tolerant of other cats as adults (Kessler and Turner, 1999).
- Previous experience of living with other cats. Positive previous experience may also increase the chances of being able to share a core territory with another cat. However, if a cat is grieving following the loss of a close feline companion, it may be less able to cope with the introduction of another cat or kitten

Fighting or playing?

Social play and fighting between cats can appear very similar and it is not uncommon for owners to misinterpret one for the other. Asking the following questions can help to identify if cats are fighting or playing.

- Do the cats make any sound? Other than an occasional chirrup vocalization, play is normally silent. Fighting usually involves vocalizations such as growling or high-pitched defensive cries.
- Does the interaction appear to be one-sided or do the cats 'swap roles'? Alternating roles is more often seen during play. If it appears that one cat is predominantly victimizing the other, this is more likely to be aggression, or unwanted play directed towards a cat that may feel intimidated by the attentions of the other.
- Following the interaction, do the cats want to stay close together, or do they try to maintain a distance from each other? Cats that play together and enjoy each other's company will usually remain together after a play session. Cats that fight will usually try to avoid each other.

If the cats are fighting, first aid behavioural advice is as follows.

- Do not attempt punishment. If a cat experiences an unpleasant event it will most likely associate that event with its main focus of attention at the time. If the cat's focus of attention is towards another cat the unpleasant event will therefore be associated with that other cat. The effect of this can be to increase rather than decrease the cat's perception of the other cat as a threat and in turn increase rather than decrease defensive and aggressive behaviour.
- Consider and investigate possible underlying medical issues. If there has been a recent breakdown in a previously good relationship between cats living in the same household it is possible that a disease process causing pain or discomfort could be a contributory or causative factor.

If fighting is frequent or severe:

- Keep the cats in separate rooms or areas of the house until a full behavioural consultation can be conducted.

If fighting is infrequent or mild:

- Ensure that the cats have easy and always available escape routes away from each other.
- Ensure that the victim has easy and always available access to safe places, such as:
  - Places to hide, for example under furniture, cardboard boxes, etc. (making sure that there is more than one entrance/exit so that the aggressor cannot trap the victim).
  - Elevated areas: tops of furniture, cat trees, wall-mounted shelves and cat beds, etc.
  - Access to a room that the aggressor cannot access or is less likely to frequent.

Watch for signs of impending fighting or chasing:

- Staring, either one at the other or both at each other
- Stalking
- Growling
- Dilated pupils
- Swishing tail

If fighting or chasing appears imminent place a physical barrier, for example a large cushion or flattened cardboard box, between the cats to block their view of each other. Then try to redirect the aggressor away with a toy or with dry food scattered on the ground a good distance away from the other cat. This should allow time for the victim to move away.

- Are the owners aware of any trigger factors? Are the fights more likely to occur at certain times or in certain places, or when the owners are doing anything in particular? For example, fights can sometimes occur during food preparation. If this is the case, separating the cats prior to preparing their food can help to avoid fighting.
- Try to reduce competition between the cats by increasing resources or ensuring that access to important resources is ample and in safe locations.
- Feed the cats in separate areas, preferably separate rooms. If this is not possible, make use of different levels.
- Provide extra feeding dishes or puzzle feeders, plus water dishes located away from the food dishes.
- Provide sufficient elimination areas (one per cat, plus one).
- Ensure a good choice of safe, warm and comfortable resting places.

The use of pheromonatherapy in cases of cat conflict

Use of a diffuser containing synthetic feline appeasement pheromone (e.g. *Feliway™ Friends* or *Feliway™ Optimum* Ceva), positioned in an area where all the cats are most likely to rest at the same time, may help to reduce tension and conflict between them. This is less likely to have a significant effect unless all other factors influencing the relationship between the cats have been fully investigated, correctly identified and addressed.

Human-directed Aggression

Cat bites and scratches can be very painful and carry a risk of infection with potentially serious complications (Oehler *et al.*, 2009). Because of the potential risks, full assessment and behaviour consultation for such cases should therefore only be conducted by a fully qualified and experienced clinical behaviourist. However, providing immediate and correct behavioural first aid advice can be of great importance to help prevent injury or further injury to the owner or other people.

The owner should be made aware of potential warning signs. A cat that is about to become aggressive may demonstrate any one or a combination of the following:

- Dilated pupils
- Ears rotated back and/or flattened to the side
- 'Swishing' tail
- Growling or hissing
- Piloerection (hair standing on end).

If the cat has a history of aggression, or is known to act aggressively, the owner should be given the following advice whenever they notice any of the above signs.

- **Do not** approach or attempt to handle the cat, even if the aggression does not appear to be directed towards you.
- **Do not** stare at the cat. Avoid direct eye contact.
- If you are already near the cat, try to move away very slowly. Avoid fast or erratic movements.
- **Never** put your face close to the cat.
- Talking gently may help to calm the cat but avoid making any sounds that might frighten the cat or cause it to react aggressively, e.g.:
  - Shouting
  - Hissing
  - 'Shushing'
  - Clapping.
- Be careful not to intentionally or unintentionally block the cat's escape route away from you or from anything it might be fearful of.
- If you can, place a physical barrier between yourself and the cat.
- If the cat bites or holds on to you with its claws, try to keep still. If teeth or claws have broken the skin, movement may cause tearing and so increase injury. Movement, especially fast movement, can also trigger further defensive aggression or a predatory reaction. When the cat releases its grip, move away slowly.

Advice should also be offered as to how the owner may avoid potential situations that might result in the cat becoming aggressive.

- **Never** attempt to punish, goad or threaten the cat.

If the owner has concerns about other aspects of the cat's behaviour, advise them that professional help is available to help them to address these concerns in a way that will not increase the risk that the cat may become aggressive.

- **Do not** encourage the cat to play with hands or feet (Fig. 18.6). Only engage the cat in play using toys that keep teeth and claws well away from human hands.

Misdirected predatory behaviour is common, especially in young cats, and can often result in injury to people, especially the cat's owners. This behaviour can develop or be reinforced when the cat is encouraged to play with and 'attack' human hands or feet. The amount of force that is applied with each bite or scratch can vary and can increase as a kitten gets bigger and stronger, causing pain and defensive reactions by the 'victim', such as yelling, erratic movement and attempts to push or kick the cat away. All of which are likely to exacerbate the cat's behaviour. Attempts may also be made to directly 'punish' the behaviour by, for example, shouting at the cat, grabbing it by the scruff or squirting it with water. These reactions can cause the cat to feel fearful and defensive but might not discourage the cat's strong and instinctive need to engage in predatory behaviour, resulting in the cat demonstrating a combination of misdirected predatory/play behaviour and defensive aggression at the same time.

Fig. 18.6. Young cats and kittens should never be encouraged to play with human hands or feet, as this can often lead to misdirected predatory behaviour and even long-term human-directed aggression. (Photo courtesy of Lucy Hoile.)

- Are the owners aware of any triggers for the aggression?
    - Is the cat more likely to become aggressive at certain times, in certain contexts or when the owner performs certain actions? If so, are they able to avoid or alter these triggers at least until professional help is available? Common trigger actions can include attempting to pick the cat up, petting the cat, attempting to groom the cat, or attempting to get the cat into a carrier.
- Owners should also be made aware of the importance of seeking medical advice as soon as possible if badly bitten or scratched.

Veterinary investigation in cases of human-directed aggression

Disease processes can influence temperament and a cat may become aggressive if it experiences or anticipates pain when touched. Physical examination and further investigation as indicated are therefore essential prior to or alongside behavioural investigation.

Over-grooming

Self-oral grooming (using the tongue and teeth on the coat and skin) is a normal feline behaviour. Over-grooming is when the frequency or intensity of self-oral grooming increases sufficiently to cause hair loss and possible injury to the underlying skin. Regular regurgitation of hairballs can also be a sign of over-grooming.

Owners may not always notice when their cat is over-grooming. Distinguishing between over-grooming and normal grooming is not easy by observation alone, as both involve licking, nibbling, and plucking of the hair and skin with the teeth. The problem might only come to light when patches of thinning hair or bald patches are discovered on examination. However, not all hair loss is due to over-grooming.

The feel of the cat's skin can provide some idea if the hair loss is spontaneous or self-inflicted. Hair that has fallen out will usually leave the skin feeling smooth to the touch, whereas over-grooming will normally leave the skin feeling slightly 'stubbly' from where the hair has been bitten off close to the root and from hair regrowth. Microscopic examination of hairs plucked from the area will also reveal angular and ragged distal tips (Griffiths, 2016).

The majority of cases of feline over-grooming have an underlying medical cause (Waisglass *et al.*, 2006; Hobi *et al.*, 2011). Veterinary investigation is therefore essential. However, stress can also be a factor that can trigger episodes, maintain the behaviour or exacerbate an underlying disease. Therefore, a full behavioural assessment may also be required to identify and address potential stressors.

Common medical causes of over-grooming

- **Pruritus**. Parasitic, fungal, or bacterial infections causing skin irritation can cause a cat to groom excessively. An 'itch–scratch' cycle may also develop whereby grooming to relieve an initial irritation causes damage and increased irritation.
- **Underlying pain or discomfort**. A cat may groom excessively over an area where it experiences pain or discomfort. Musculoskeletal disease and neuropathic pain should therefore also be considered and investigated, as well as possible visceral pain. For example, lower urinary tract disease may cause a cat to over-groom the ventrocaudal abdomen and medial aspects of the hind legs (Fig. 18.7).

First aid behavioural advice for over-grooming

As well as explaining the importance of veterinary investigation, first aid advice in cases of over-grooming should aim to help the owner to recognize potential stressors and to avoid

Fig. 18.7. Although psychological stressors may trigger, maintain or exacerbate the behaviour, most cases of feline over-grooming have an underlying medical cause. Veterinary investigation is therefore essential. (Photo courtesy of Celia Haddon.)

actions that might exacerbate any stress experienced by the cat.

- Do not try to punish or physically prevent the cat from grooming. Distraction may be attempted, for example by calling the cat's name or by using a moving toy. But because the behaviour is self-rewarding, it might not be possible to stop or prevent the cat from self-grooming and owners should be advised that attempting to do so might even increase the behaviour by causing additional distress or frustration or by inadvertently reinforcing the behaviour.
- Physical barriers to self-grooming, such as Elizabethan collars or bandaging the area, can help injured areas of skin to heal and will allow the hair to grow back. However, this might also cause the cat to experience increased distress and if the source of the over-grooming has not been effectively addressed the over-grooming behaviour is likely to return with increased vigour when the physical barrier is removed.
- To help identify possible trigger factors, advise the owner to keep a diary noting when over-grooming occurs and if episodes occur at the same time or are preceded by potential stressors.
- If the cat lives in a multi-cat household, address any issues of cat conflict as advised earlier in this chapter.
- The stressors that might trigger or exacerbate over-grooming can also be the same as those possibly underlying indoor urine marking. Following the same advice as described earlier in this chapter to reduce stressors related to urine marking may therefore also be relevant and helpful in cases of over-grooming.

References

Atkinson, T. (2018) *Practical Feline Behaviour: Understanding Cat Behaviour and Improving Welfare*. CAB International, Wallingford, UK.

Bradshaw, J.W. and Hall, S.L. (1999) Affiliative behaviour of related and unrelated pairs of cats in catteries: a preliminary report. *Applied Animal Behaviour Science* 63(3), 251–255.

Bradshaw, J.W.S., Casey, R.A. and Brown, S.L. (2012) *The Behaviour of the Domestic Cat*, 2nd edn. CAB International, Wallingford, UK.

Carney, H.C., Sadek, T.P., Curtis, T.M., Halls, V., Heath, S. *et al.* (2014) AAFP and ISFM guidelines for diagnosing and solving house-soiling behavior in cats. *Journal of Feline Medicine and Surgery* 16(7), 579–598.

Griffiths, K. (2016) How to approach overgrooming in cats. *Veterinary Focus* 26(2), 32–39.

Guy, N.C., Hopson, M. and Vanderstichel, R. (2014) Litterbox size preference in domestic cats (*Felis catus*). *Journal of Veterinary Behavior: Clinical Applications and Research* 9(2), 78–82.

Hart, B.L. and Cooper, L. (1984) Factors relating to urine spraying and fighting in prepubertally gonadectomized cats. *Journal of the American Veterinary Medical Association* 184(10), 1255–1258.

Heath, S. (2007) Behaviour problems and welfare. In: Rochlitz, I. (ed.). *The Welfare of Cats*. Springer, Dordrecht, pp. 91–118.

Hobi, S., Linek, M., Marignac, G., Olivry, T., Beco, L. *et al.* (2011) Clinical characteristics and causes of pruritus in cats: a multicentre study on feline hypersensitivity-associated dermatoses. *Veterinary Dermatology* 22(5), 406–413.

Kessler, M.R. and Turner, D.C. (1999) Socialization and stress in cats (*Felis silvestris catus*) housed singly and in groups in animal shelters. *Animal Welfare* 8(1), 15–26.

Oehler, R.L., Velez, A.P., Mizrachi, M., Lamarche, J. and Gompf, S. (2009) Bite-related and septic syndromes caused by cats and dogs. *The Lancet Infectious Diseases* 9(7), 439–447.

Pryor, P.A., Hart, B.L., Bain, M.J. and Cliff, K.D. (2001) Causes of urine marking in cats and effects of environmental management on frequency of marking. *Journal of the American Veterinary Medical Association* 219(12), 1709–1713.

Villeneuve-Beugnet, V. and Beugnet, F. (2018) Field assessment of cats' litter box substrate preferences. *Journal of Veterinary Behavior* 25, 65–70.

Waisglass, S.E., Landsberg, G.M., Yager, J.A. and Hall, J.A. (2006) Underlying medical conditions in cats with presumptive psychogenic alopecia. *Journal of the American Veterinary Medical Association* 228(11), 1705–1709.

19 First Aid Advice for Common Behavioural Signs: Rabbits and Rodents

Emma Magnus and Anne McBride

Many behaviour problems commonly seen in rabbits and rodents can be prevented relatively easily, as discussed in Chapters 4 and 10. This requires species-specific knowledge regarding normal communication signals and how to meet welfare needs. This includes ensuring that the environment is safe, provides adequate space for normal movement, slow and fast, and has appropriate cage furniture, to provide for resting, climbing, hiding and digging. Further, the animal must be provided with: a varied, appropriate diet; toys and other manipulanda to encourage physical and mental stimulation; and suitable compatible cage-mates and, if kept alone, regular and frequent contact with trusted humans (see Box 10.1 in Chapter 10).

The checklist below highlights key areas of husbandry that can help prevent or address problem behaviour. Each animal should have the following.

- Refuges to retreat to if it is scared or wants to be left alone.
- An environment free from predators, unusual or loud noises, strong smells and bright lights.
- The opportunity to dig, work to gain food, or undertake other species-specific activities.
- The opportunity to mark its territory effectively with body secretions, urine and faecal pellets.
- Correct handling so that it feels secure and relaxed when picked up and interacted with by people.
- An appropriate diet that encourages normal digestive function, dental abrasion and prolonged feeding time.
- The opportunity to engage in regular and frequent physical and mental exercise and appropriate social contact with another individual; ideally at times of the animal's own choosing.
- Freedom from reprimands or punishment from people.

Both behavioural and physical problems can arise when these conditions are not met. This is not only the responsibility of the owner. How animals are handled and managed in veterinary practice, even on routine visits, can exacerbate current behaviour problems or be causal in the development of others (McBride, 2017a,b).

As in previous chapters, the general advice given here applies to captive small mammals kept for any purpose, be they rabbits, rodents, or from other orders such as the insectivorous Pigmy hedgehogs or carnivorous ferrets.

Pain, Stress and Behaviour

Underlying physiological or pathological changes can lead to alterations in behaviour,

 Companion Animal Behaviour Problems: Prevention and Management of Behaviour Problems in Veterinary Practice (eds. R. Casey, S. Heath and H. Zulch)
DOI: 10.1079/9781800621312.0019

which may be noticed before any clinical signs are identified. Pain can be indicated by various behaviours, including changes in facial expression (Chapter 4), restlessness, apathy, irritability and aggression (Whittaker and Howarth, 2014; NC3Rs, n.d.). It is important to heed the owner's descriptions of the animal's behaviour at home as part of any examination, because the need for vigilance means that these animals mask their symptoms if they detect a threat. So even painful conditions may not be detected easily in the practice environment, and X-rays may be needed to discover an underlying fracture, arthritis or dental problem (Meredith and Delany, 2010).

In the main, behaviour problems are associated with aversive emotions of anxiety, fear and frustration. Box 19.1 gives generic descriptions of behaviours indicative of these emotions. Readers should research species-specific details (Yeates, 2018; McBride and Hinde, 2022). Suggested sources are given at the end of Chapters 4 and 10 and major UK animal charities have freely downloadable advice leaflets.

The term stress is often used non-specifically, but here it is used analogously with 'distress' and describes situations in which environmental conditions are having an adverse effect on an individual. Stress is therefore an environment-related state in which the animal finds itself, which will have various effects on its physiology and emotions, the latter being expressed through behaviour (Broom and Fraser, 2015). In this chapter, only psychological distress is considered unless otherwise stated.

Environmental factors that cause stress are called *stressors*, to which the individual

Box 19.1. Behavioural indicators of fear, anxiety and frustration in rabbits and rodents.

Fear-related behaviour. The first response of a small prey animal to a fear-inducing stimulus is to freeze and crouch low to the ground to avoid detection. This is usually associated with a decreased heart rate and rapid breathing.

If the stressor does not go away, this will be followed by trying to hide or flee. If this is not possible, they will use aggression to defend themselves.

Occasionally displacement activities are seen in the presence of a fear-provoking stimulus, which may be redirected escape attempts, such as chewing cage bars.

Anxiety-related behaviour. Anxiety can last longer than fear and is usually associated with the anticipation of an unpleasant event or interaction, or exposure to an unknown (novel) situation.

Behavioural signs in small animals include jumpiness (reactivity), irritability, hiding, frequent urination and defecation, increased or decreased eating or drinking. Chronically or extremely anxious animals may show apathy, appearing relatively unresponsive or lethargic.

Adult females and youngsters may show increased apathy and decreased social behaviours associated with sudden forced weaning, which is likely to be associated with an increased level of anxiety.

Frustration-related behaviour is due to lack of opportunity, or of sufficient opportunity, to show normal behaviour. Examples of frustration behaviours include: changes in maintenance behaviours; over- or under-eating/drinking; lack of grooming; over-grooming of self or a cage mate (barbering); self-directed injurious behaviour, for example biting or gnawing own body; and aggression to cage mates and other animals.

Chronic frustration can lead to apathy, learned helplessness and depression. This is common in animals kept in physically barren environments and can be exacerbated in social species if there is no or inappropriate social contact.

Anxiety and frustration can also be displayed through repetitive behaviours. Where these interfere with the animal's normal behaviour they are considered as 'excessive' and 'stereotypical', by which stage the animal likely is in a state of very chronic stress.

Repetitive behaviours include: repeated digging in cage corners and bar-biting for minutes or hours at a time; and persistent jumping in loops off cage walls, as is seen in chronically frustrated chinchillas (Würbel, 2006). It has been shown that rats and mice in barren environments containing only a wheel for enrichment can show repetitive wheel running even at the expense of eating properly (Sherwin, 1988). These repetitive behaviours occur even if the animals are socially housed, suggesting that lack of resources or incompatible social groupings lead to frustration and anxiety in animals.

shows *stress responses*. Stressors lead to acute (of short duration) or chronic states of stress and accompanying aversive emotions of anxiety, fear or frustration. Particularly stressful are situations where individuals lack control within their environment or are unable to predict events (Broom and Johnson, 1994), such as having their cage moved across a room (Dallman *et al.*, 2006). Lack of control includes not being able to avoid or escape from potentially dangerous stimuli and thus avoid anxiety and fear. There are many stressors in captivity; some examples are given in Box 19.2. Considering causes of stress is of paramount importance in behaviour work.

Some circumstances, such as veterinary examination and treatment, will inevitably have associated stressors (Fig. 19.1). These can be minimized with appropriate clinical facilities and clinical management, as summarized in Table 19.1 (Varga, 2014; McBride, 2017a,b; Rendle and Hinde, 2022).

Responses to stress reflect species-typical behaviours, though there will be some individual variation. Influences on how an individual will respond include: genetic predisposition; the individual's susceptibility (its physical and psychological health); age; and previous experience – what it has learnt about the stressor.

Depending on the actual circumstances and the individual animal, the same stressor may affect individuals quite differently. For example, social stress occurs in groups housed with limited resources, for example too few places to retreat, eat or drink. Some individuals may find this highly frustrating and respond with increased levels of aggression towards their cage mates. Others may also find it frustrating, but this is accompanied and superseded by

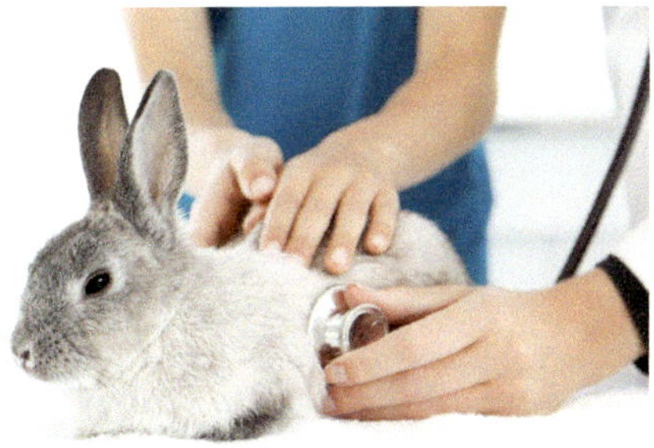

Fig. 19.1. A well-handled, relaxed rabbit at the veterinary surgery. (Image: iStock.)

Box 19.2. Common sources of stress for rabbits and rodents leading to aversive emotions of anxiety, fear or frustration.

A stressor can cause different emotions, depending on the individual and circumstances. For example, the scent of unfamiliar animals may cause anxiety, or feelings of frustration at not being able to reach them, especially if they may be a potential rival or a potential mate.

- Novelty – examples include the first trip in a car, the first visit to a show, handling by a 'stranger', new smells or objects in the environment.
- Stimuli causing an automatic or species-specific fear response – examples include sudden noises or lights, rapid movement, scent of predators or unknown animals or people, or poor handling.
- Social stress – examples include a lack of social contact or interactions with many individuals in a limited space (overcrowding), or housing with incompatible others.
- Inability to perform normal behaviour patterns – examples include a lack of social contact, lack of physical or mental exercise opportunities or an inability to retreat from a stressor.
- Pain, discomfort or illness.
- Anticipation of pain or discomfort – examples include poor or excessive handling. This is a classically learnt response.
- Inability to control environmental factors – examples include poor ventilation, travelling in a car on a hot day, having a brightly lit environment, exposure to loud sonic or ultrasonic noise.
- Lack of space – this may be for a 'short' period, such as confinement in a show pen or travel cage; or may be a chronic condition of the animal's living accommodation, such as hutches and indoor cages that are too small.
- Unavailable food or water – for example when travelling, or if the animal lives outside and water in bottles or bowls has frozen.

Table 19.1. Summary of key points to improve small prey mammal (SPM) behavioural welfare in veterinary practice. (From McBride (2017a).)

	Action
1	Have a separate part of the waiting room for SPM, to reduce visual and auditory disturbance from predators.
2	Have a separate recovery room for SPM.
3	Have low light in recovery rooms, and test for and minimize ultrasound and sound levels, including baffling on doors to promote quiet closure.
4	Provide suitable cage furniture to offer retreat from lights and ultrasound in both recovery and transport cages.
5	Provide shelters, lookout places and light-proof nest boxes with access via a dark tunnel.
6	Avoid transmission of scents from predators into areas where SPM are held and before interacting with SPM.
7	Avoid transmission of scent between non-group members, especially in territorial species.
8	Leave some used bedding in cleaned cages.
9	Advise owners to bring all cage mates to all visits to the surgery.
10	Ensure group scent profiles remain similar should any animal need to be temporarily removed from the group.
11	Consider effects of stress on masking behavioural signs of ill-health.
12	Gently tap cages before moving and carry them steadily. Wherever possible, allow recovery time after moving of at least an hour before examination or treatment.
13	Ask owners about the animal's normal behaviour, activity cycles, cage and wider environment, management and how handled, and changes in any of these aspects.
14	Ask owners if patient has been trained to any cues, including relating to handling and entering carry cages. These should be used if the animal needs hospitalizing.
15	When holding SPM ensure that the body weight is supported and the head remains upright, to prevent perceived loss of support.
16	Where appropriate to the individual's health status in hospital, provide free access to enough space to facilitate the full locomotor repertoire, including climbing and digging opportunities. Ensure that surfaces are not slippery.
17	Provide variety in a species-suitable diet to encourage eating and natural movement.
18	Ensure that all staff are knowledgeable in SPM behaviour and Facial Grimace Scales.
19	Ensure that all staff are knowledgeable in SPM management and appropriate handling, and individual particulars such as cue words used by the owner.
20	Advise all current SPM owners how to meet their animal's welfare needs: its diet; a suitable environment; how to keep it physically healthy and mentally active and happy; and how to detect if it is unwell or in pain, including normal behaviour and Facial Grimace Scales.
21	Encourage practice participation in outreach education to current and potential owners (including children), breeders, rescue and pet shop staff regarding SPM ethology, behaviour, Facial Grimace Scales, and how to meet the animals' physical and behavioural welfare needs.
22	Design in-house or use information leaflets regarding management available from credible sources such as the RSPCA, Rabbit Welfare Fund and freely distribute to all SPM owners; perhaps as species-specific packs.
23	Encourage companies to supply better designed, species-appropriate housing to meet welfare needs.
24	Encourage all companies to use images in their publicity material of breeds with natural conformations (head shape, upright ears, and appropriate fur length) in suitable environments. In this way welfare can be indirectly improved by influencing owner perception, attitudes and expectations.

anxiety. These are likely to be victims of the aggression, showing fear when threatened, but also becoming depressed, hiding a lot and losing condition. Thus, it is important to regard each behaviour case as unique (McBride and Hinde, 2022).

Raising Awareness of Behaviour Problems

Whilst many animals in this group have issues that would benefit from both veterinary and behaviour intervention, they are not frequently presented to veterinary or behaviour practitioners. Often owners are not aware that help is available or do not realize that their pet is suffering. This lack of awareness means many pets may suffer, or are re-homed, euthanized or simply 'effectively abandoned'.

Thus, it is important that owners are advised of behaviour signs that indicate that their animal is distressed or in pain (NC3Rs, n.d.) and that veterinary or remedial behaviour therapy may be needed and is available. The sooner such help is provided, the better is the prognosis. Veterinarians should ask questions when animals come for routine appointments (Fig. 19.2) or treatment for a physical condition, and should have advice leaflets in the waiting room (Roshier and McBride, 2012).

As discussed in Chapter 10, most small mammals will never see a veterinarian, or only rarely. With a little effort, many of these can still be helped. Owners of small mammals often own other species, especially cats and dogs. Veterinary practices could routinely ask (and record) if clients own small mammals. Such records would alert the veterinarian to ask about all pets when the owner visits with any one of them. Additionally, advice leaflets and posters can be put in the waiting room. All will increase awareness in other owners through general social osmosis, such as interactions between pet-owning school children.

History Taking

Problem behaviour in animals can be classified into three broad categories. The first category is normal behaviours that are simply not understood by the owner but are not indicative of any stress in the animal. Once explained, these may no longer be perceived as problematic. The remaining two categories comprise behaviours that may not be a problem to the owner but are of welfare concern. These are normal behaviours indicative of stress, such as aggression or breakdown in toileting behaviour, and 'abnormal' behaviour, including the repetitive behaviours described in Box 19.1.

Causes of problematic behaviour include: inappropriate or lack of experiences during development (see Chapter 10); lack of socialization; fear-provoking experiences at any time of life; and a lack of appropriate mental and physical stimulation. As with other species, there are possible medical reasons underlying problem behaviour. These must be eliminated as the first part of any diagnosis (Meredith and

Fig. 19.2. Veterinarian clipping a rabbit's nails (note the good support of its body weight): an opportunity to raise awareness about suitable exercise and substrate and ask about behaviour. (Image: iStock.)

Delany, 2010; Buseth and Saunders, 2014; McBride, 2014; Yeates, 2018).

Behaviours can have more than one motivation or cause, and a correct diagnosis is essential to developing an appropriate behaviour modification programme. This requires taking a full history of the animal's life, environment, management, handling and interactions with people and other animals. It is important to obtain clear descriptions of what is happening, as these are imperative to understanding both the animal's motivation and the factors that predispose, initiate and maintain the behaviour. Information is needed about the behaviour both of the individual animal and of other animals and humans present immediately before, during and after the display of the problematic behaviour.

For small mammals, environmental factors are easier to identify by visiting the animal *in situ*. Where this is not feasible, asking owners to provide video footage of the environment (physical and social) *and* of the problem behaviour is strongly advised. Given the subtle nature of visual displays in these species and the rapidity with which behaviour, such as aggression, can occur, video has the advantages of repeatability and slow-motion replay. This enables more accurate determination of the sequence of events and of potential triggers, including inappropriate handling technique. Further, the video can be used to help educate the owners.

It must be emphasized that a full veterinary examination is essential, especially where behaviour has changed, and in all cases of aggression. In many cases, pain will be a factor, and as vigilance may mask signs in the practice environment, owner reports and video footage are important sources of corroborative evidence.

Resolving Behaviour Problems by Neutering

For many species the best groupings comprise members of both sexes (see Table 4.1 in Chapter 4). Neutering is advised to avoid unwanted young and can help prevent the development of most hormonally driven anti-social behaviours.

Where behaviour problems have already developed, neutering may not be appropriate. Even where behaviour is influenced by sexual hormones, such as seasonal territorial aggression and urine spraying by rabbits, neutering will only be part of the resolution process. Aggression is likely to have an emotional root in anxiety, fear or frustration, and to have additional learned components. All these aspects will need addressing in order to resolve the problem as fully as possible.

Common Behaviour Problems: Aggression, Destruction, Toileting

For owners of rabbits and rodents, three areas are commonly perceived as problematic: aggression (inter-specific or intra-specific); destructive behaviour; and excretion (loss of toileting in places the owner considers appropriate). As all can have several motivations, the full range of possible differential diagnoses must be considered in every case (McBride *et al.*, 2016). The following sections describe some differentials and potential resolution techniques; these are not claimed to be definitive or sufficient for every case. It is presumed that there are no medical differentials, or that these have been resolved.

Aggression to People

Aggression in these small species is sometimes considered an irrelevance and that owners are making a 'fuss about nothing'. However, the chisel teeth of rodents or the rabbit's slicing incisors can cause a lot of pain and damage to faces and wrists. Furthermore, rabbits can kick and scratch with their powerful back legs. Even though some rabbit breeds weigh only a couple of pounds (1 kg), medium-sized breeds are around 7–12 lb (3–5 kg) and all can pack a lot of punch or leave long open wounds.

Aggressive behaviour towards people, particularly children, is a major reason for seeking behavioural advice. If owners are advised appropriately, then a lot can be done to turn such situations around and restore a harmonious relationship.

Animals can be aggressive towards people for a variety of reasons, including learnt attention seeking and sexually driven behaviour. However, the most common reason is fear

(McBride, 2000). Fear is also the basis of territorial aggression to people, commonly reported in rabbits. Typically, this aggression is concentrated around times when the owner tries to feed the animal or clean out the cage. It may be due to lack of space and resources, or a fear of people and being touched.

Shouting, flicking the animal's nose, shaking, hitting, etc., are not appropriate human responses. Punishment simply increases fear and future aggression and further ruins the relationship between animal and owner.

Fear of being handled

When an animal has not been well handled as a youngster, or has had a bad experience, it is quite normal for it to view humans as threats throughout the rest of its life. Unlike a dog, who might whimper when scared, small prey animals try to be inconspicuous by being quiet, flatten their body to make themselves as small as possible, and go stiff and still (the 'freeze' response). If the threat does not go away, they will run and try to hide. Only if there is no escape will the fear be displayed as aggression; and if terrified they will cry or scream, usually when caught, in an attempt to shock the predator into letting them go. All these behaviours can lead to a breakdown in the human–animal relationship.

Unfortunately, the initial signs of fear are often unrecognized and grabbing or 'chasing' the pet around the cage becomes the norm. Eventually the animal uses aggression – in the form of bites, scratches and kicks – to defend itself and prevent it being picked up. It quickly learns that aggression 'works' and the earlier warning signals do not, and which stimuli predict the scary situation. Thus, it will show fewer warning signals and resort to aggression earlier; possibly even as soon as a face or hand appears (McBride and Hinde, 2022).

Prevention

Ideally, potential owners will know about care needs and the species' communication signals. They should be advised to not choose fearful individuals, i.e. not taking home the rabbit or rodent that is crouching at the back or fleeing around the cage to avoid capture. Recognizing anxious and fearful behaviour at home is also important, as remedial steps can be taken.

When first brought home, pets need time to become accustomed to their new surroundings. Apart from feeding, they should be left quiet and undisturbed for a few days before trying to touch them. Products such as Pet Remedy (Unwin *et al.*, 2020) or the synthesized rabbit pheromone (Bouvier and Jacquinet, 2008) may help lower stress levels in the new home, the veterinary surgery or as part of a behaviour modification programme.

An animal should be approached, and picked up, in ways that help it feel secure (Oxley *et al.*, 2019). Considering the animal's perspective can help people realize that some things they do are very like the behaviours of natural predators. Staring, bending (looming) over animals or swooping a hand down from above are all actions likely to cause anxiety, as is moving quickly or being noisy.

Talking or whistling softly to warn them of human arrival (Hawkins *et al.*, 2016), approaching from a lower height, with head turned to one side and with a tasty titbit to offer, can often make a lot of difference to an animal's interest in being stroked or handled (Fig. 19.3). They can also learn that a repeatedly used word, whistle or signal, such as a hand placed on the cage floor, is a cue for it to approach. Veterinary professionals should ask owners what cues the animal knows and use them. Noting them on the animal's hospital cage means that all involved in its care will act accordingly. This will decrease anxiety and assist recovery.

All these species rely heavily on scent for identifying individuals. Thus initial handling of a new pet, whether it is friendly or fearful, should promote a group 'friendly' scent. Although quite distasteful, it helps to roll hands in some of its used bedding before attempting to interact with the animal. Likewise, do not stroke a dog or cat just before interacting with these prey species. In the veterinary surgery, steps should be taken to keep predators and their scent away from prey species (Varga, 2014; McBride, 2017a).

Treatment

Animals who are fearful of humans can take a lot of hard, slow work to rehabilitate and the fear

Fig. 19.3. Hand-feeding a rabbit can help them become relaxed around people. (Images: iStock.)

is never completely forgotten. Behaviour modification is based on de-sensitizing and counter-conditioning (DS&CC). The basis of this is making new, pleasant associations initially with the presence of humans, then with human touch and ultimately with being picked up (Figs 19.3 and 19.4). The easiest way to make such associations is using favourite food items. Providing a selection of appropriate foodstuffs and observing the pet from a distance will indicate which are the most preferred. It is these that should be used in the training. Unless otherwise stated, these should be offered in single mouthful-sized pieces.

In all cases it is important that all attempts to pick up the pet are stopped for a period. For the small rodents this may only be for a week or two; for rabbits and chinchilla a period of at least 6 weeks is often required. This is the first part of the DS&CC process, altering the way the animal views the owner and enabling it to gain confidence in other day-to-day interactions.

It is also essential that the owner is taught the correct method of handling during this period (Hurst and West, 2010; McBride, 2010, 2014). This should be demonstrated and owners allowed to practise on another confident and friendly hamster, rat, rabbit, chinchilla, etc., or even a life-sized cuddly toy.

Fig. 19.4. When holding rabbit or rodent, remember to make it feel safe by supporting its body weight. (Image: iStock.)

Rehabilitating a frightened animal is essentially the same process as initially training a confident one to be picked up (see Chapter 10). Several short sessions a day are advised and progress must be at the animal's own pace. To force the issue is likely to cause fear and set the proceedings back by days or weeks.

Smaller species

Start by placing a few pieces of food near the hand, which is resting, palm upwards, just inside on the cage floor. The hand *must* remain motionless. Owners should be sitting and they should not be staring into the cage, nor at the animal. As each session can take many minutes, sitting is more comfortable as well as being less intimidating for the pet.

Wait for the pet to approach, take the food, retreat somewhere to eat it and come back for more. If the pet does not approach, it is likely the hand is too close to the food, so increase the distance. For very anxious individuals, start at a greater distance. After placing the food in the cage, the person sits at a distance, where they can be seen but are far enough away that the animal will venture to investigate the food.

Only when the pet is reliably coming to take the food, and ideally remaining in proximity to eat it, should the food be placed nearer the hand or held in the fingertips. Gradually, work up to the food being in the open palm so the animal climbs on the hand to get it. When the pet is reliably coming on the hand, a cue word can be introduced.

When it is happy to sit on the open palm to eat, the next stage is to curl the fingers slowly over the animal's back. Again, this may have to happen in several stages so that the pet does not become frightened. Eventually, the pet will accept being enclosed by the fingers and more pressure can be applied so that it is gently but firmly held and cannot wriggle free. Do not take the hand out of the cage, but hold the animal just for a second or two, releasing it on the cage floor, being careful it does not jump away and injure itself. Slowly increase the time it is held, still with the hand on the cage floor. Once it can be held for a few seconds, start to lift it gently and slowly, ensuring that it is placed back on the cage floor and released before it starts to get frightened. With a fearful animal this can be a slow process taking many tens of training sessions, but the result is worthwhile.

Incorporating clicker training can be very successful, using it to mark approach behaviour. Given the sensitivity to loud and ultrasonic noise, the clickers normally used by dog trainers are unsuitable. Use a small plastic 'bug' clicker or the click of a biro kept especially for training. It can be helpful to have a clearly identified 'reward station' for delivering the food reinforcers and only used during training. A jar lid will do and can be used in other training locations. Later, when the pet is happy to approach the hand, then single mouthfuls of reinforcer can be offered from the fingers.

Note that it may take several months for an individual to learn to confidently tolerate or enjoy being handled. This is especially true if it is a wild species, including chinchilla (DEFRA, 2012), or was caught from the wild or has bad experiences along the supply chain (Blue Cross and Born Free Foundation, 2016).

Rabbits

The programme for teaching a rabbit to accept being picked up (McBride, 2000) is appropriate for similar-sized animals like guinea pig and chinchilla. It is an adaptation of that described above for the smaller species.

1. The rabbit's favourite treats should be identified and initially offered from a stationary hand, with no attempt to stroke the rabbit. The treat may be long strands of herbs such as parsley. As with the small rodents, it can be useful to introduce a clicker and reward station at this point, using small mouthful-sized treats, such as a parsley leaf rather than whole stalk. A calm approach to the person is reinforced by click/treat.
2. Hands should be rolled in some of the rabbit's used bedding first to remove any unusual smells. Again, the person should be sitting at rabbit level, ideally on the floor in the rabbit's pen or next to the hutch, and not staring at the animal.
3. The next stage involves touching the rabbit, initially around the head and neck. Once the rabbit is happily taking the treat from the person's fingers (this may take a few days) then the key (cue) word is introduced just before stroking the rabbit on the top of its head. The stroking should be done whilst the rabbit is offered and eating the food. By linking food, the cue word and small amounts of stroking, the rabbit will make pleasant associations with touch and eventually with anticipation of being lifted.
4. Once the rabbit accepts being stroked on the head, gradually stroke over the body, using short gentle movements.

5. For very fearful, aggressive rabbits, stages 2–4 should first be accomplished using a soft baby brush attached to a handle of 8–12 inches (20–30cm) in length and an inch (2.5 cm) in diameter. This too should be rolled in some of the rabbit's bedding first. Remember frequent, short sessions are best.

- Place a few mouthfuls of food near the brush. If the rabbit bites the brush, hold the brush still. This helps the rabbit learn that the brush is harmless. Once the rabbit has relaxed and is happy to eat next to the brush, start to gently brush for a second around the top of the head. If the rabbit reacts, stop brushing and only start again when it has resumed eating the treats.
- Only when the rabbit is consistently relaxed about being brushed is it time to start to replace the brush with the hand. This is done in stages over several sessions, gradually shortening the length of the handle, eventually holding the hand over the main head of the brush and finally removing the brush altogether, as in stage 4.

6. Next is to introduce some of the more invasive components of handling, one by one, whilst the rabbit is eating. These include laying a hand over the rabbit's shoulders, placing a hand lightly on the rabbit's rump and leaning over the rabbit.

7. Picking up the rabbit only happens once it is totally confident in all of the above stages. At this point sitting so that the owner's lap is at the same height as the floor of the rabbit's home will help to scoop the rabbit on to the lap where hand-feeding can continue (Fig. 19.5).

Fig. 19.5. Rabbit on lap. (Image: iStock.)

8. Finally, short lifts can be performed – always ending in a reward. The time spent on the owner's lap or in the owner's arms can be increased. Care must be taken when returning the rabbit to the floor or into its hutch to avoid injury.

Aggression to people due to territoriality (resource-guarding)

Territorial aggression is where the animal defends valuable resources which it fears it will lose, such as the food bowl or hay rack. Some owners will try to deflect the rabbit's aggressive behaviour on to a different and, from the rabbit's view, still aversive or threatening stimulus, such as a large gardening glove, dustpan/brush or even a bar of soap. This may enable them to manage the problem in the short term, but is not a solution, as it is simply dealing with the symptom and not the cause. It is likely to exacerbate the issue by increasing the animal's general fear of people.

Prevention and rehabilitation include providing food in several locations, several bowls or hay racks, activity toys or scattering food throughout the home environment, as this reduces the perceived 'value' of each bowl of food. Rebuilding trust, as described above, may also be required.

Neutering is worth considering, particularly if the behaviour coincided with the start of puberty. This will not be sufficient by itself, as learning will also have taken place. Where animals are group-housed, puberty and breeding seasons can be triggers for intra-specific aggression based on resource-guarding.

Intra-specific Aggression

There are several possible causes of aggression between cage mates. These include re-direction of fear-related aggression, which occurs when the animal is frightened but cannot escape the perceived threat, be that a loud noise, the presence or scent of a predator, or someone trying to pick it up; frustration due to a barren environment; or competition over limited resources. For some species, such as rabbits, chinchillas and degus, the latter is particularly evident between members of the same sex, starting when they reach puberty or during the breeding season (McBride, 2000).

Under natural circumstances the levels of aggression would be reduced by the loser of a dispute moving out of the victor's territory and few injuries would be sustained. However, in the cage situation there is nowhere to go and this causes prolongation and escalation of aggression. In such situations serious injuries are likely and can be fatal.

It is therefore of great importance that owners are advised on suitable pairing or groupings of animals and the need to provide sufficient space and resources. A general rule is to provide at least one of each resource per animal and one extra. Owners also need to be educated in good time about neutering; how to reintroduce animals that have been separated for any reason; and how to introduce a new animal to an already established group. It may all be too late if aggression has already occurred.

Prevention

The animals' home environment should be large and provide lots of boltholes and resting places, whilst also enabling free movement at both slow and fast speeds (see Chapter 10). There should be at least one resting place per animal, enabling them to avoid visual and physical contact with each other if they so choose, and one where they can all rest together. Owners should be advised that more than one water source should be provided, in different areas, and several feeding sites distributed around the accommodation. Scatter feeding is useful, as is the use of puzzle feeders, though again several of the latter need to be provided to avoid monopolization and aggression (Shepherdson *et al.*, 1999; Young, 2003). Likewise, if running wheels are provided, there should be at least two, and a variety of toys, ideally at least as many as there are animals in the group (Figs 19.6 and 19.7).

When obtaining littermates or another companion for an existing pet, an individual of the opposite sex makes for a more likely union (see Table 4.1 in Chapter 4). At the appropriate age, around puberty, the owner should have one or both animals neutered, preventing pregnancy and reducing intra-specific aggression.

When introducing a new animal, time should be taken to introduce them gradually, particularly for the territorial chinchillas and rabbits. A male rabbit should never be put into a female's hutch; to do so will cause enormous problems and probably quite severe injuries. Campbell (2010) and others recommended that introductions take place on neutral territory, following a period of scent swapping (wiping a clean cloth – one for each individual – over the individual and then on all the other individuals in the group).

Not infrequently, animals that have lived together quite happily are aggressive to each other when reintroduced after they have been separated to go to the vet, or elsewhere. This most

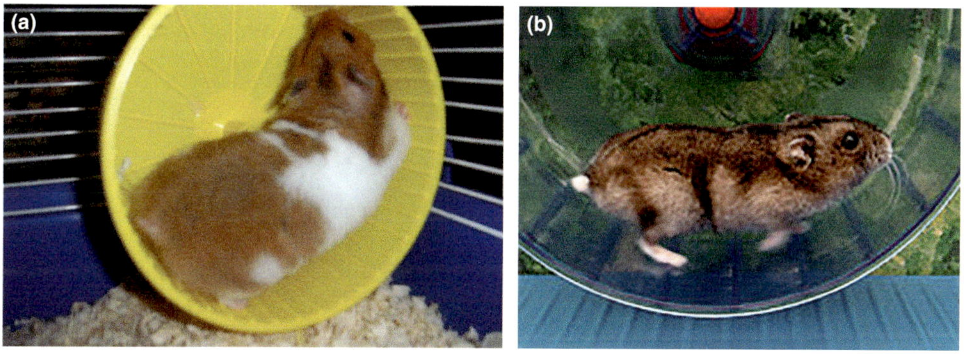

Fig. 19.6. **(a)** Small mammal wheel: a good design has a solid floor, which prevents toes being broken, but this is too small for this animal, as the back is curved when running, which will cause pain. (Image: Wikipedia Commons.) **(b)** A better size, as the animal's back is straight.

Fig. 19.7. Small mammal wheel. With this saucer-shaped design the animal will not be curving its spine when running. (Image: iStock.)

likely occurs because they no longer share a common scent (see Chapter 10).

Where relationships have broken down, they can be re-established in some circumstances. However, warn owners that some breakdowns are irrevocable. The prognosis for successful reintroduction will be determined by various factors, including the duration of the problem and intensity of fighting.

Taking all precautions to prevent relationship breakdown is strongly advised. Prevention entails keeping the group scent. Keep the group members together as much as is possible, including taking them all to the vets either in the same travel basket or keeping the baskets close to each other, and swapping bedding and scent cloths (McBride, 2017a and Table 19.1).

The process described below is for rabbits but can be applied to introductions and reintroductions in other species, such as chinchillas.

Please note that introductions and reintroductions have a better chance of being successful if they are made *out* of the breeding season (Magnus, 2007).

Treatment

Neutering can help with some situations, particularly in the case of same-sex individuals that are fighting around puberty or the breeding season. All animals should be taken to the veterinary surgery and scent profiles kept as similar as possible to prevent the problem worsening.

The home accommodation and management should be assessed. Often a total revamp of the environment is required to make it larger and more complex, with multiple resources, more to do and areas that allow the rabbits to avoid visual contact with each other. The steps described above to reduce resource (territorial) aggression should be taken. If the accommodation was already the home of one of the rabbits, it should be washed down to get rid of territorial scent marks (see Excretion, below).

When introducing or reintroducing two rabbits to each other, simply putting them into an enclosure and 'letting them get on with it' often causes huge problems. Invariably the damage in some rabbit relationships is done during such first meetings. It can be very hard to ever reintroduce some individuals after a bad start (Campbell, 2010). There is some evidence that when introducing two unfamiliar rabbits into an already established group, this may be better done in the group's home cage (Graf *et al.*, 2011). This may not be appropriate for introducing a single new animal to a group.

The programme below is used both for initial introductions and for rehabilitating individuals whose relationship has broken down. It is one of several possible methods (Campbell, 2010). Fear-provoking methods, as in putting animals in the same carrier and driving them around, are not recommended. Such methods are unethical and the forced 'relationship' may quickly break down. Advice should be taken only from an appropriately knowledgeable and experienced person or rescue centre (RWAF, n.d.).

- A neutral, safe, enclosed environment should be used. Ideally this should be an area where none of the animals have been before, like a garden shed or bathroom. Ensure that there are safe hiding places for the animals to retreat to.
- Initially, the rabbits should be brought together using animal travel baskets, with water and hay available. Baskets need to be big enough for the animal to be able to stretch fully, and ideally allow it to see out of the front and the sides. Place the baskets at a distance apart that allows the rabbits to relax. This may be several metres to begin with. Feeding the rabbits some favourite treats, by hand, through the sides of the baskets will help them make pleasant associations with the presence of the other rabbit.

- If the rabbits are not distressed or showing aggressive behaviour, reduce the distance on a little-and-often basis over a period of several days. If either individual reacts adversely, temporarily increase the distance between them until both are relaxed.
- At the end of each training session, the rabbits should be returned to their own separate home cages.
- Assuming that the rabbits appear relaxed once they are close together, the next time they come together will be without the baskets. Before this happens, ensure that the environment where this is going to take place has lots of hiding places and nothing dangerous such as sharp edges. Scattering lots of hay, green vegetables and healthy treats around the area will provide plenty of distractions.
- The animals should be supervised throughout, but not interacted with, unless the session needs to be curtailed if any serious chasing occurs. (This may indicate that this is not going to be a successful bonding.)
- If this meeting is successful, then it should be repeated several times. The rabbits should be separated at the end of each session.
- When the rabbits show relaxed behaviour, ideally grooming each other, then a long (overnight) session takes place. The rabbits must still be monitored. Live-stream video is helpful here, so the rabbits are watched both when a person is present and when not.
- If all is well, then rabbits can then be introduced to their new spacious and interesting home.

As suggested earlier, Pet Remedy may be a helpful adjunct for all species (www.petremedy.co.uk).

Chewing, Digging or Other Destructive Behaviour

Destructive behaviour is usually only seen as a problem by owners if it affects belongings they value. This may be normal inquisitive behaviour deemed inappropriate, such as biting through television wires, or chewing books. Where chewing or digging is repetitive and occurs for minutes at a time on inappropriate items, such as cage bars and floors, it is abnormal and indicative of stress (Würbel, 2006). This may not be recognized as a problem but regarded as 'normal', simply as 'what these small animals do'. This again underlines the need for prophylactic education and to ask owners about the behaviour of all their pets.

Prevention and treatment of destruction problems are the same and are considered together below.

Neutering is not a treatment for destruction problems. It may be helpful for female rabbits whose digging is related *solely* to the breeding season and pregnancy/pseudo-pregnancy cycle (Elliott and Lord, 2014).

Prevention and treatment

Providing a natural diet helps prevent dental problems that can lead to repetitive chewing. Provision of the diet in ways that increase foraging and eating gives a natural use of the activity budget and prevents many problems associated with chewing or destruction.

Chapter 10 describes ways of encouraging the animals to work a little harder for their food. Other items appropriate for chewing include untreated fruit-tree branches, cardboard tubes or boxes. These also act as mental stimulation. Owners should provide toys and areas/objects to investigate. Alternating toys and making small changes around the environment provides novelty and stimulation in a non-threatening manner.

Where a social species live alone, compatible company of its own kind can help to relieve the boredom. Additionally, or alternatively, daily stimulating interaction with owners is highly beneficial. If animals live outdoors, facilitate owner interaction by making part of the accommodation comfortably accessible in all weathers, e.g. a shed. For all animals, a range of 'tricks' can be clicker trained, e.g. for heath checks (Mackie and Patel, 2022), retrieving objects, agility courses and even playing basketball (Reid, 2009).

Areas to which the animal has access must be 'proofed' to ensure that there are no dangerous or valuable items that could be climbed, chewed or dug. Even when supervised, it is the work of an instant to chew a wire, strip wallpaper or dig a hole in the carpet, cushion or back of the sofa. For the smaller species, exercise pens are available where interactions can occur. For the larger species, more extensive steps need to be taken, such as encasing wires in thick plastic or metal trunking (Dykes and Flack, 2003).

Excretion

Excretion problems involve both faeces and urine. Owners of chinchillas, degus, rabbits and guinea pigs who notice soft faecal pellets (caecotrophs) on the floor of the cage should seek veterinary help. This indicates a physical problem that needs addressing, as well as possible behavioural issues (McBride, 2014). As with any problem, the motivation for behaviour change may be quite subtle and require good history-taking skills to uncover the underlying cause (McBride *et al.*, 2016).

Though all species will deposit faeces and urine as they traverse their environment, rabbits and chinchillas predominantly use latrines and can be litter (toilet) trained. Urine deposition may be inappropriate urination, i.e. loss of toilet training, or it may be urine spraying. Both species deposit urine to mark territory. Chinchillas spray urine when fighting and as a defence against a threat. If the latter, and the human is the threat, then rehabilitation to people as described earlier will be needed as part of any programme.

Rabbits also spray urine in territorial disputes and in courtship (McBride, 2000). A perfectly toilet-trained house-rabbit can start to deposit faeces and urine in locations other than the tray as it reaches puberty. Owners may also describe the rabbit running in circles around their legs, possibly grunting and depositing faeces. On occasion a spray of urine may follow this.

Animals can lose toilet training if there has been a sudden change in the type of litter used or if the tray has become aversive. It may have been moved to a place where the animal feels unsafe, or accessing and using it has become associated with pain from a fracture, arthritis or bladder infection.

Once a breakdown in training has occurred, a secondary problem of over-marking can develop as the animal continues to scent mark in the new toileting place.

The advice below describes how to initially train or re-instate litter tray use. It is based on the rabbit but can be applied to chinchilla.

Treatment

In all cases, possible medical causes should be investigated and resolved first.

Neutering should not be an action of first resort. It will only be effective if the toileting problem is associated specifically with puberty or the breeding season (McBride, 2000).

If problems develop after a change of litter, then the original litter should be used. If the tray has been moved, it should be replaced in its previous location. Trays in busy areas of the home, or where the animal using it can be easily seen by the pet dog or cat, can be threatening. The tray should be put within an indoor hutch or other undisturbed, visually screened place (Buseth and Saunders, 2014).

Once an incorrect association has been formed – for example the rabbit is routinely urinating on the sofa – it can be a very hard habit to break. Retraining, as with initial training, means confinement into a smaller area, an indoor cage with its food, water, toys and tray, for at least 3–4 days to reintroduce toileting on the desired substrate in the desired location.

Whilst the rabbit is confined, the owner needs to thoroughly clean all the inappropriately soiled areas (urine or faeces). It is not helpful to use disinfectants and 'floral' products, as they only mask the smell to humans. Likewise, using products containing ammonia (as in bleach) can be counter-productive and encourage the rabbit to over-mark the area. A warm (but not hot) solution of biological washing powder or liquid should be used. Several washes are probably needed to ensure that the area is clean enough to allow the rabbit supervised access in the future. Some soft furnishings may need replacing.

Any mishaps should always be cleaned in the above manner. If the problem is quite severe, then access to certain areas may have to be denied for much longer periods of time to truly break the habit.

Summary

The three linked chapters on rabbits and rodents have provided an overview of normal behaviour (Chapter 4), means by which welfare needs can be met and problems prevented (Chapter 10) and, in this chapter, advice on causes of common problem behaviours and tips for implementing appropriate behaviour modification programmes. Members of the veterinary practice team are encouraged to enable owners to recognize potential problems and to provide opportunities for them to express concerns about their pet's behaviour. Both potential medical issues and the physical and social environment should be investigated. Problem behaviour should be dealt with by a knowledgeable person (McBride and Montgomery, 2018), either in-house or by referral (see Chapter 7).

References

Blue Cross and Born Free Foundation (2016) One Click Away: an investigation into the online sale of exotic animals as pets. Available at: https://www.bornfree.org.uk/storage/media/content/files/Publications/ONE_CLICK_AWAY_REPORT_SPREADS.pdf (accessed 30 June 2022).

Bouvier, A.C. and Jacquinet, C. (2008) Pheromone in rabbits: preliminary technical results on farm use in France. In: *Proceedings of the 9th World Rabbit Congress*, Verona, Italy, 10–13 June 2008. World Rabbit Science Association, pp. 303–307.

Broom, D.M. and Fraser, A.F. (2015) *Domestic Animal Behaviour and Welfare*, 5th edn. CAB International, Wallingford, UK.

Broom, D.M. and Johnson, K.G. (1994) *Stress and Animal Welfare*. Springer, Heidelberg, Germany.

Buseth, M.E. and Saunders, R.A. (2014) *Rabbit Behaviour, Health and Care*. CAB International, Wallingford, UK.

Campbell, F. (2010) *Bonding Rabbits*. Coney Publications, Ipswich, UK.

Dallmann, R., Steinlechner, S., von Hörsten, S. *et al.* (2006) Stress-induced hyperthermia in the rat: comparison of classical and novel recording methods. *Laboratory Animals* 40, 186–193.

DEFRA (2012) *Zoo Licensing Act 1981 Guide to the Act's Provisions*. Department for the Environment, Food and Rural Affairs, London.

Dykes, L. and Flack, H. (2003) *Living with a Houserabbit*. Interpet Publishing, Dorking, UK.

Elliott, S. and Lord, B. (2014) Reproduction. In: Meredith, A. and Lord, B. (eds) *BSAVA Manual of Rabbit Medicine*. BSAVA Publications, Gloucester, UK.

Graf, S., Bigler, L., Failing, K., Würbel, H. and Buchwalder, T (2011) Regrouping rabbit does in a familiar or novel pen: effects on agonistic behaviour, injuries and core body temperature. *Applied Animal Behaviour Science* 135, 121–127.

Hawkins, P., Atkinson, J., Birt, R., Cruden, J., Duran, A. *et al.* (2016) Report of the 2015 RSPCA/UFAW rodent and rabbit welfare group meeting. *Animal Technology and Welfare* 15(1), 9–22.

Hurst, J.L. and West, R.S. (2010) Taming anxiety in laboratory mice. *Nature Methods* 7(10), 825–826.

Mackie, J. and Patel, C. (2022) Welfare-focused animal training. In: Rendle, M. and Hinde, J. (eds) *BSAVA Manual of Practical Veterinary Welfare*. BSAVA Publications, Gloucester, UK.

Magnus, E. (2007) *Teach Yourself: Keeping a Rabbit*. Hodder Publications, London.

McBride, A. (2000) *Why Does My Rabbit ...?* Souvenir Press, London.

McBride, A. (2010) *Hamsters: Understanding and Caring for Your Pet*. Magnet and Steel, Llandow, Cardiff, UK.

McBride, A. (2014) Normal behaviour and behaviour problems. In: Meredith, A. and Lord, B. (eds) *BSAVA Manual of Rabbit Medicine*. BSAVA Publications, Gloucester, UK.

McBride, E.A. (2017a) Small prey species' behaviour and welfare: implications for veterinary professionals. *Journal of Small Animal Practice* 58(8), 423–436. doi: 10.1111/jsap.12681

McBride, E.A. (2017b) Stress in the veterinary surgery – small mammals. *The Veterinary Nurse* 8, 376–381.

McBride, E.A. and Hinde, J. (2022) Behaviour. In: Rendle, M. and Hinde, J. (eds) *BSAVA Manual of Practical Veterinary Welfare*. BSAVA Publications, Gloucester, UK

McBride, E.A., Magnus E. and Hearne, G. (2016) Behaviour problems in the domestic rabbit. In: Appleby, D. (ed.) *The APBC Book of Companion Animal Behaviour*. Souvenir Press, London, pp. 164–182.

McBride, E.A. and Montgomery, D.J. (2018) Animal welfare: a contemporary understanding demands a contemporary approach to behavior and training. *People and Animals: The International Journal of Research and Practice* 1(1), Article 4.

Meredith, A. and Delany, C.J. (2010) *BSAVA Manual of Exotic Pets*, 5th edn. Wiley, Oxford.

NC3Rs (n.d.) Grimace scales. National Centre for the Replacement Refinement & Reduction of Animals in Research. Available at: https://nc3rs.org.uk/grimacescales (accessed 18 June 2022).

Oxley, J.A., Ellis, C.F., McBride, E.A. and McCormick, W.D. (2019) A survey of rabbit handling methods within the United Kingdom and the Republic of Ireland. *Journal of Applied Animal Welfare Science* 22(3), 207–218.

Reid, A. (2009) Rat basketball at Wofford College. Available at: http://www.youtube.com/watch?v=jAQSEO25fa4 (accessed 18 June 2022).

Rendle, M. and Hinde, J. (eds) (in press) *BSAVA Manual of Practical Veterinary Welfare*. BSAVA Publications, Gloucester, UK.

Roshier, A.L. and McBride, E.A. (2012) Canine behaviour problems: discussions between veterinarians and dog owners during annual booster consultations *Veterinary Record* 172(9), 235. doi:10.1136/vr.101125 (open access).

RWAF (n.d.) Rabbit Welfare Association and Fund advice sheets and helpline. Available at: www.rabbitwelfare.co.uk (accessed 30 June 2022).

Shepherdson, D.J., Mellon, J.G. and Hutchins, M. (eds) (1999) *Second Nature: Environmental Enrichment for Captive Animals*. Zoo and Aquarium Biology and Conservation Series. Smithsonian Books, Washington DC.

Sherwin, C.M. (1988) Voluntary wheel running: a review and novel interpretation. *Animal Behaviour* 56, 11–27.

Unwin, S.L., Saunders, R.A., Blackwell, E.J. and Rooney, N.J. (2020) A double-blind, placebo-controlled trial investigating the value of Pet Remedy in ameliorating fear of handling of companion rabbits. *Journal of Veterinary Behavior* 36, 54–64.

Whittaker, A.L. and Howarth, G.S. (2014) Use of spontaneous behaviour measures to assess pain in laboratory rats and mice: How are we progressing? *Applied Animal Behaviour Science* 151, 1–12.

Würbel, H. (2006) The motivational basis of caged rodents' stereotypies. In: Mason, G. and Rushen, J. (eds) *Stereotypic Animal Behaviour: Fundamentals and Application to Welfare*, 2nd edn. CAB International, Wallingford, UK, pp. 86–120.

Varga, M. (2014) The rabbit-friendly practice. In: Meredith, A. and Lord, B. (eds) *BSAVA Manual of Rabbit Medicine*. BSAVA Publications, Gloucester, UK.

Yeates, J. (2018) *Companion Animal Care and Welfare: The UFAW Companion Animal Handbook*. Wiley, Oxford.

Young, R. (2003) *Environmental Enrichment for Captive Animals*. Blackwell Scientific, Oxford.

20 First Aid Advice for Common Behavioural Signs: Parrots

Clare Wilson

It is common for behavioural problems to have been present for some time before appropriate professional help is sought and, by the time this happens, owners are often in a stressed state themselves. There may be further time delay involved with organizing referral to an avian behaviourist and it can therefore be extremely beneficial to owners and their pets if the veterinary practice is able to provide some first aid advice. The aim of such advice is to prevent deterioration and, in cases where owners do present a problem early enough in its development, it may be sufficient to resolve the situation. Parrots presenting with behavioural problems should have a thorough physical work-up from a specialist avian veterinarian prior to behavioural referral.

Providing possible motivational explanations for behavioural problems and educating owners about normal parrot behaviour can significantly improve their perception of the situation and their willingness to work with their bird. For example, in the case of a bird that is presented for biting people, the owner is often scared to handle the bird. Explaining that the bird is likely to be frightened as well, and showing defensive aggression as a result of its fear, can help the owner to put the problem behaviour in context and start to feel empathy with the bird's emotional state. Another example where better human understanding can help to repair a damaged relationship between a bird and its owner is explaining the use of contact calls to an owner of a parrot showing excessive vocalization.

Management to Prevent Practising of Problem Behaviours

One of the main aims of first aid advice is to prevent unintentional reinforcement of unwanted behaviours. This may occur via external responses to the behaviour, such as the act of rapidly withdrawing a hand when the bird attempts to bite, or via internal responses, such as the emotional state of relief when the bird feels safer again. When a behaviour is reinforced, it is more likely to occur again in the future. It is therefore essential to try to limit displays of inappropriate behaviours as far as possible and discussion with the owners about specific circumstances in which the behaviour occurs can help to decide management strategies to reduce the incidence. Inappropriate behaviour is often associated with underlying stress for the bird as well as the owner. First aid advice to help reduce occurrences of incidents will therefore not only assist in the long-term success of a behavioural modification programme, but will also immediately improve the welfare of the parrot and its caregivers. Some examples are discussed below but it is crucial that the clinician treats every

 Companion Animal Behaviour Problems: Prevention and Management of Behaviour Problems in Veterinary Practice (eds. R. Casey, S. Heath and H. Zulch)
DOI: 10.1079/9781800621312.0020

case on an individual basis, as the exact circumstances in each case will vary.

History Taking

A thorough history is essential in order to establish an accurate diagnosis and thereby prescribe appropriate behavioural modification. Referral-level behavioural work is therefore time consuming. In the first aid context of a general veterinary practice where a client discusses a behavioural issue during a routine consultation, time is limited but even a brief history can be beneficial in providing suitable initial advice.

Knowledge of the origins of the bird (captive-bred or wild-caught) and its rearing history (parent-reared or hand-reared, parent-weaned or hand-weaned) can give important clues as to the risk of problematic emotional responses such as fear, which would be more common in wild-caught birds and in inappropriately hand-reared birds who lacked early environmental and social stimulation.

Knowledge of the normal daily routine can identify management issues that may be affecting behaviour, for example insufficient enrichment or insufficient sleep. General advice about a sensible daily routine for the bird, which is based on an understanding of normal parrot behaviour, can be invaluable in helping owners cope whilst awaiting a referral for a full behavioural consultation. Chapter 11 discusses what an appropriate routine should be in the section headed 'Husbandry requirements as related to behavioural needs'. Monitoring and altering the amount of sleep that a bird is getting is an easy and extremely beneficial component of the first aid approach. Serotonin is produced in significant amounts overnight. When a bird is lacking the required 9–12 hours of sleep in a quiet environment, low serotonin levels may result in irritability, increased aggression and other fear- and anxiety-based behavioural responses. However, in contrast with this, Jayson *et al.* (2014) found that African Grey parrots who were shut away in the dark and isolated for over 12 hours had a significantly higher risk of feather plucking. This was likely due to such birds experiencing excessive levels of isolation and lack of stimulation.

General Approaches to Alter Inappropriate Behaviour in Parrots

Parrots establish and maintain behaviours using the same learning processes as those applied in the fields of canine and feline behaviour. Parrots respond very well to reward-based training, but as a prey species they are extremely sensitive to aversive or punitive techniques and such strategies are strongly discouraged, as discussed in Chapter 11. Identifying rewards that motivate the individual parrot is crucial. Parrots are unable to preen the feathers on their own heads and therefore allopreening (the act of preening one another) not only serves to cement and maintain bonds but also serves a functional purpose. A head rub can therefore be a very useful tool to reward appropriate behaviour in parrots and induce a positive emotional state. Food and toys can also be used but it is important to identify differing values for these rewards, as some situations will require the use of a higher-value intervention than others. The contexts in which different rewards are valued should also be assessed, because what is rewarding to a parrot in one context may not be in another.

Desensitization and counter-conditioning are extremely useful techniques for addressing fear responses. Desensitization is the process whereby a stimulus that concerns the bird is introduced at a low intensity, ensuring that the bird remains relaxed and establishing a neutral emotional reaction to its presence. The intensity of the stimulus is then gradually increased, provided the bird remains relaxed. An example of how this might be used would be introducing a novel household object such as a new clock at a distance from the bird's cage or play stand so that the presence of the clock does not create a significant emotional response. The clock can then be gradually moved to its permanent position, ensuring that the bird remains in a neutral emotional state at each stage of the process. In this example, desensitization results in a neutral response to a stimulus that had the potential to induce fear or anxiety. If a stimulus is already generating a negative emotional response, it would be introduced in a sufficiently diluted form as to allow the bird to remain in a neutral emotional state. The dilution may be achieved in various ways, for example by introducing the stimulus at an increased distance or at a reduced volume. Counter-conditioning takes the learning process a step further and is extremely important in situations where the parrot is showing overt fear responses. This process changes the neutral emotional state achieved during desensitization into a positive emotional response. For example, if a bird has been showing territorial behaviour around its cage, the process of desensitization can be used to neutralize its emotional state in response to the proximity of people and counter-conditioning can then be used to teach it to expect a favourite food treat when people approach. Replacing the problematic emotional reaction of feeling threatened with a positive emotional expectation will be associated with an alteration of the behavioural expression from one of defensive aggression to positive social interaction. It will be beneficial to associate the new positive emotional expectation with a verbal cue, such as 'hello', as this will enable the cue to become associated with the desired behavioural response. Indeed, the teaching of verbal cues in association with positive emotional states and with desired behaviours is beneficial in all behavioural cases, as it provides the owner with more control and creates a more predictable relationship between the bird and owner.

Specific Problem Presentations

Excessive vocal behaviour

Concerns over noise levels associated with vocal behaviour in parrots are one of the most common reasons why owners seek behavioural advice and even relinquish their avian companion. As with all behaviour problems, the underlying motivation for vocalization must be identified. Parrots vocalize for many different reasons. They may be motivated by play, fear, distress, a desire to contact their owner, territorial defence, pain, lack of enrichment and so on. In order to be able to give appropriate first aid advice it is essential to investigate the underlying causes, because inappropriate advice could aggravate problems and potentially increase stress for the bird and the owner. Table 20.1 describes the different types of parrot vocalization.

It is important that owners are aware of the natural activity pattern expected of their parrot species in order that they can have realistic expectations of the potential to effect any change in the vocal behaviour. The majority of parrot species kept as companions have highly vocal periods as they gather in flocks ready to forage first thing in the morning and again as they flock to roost at sunset. They are relatively quiet during the day and usually silent during the hours of darkness. It is therefore important to discuss the timings of reported vocal problems and to establish whether the behaviour is motivated by this innate pattern of activity. Although management strategies may reduce the vocal disturbance in this instance, it would be crucial to educate the owner about their parrot's natural requirements and emphasize the need for them to tolerate a certain degree of noise in the morning and evening. For the species that live in equatorial regions in the wild, a simple measure such as a blackout cage cover to mimic the 12-hour daylight, 12-hour darkness cycle can reduce the chances of parrots vocalizing very early in the morning over the summer. However, such a measure would be both ineffective and inhumane at other times of the day or if there is a different motivation for morning/evening vocalization. If excessive vocalization continues when the cover is removed it can lead to complaints from neighbours and significant stress for owners. In these cases, it may be appropriate to establish alternative behaviours that are incompatible with vocalization, such as foraging and playing, through the provision of foraging enrichment and novel toys. Providing this alternative behavioural outlet at specific times of day may help to reduce the severity of this time-dependent vocalization. This time of vocal activity can also be used productively to teach the bird new more appropriate vocal sounds for owners who wish to teach their bird to imitate human language or whistling tunes and so on.

Contact calls

These calls are often misunderstood by owners and this can result in the onset of a screaming problem. In wild parrots, contact calls are used to ensure that bonded pairs remain in contact with each other, to keep the flock together and to help parrots locate each other within the

Table 20.1. Vocalization types in parrots. (Data: author's own.)

Type of vocalization	Possible motivations	Advice for owners
Morning and evening flock gathering	Natural activity pattern	Use cage cover to create 12–15-hour day, 9–12-hour night cycle Provide enrichments at these specific times of day to redirect parrot into quieter activities Use the opportunity to teach the parrot new sounds or words Reward the bird for low-volume vocalization
Contact calls at other times	Attempting to make contact with owner	Owner to respond quietly if parrot calls quietly Owner to initiate the contact by talking or whistling before the parrot starts to vocalize
Territorial declarations	Onset of breeding condition, which may be seasonal or may be related to exposure to nesting materials or high-fat diet	Address any inappropriate behaviour of owner and environmental triggers (see details in section 'Reproduction-related aggression')
Alarm calls	Presence of a potential threat. This is a distress call and, if occurring frequently, indicates a significant welfare issue	Identify stimulus causing alarm and limit exposure or remove. Treat by desensitization and counter-conditioning if appropriate
Screaming	Frustration associated with inability to achieve expected outcome in any of the contexts above	Increase enrichment of all types – social, physical, foraging. Ensure 12–15-hour light, 9–12-hour dark cycle. Do not use aversive responses to the screaming and ensure that quiet behaviour is rewarded

flock. These calls are a critical natural behaviour and owners must accept that their companion parrot has an innate need to be vocal and to be responded to. Individual parrots will vary in their use of contact calls depending on their ability to cope with being alone and with the type and strength of the bond they have with their owners. If the parrot has pair-bonded with the owner, contact calls are likely to occur with minimal separation, such as the owner going to a different room in the home, and increase significantly if the bird comes into breeding condition. If the owner does not respond appropriately by replying to these contact calls, then the parrot may become frustrated, which will intensify and accelerate the behaviour. In severe cases parrots may develop an extreme negative emotional response to being separated from their owner, particularly when they are out of auditory contact. Behavioural problems in this context may be referred to as separation anxiety and these issues are discussed separately below.

Alarm calls

These can occur if the parrot is feeling threatened and in this case it is vital to identify the trigger for the fear reaction. For example, it may be a neighbouring cat passing a window, a buzzard flying by, the pet dog entering the room and so on. Parrots are prey species that are naturally neophobic and apparently innocuous novel inanimate objects, such as a new toy or a new item of furniture in the home, may trigger a fear response. Alarm calls can be particularly problematic on two fronts. Firstly, they tend to be very loud, because their aim is to alert the flock to a dangerous situation. The volume of the call can cause serious problems for owners and also result in complaints from their neighbours. Secondly, and more importantly from the bird's perspective, these calls are used when a bird is genuinely distressed and therefore frequent occurrence of these calls is highly significant in terms of compromised welfare. Once the source of alarm has been identified the owners can work out management strategies to reduce or eliminate the parrot's exposure to that stimulus. It is also important to bear in mind the level of distress indicated by the bird. In some situations, a bird may give a brief alarm call and then adapt to the situation, for example if a visitor comes to the house that the bird is unsure of or if the household dog approaches the bird's cage. In this case the owner should remain calm and relaxed so that the bird can see that there is not a genuine threat and the stimulus (e.g. dog, visitor) should be prevented from approaching further. Alternatively, the parrot may be given the opportunity to retreat to a higher perch where it feels safer. As long as these brief alarm calls do not occur frequently and the bird adapts to the situation, it is best for the owner just to ignore the calls and ensure that the bird can see that their owner is unconcerned. It can also be beneficial to reward the bird once it settles, either with a food treat or with praise. Parrots are excellent at learning from observation, so if the parrot observes the owner approaching and interacting with a novel object this can help to settle the parrot and may encourage it to approach the object to investigate for itself. Using observational tactics can also work if there is an issue with a visitor to the home, with a quiet calm conversation between the owner and the visitor helping the parrot to see that there is no real threat. It is important that the owner correctly identifies the motivation for vocalization in association with visitors, since it is possible for it to be motivated by inappropriate bonding where the parrot is attempting to drive away a rival; in this case interaction between the owner and the visitor is likely to induce frustration and intensify the vocal response.

If there is an external trigger for alarm-motivated vocalization, such as aeroplanes flying overhead or dogs passing by for a walk, and the bird can see these triggers through a window, then moving the cage to a new location in the house may be a very beneficial first aid intervention. If this is not possible, partly covering the cage or the window can offer the bird some level of visual protection. If the introduction of a new toy has caused alarm, then simply removing the toy and later reintroducing it at a distance can be helpful. When the bird is out of its cage, it may then choose to approach and investigate the toy itself or, as it starts to get used to it, the owner can gradually bring the toy closer to the cage. In cases where parrots perceive the stimulus as a continuing threat, such as a dog living in the same household, a behaviour modification programme can then be used to desensitize and counter-condition the bird to that stimulus and teach him to cope with exposure. It is also important to bear in mind that training, behaviour modification and appropriate management of the dog in this scenario may also be required for a successful outcome.

Inappropriate mimicking

Parrots can sometimes learn inappropriate words, such as swear words, or inconvenient sounds, such as oven timers and phone ringtones. These can cause embarrassment for owners or result in annoyance when they keep jumping up to check if their dinner is ready or the phone is ringing. Unfortunately, there is no quick solution for this problem and the best way to deal with it is for the owner and visitors to completely ignore the unwanted sounds in order to avoid unintentional reinforcement. Rewarding appropriate or acceptable mimicking can also be helpful. Parrots will very often learn certain vocalizations in order to gain a response from their owners and can find it very entertaining to pretend to be a text message alert or a dripping tap. The owner must therefore make efforts to give attention to the bird for desirable sounds if the undesirable ones are to become extinguished. If ignoring and replacing alone are not proving successful at decreasing the inappropriate vocalization after a few weeks of consistent owner response, then a more active approach of using social isolation may be warranted. Parrots are highly social and therefore the removal of company can be a powerful tool in encouraging a change in behaviour. If the parrot makes the undesired sound, the owner can turn their back on the parrot or walk out the room as a clear signal that this behaviour is not going to be reinforced. Owners must be told that this social isolation only needs to be momentary, as many old-fashioned texts advise prolonged periods of isolation and this can result in unnecessary stress for the bird. As soon as the parrot is showing any more appropriate behaviour, such as making more desired vocalizations or playing with a toy or feeding, the owner can return to the room and interact with the bird to reward it. In addition to addressing the vocalization, one must also consider why the parrot is attempting to gain attention in this way and ensure that he has sufficient social, physical and mental stimulation.

Attention-seeking vocalization

Social interaction is such an important part of natural parrot behaviour that they will often learn inappropriate behaviours as a means of gaining the attention of their owners. Any of the different types of vocalization can be inadvertently rewarded and develop a learned component whereby the parrot uses them to encourage their owner to respond to them. In this case the owner responses are crucial to resolving the excessive noise but the underlying motivations must still be addressed and the owner must ensure that the parrot is receiving sufficient social interaction and environmental enrichment.

Any of the vocalizations in Table 20.1 can become learned behaviours as a means of getting attention from the owner, but it is still crucial to address the original underlying motivation.

Aggression problems

There are several possible emotional motivations for aggression, including fear (defensive aggression), frustration (which can result in redirected aggression, i.e. the victim of the aggression is not the inciting stimulus), pain, excitement, inappropriate play or inappropriate exploration. Contextual motivations such as lack of sleep and possession-related aggression can also be identified and any aggressive behavioural response can be associated with a learned component. It is essential that the clinician accurately identifies the underlying motivation for any behavioural problem in order for appropriate treatment to be recommended. It is also important to rule out normal use of the beak as a supporting 'limb' where, for example the parrot might reach forward to grip the owner's arm or hand with its beak as it climbs on. This can be misinterpreted as 'biting' but careful questioning and observation can easily rule this out. When the bird is climbing on to the hand or arm, it will often test it for stability with its beak, so if the owner pulls away or wobbles their hand the parrot will not be keen to step on. Chapter 11 discusses bite inhibition training and the importance of owners teaching their bird to hold gently with their beak when using it as a third hand.

First aid advice for all cases of aggressive behaviour must involve educating the owner about their bird's communication strategies and helping them to avoid situations where the bird feels the need to bite. Safety advice is absolutely crucial in the case of large parrots, as they can inflict severe damage that is not only an issue in itself but also has severe consequences for the owner–pet relationship and may affect whether the owner is compliant and enthusiastic about resolving the issues. When giving advice it is important to be fully aware of the make-up of the household, because the situation will be more complicated and potentially more urgent if there are children, elderly or infirm adults or other pets involved. It is also crucial to be aware of whether the bird is flighted or has clipped wings, as this will make a difference to safety issues. However, regardless of whether the parrot can fly or not, it is essential in all cases to modify the social and physical environment such that the parrot does not feel any need to show aggressive responses.

In order for appropriate first aid advice to be given, a detailed discussion about the circumstances of when bites occur is required. For example, the parrot might be resting on the owner's lap or shoulder when their spouse appears suddenly and the bird bites the owner. In this case, asking the spouse to give a vocal warning that they are entering the room, and seating the owner in a chair away from the doorway, would give an opportunity to return the bird to its stand, distract it with a foraging toy, or cue another behaviour. Another example might be the parrot that bites when the owner tries to return it to its cage. As a management strategy the owner could place the bird's favourite food treat in the cage in view of the bird. This not only provides a motivation for the bird to want to re-enter his cage but also rewards him once he is in there. If children or pets are at risk, they must never be left unsupervised with the parrot, even if the parrot is caged. This is to protect the children and other pets but also to ensure that the parrot feels safe. If the parrot is fearful of these stimuli, then it can still feel threatened if they approach too close to its cage and this is not only a welfare issue but will also counteract any behaviour modification to alter the bird's perception of them.

Gloves should be used with extreme care. The author would only use gloves in an emergency situation where a bird needs urgent handling and the handler needs protection. When attempting to modify behaviour, it is crucial that the bird is not put in any situation that would result in aggression and there is a risk that handlers who are wearing gloves may feel protected and therefore continue with interactions that are detrimental to the bird's emotional health.

Fear, anxiety and phobia as a cause of aggressive behaviour

Full details regarding parrot body language associated with fear, anxiety and aggression are available in Chapter 11. Parrots that are anxious, fearful or phobic tend to use avoidance as their behavioural response if it is possible. The most common reason why parrots show aggression is because they feel threatened and avoidance is not an option. Treatment of fear aggression or avoidance behaviours involves identifying the trigger for the fear and then using the processes of desensitization and counter-conditioning to alter the parrot's perception. In order for these processes to be successful, owners must be given management advice to avoid aggressive or avoidance incidents whilst the training is carried out. First aid advice from the general veterinary practice needs to focus on helping the owner to identify the trigger(s) and reduce the parrot's exposure. In addition, educating the owner to be able to observe their bird's body language and look for subtle signs of anxiety will assist them in achieving the aim of ensuring that the parrot remains relaxed. One common sign of anxiety is the occurrence of displacement behaviours, i.e. a normal behaviour occurring out of context (see Chapter 11). Punitive or aversive techniques must be avoided, as these will exacerbate the fear response.

Territorial aggression

This contextual description relates to the expression of fear-related aggression in situations where the bird perceives its territory to be under threat. The nest site and territory are highly valued resources for a breeding bird and if they feel they are threatened they will defend these resources. First aid advice for these cases is the same as that in any case of fear-motivated aggression. Prevention of incidents is key and the location of the cage may need to be altered to reduce the exposure of the bird to perceived territorial insults. For example, if the cage is close to a doorway that people regularly walk through, moving the cage to another location may immediately reduce the problematic behaviour and the potential for the response to be reinforced each time the parrot lunges at people as they pass. Care must be taken when moving the cage to ensure that reduction of perceived threat does not lead to unacceptable restriction in the bird's access to ample social interaction and environmental enrichment. Counter-conditioning can be used to teach the parrot that people passing the cage are a pleasant experience and not a threat. General training of cues such as step-up, step-down and recall can also be very beneficial in territorial issues and signal to the bird that human approach is associated with a positive emotional experience.

Reproduction-related aggression

Aggressive behaviour that only occurs when the bird comes into breeding condition may be referred to as reproduction-related aggression. It is often displayed in the context of the bird's territory and is associated with a perception of threat to resources such as nesting sites (Fig. 20.1).

There is therefore the potential for overlap between the use of the terms territorial and reproduction-related, since one is a contextual and one a motivational label. The most serious issue that arises in pet homes is where the parrot pair-bonds with one member of the household and regards that person as a high-value resource, particularly when in breeding condition. At this time the parrot may increase contact calling to the bonded owner, be less tolerant of being handled and show

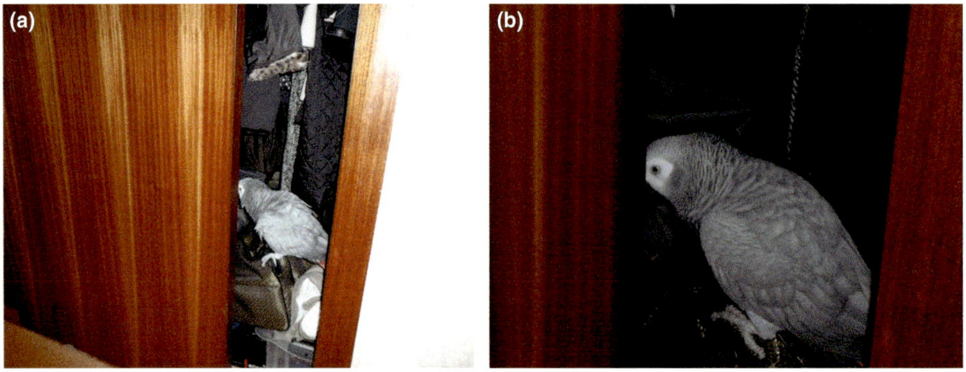

Fig. 20.1. **(a)** This African Grey aggressively defends her 'nest site'. All her feathers are raised and a bite is imminent if the person continues their approach. **(b)** Only the nape of the neck feathers are raised, showing a lower level of arousal. Access to this site must be prevented to avoid triggering defensive aggression. (Photo credit: A.G. Turner.)

sexual displays such as regurgitating to the owner, masturbation and nesting activities. This may occur seasonally or due to triggers such as the presence of potential nest sites, nesting materials or a high-fat diet. The bird may attack any people or pets approaching their pair-bonded owner, due to a perception of them as potential rivals.

Treatment involves careful management to prevent attacks on the other members of the family, which may involve reduced access to the people at risk of being attacked, short-term wing clipping as a safety measure, removal of access to potential nesting sites and reducing the fat context of the diet.

If the parrot is not already on a 12-hour day/night cycle and breeding condition has been triggered by a change in photoperiod (i.e., onset of spring), then artificially altering the photoperiod can help bring them out of breeding condition. It can be beneficial for other family members to become involved in interacting with and caring for the parrot but this may well need to occur in the absence of the bonded owner. Stroking the parrot over the back can be sexually stimulating and so this should be avoided. Counter-conditioning can be used in some situations to teach the bird that the other human members are not a threat, but this type of training may not be applicable in all cases. Hormonally driven behaviours are powerful innate responses and finding a reward of sufficiently high value to create a comparable positive association with other people may prove difficult. In extreme circumstances hormonal therapy may be required but this should only be considered in conjunction with a behavioural modification programme (Shoemaker, 2018).

Aggression around food

Wild parrots generally feed in areas of ample food and therefore competition over food is not high on their agenda and the need to use aggressive responses in defence of food resources is unusual. They tend to eat small items and are not predisposed to share an item of food once they have hold of it. In order to reduce the potential for birds to perceive threat in the context of people approaching them with food or attempting to remove food from them, it can be beneficial to teach them to share food items. This can be achieved through observation of the owner sharing and by training a 'leave' or 'drop' cue. In general terms, learned aggression can readily be unlearned by using appropriate reward-based training.

Learned aggression

In some texts, learned aggression may be referred to as a 'dominant' behaviour but the author does not feel this is a helpful way to classify aggression, as it tends to create an inappropriate atmosphere of competition between the bird and its owner. The use of the term 'dominance' in relation to avian behaviour is controversial and its use has been complicated by the assumption that hand-reared parrots, which have been imprinted on to humans, may have a greater tendency to interact with people in the same manner in which they would communicate with their own species. It is important to remember that even hand-reared parrots have the same genetics as their wild cousins and that wild parrots show very low levels of serious aggression and wounds are very rare. All social species have communication strategies to enable them to live in harmony with their conspecifics, because conflict uses up vital energy reserves and is not in the best interests of any individual. The debate about dominance is relatively academic and the more important issue is the need to obtain accurate descriptions of the behaviours that are seen in order to establish how they are motivated emotionally, how they have been learned and how best to respond to them. Behaviours should be discussed in terms of what actions precede the behaviour (antecedents) and what actions follow the behaviour (consequences), which is a far more productive and appropriate manner in which to examine interactions between parrots and their owners, as well as parrot–parrot interactions. Parrots are highly intelligent creatures and their prime aim in life is to maintain emotional stability through obtaining and keeping resources that are important to them for survival and for enjoyment.

Parrots will quickly learn to use any behaviour, including biting, as a means of communication with humans if they find it to be effective in achieving a desired outcome. For example, they may use biting as a means of retaining resources, as a means of entertainment or in order to remain on a desired perch. Some parrots delight in making their owners squeal or shriek when they bite too hard and can quickly learn this as an attention-seeking behaviour. In this case the owner must be advised to create situations whereby the parrot can gain attention through more appropriate means and ensure that the bird is rewarded and therefore reinforced when it shows more appropriate techniques for interacting with its owner.

Aggressive behaviours may develop in association with the bird adopting an elevated position in relation to the owner. For example, parrots may learn that they can stay on top of their cage or on a high perch if they bite their owner instead of coming on to the hand. Parrots feel safer when up high and therefore this is a rewarding outcome for them. Rather than assuming that the bird is attempting to 'dominate' the owner, it is more helpful to examine this behaviour in terms of the perceived reward and to resolve it by encouraging the parrot to come down to a lower height in order to feel safe (Fig. 20.2). Examining the environment in order to identify anything that is causing concern for the parrot and establish why it wants to remain high up is important.

Fig. 20.2. African Grey voluntarily coming down from top of cage for a food reward. (Photo Clare Wilson.)

Separation-related behaviours

This term is not clearly defined in the context of parrot behaviour but is sometimes used to describe behaviours that are displayed when the owner is either partially or totally absent. A wide range of behaviours can be displayed in this context, but a very common presentation is excessive vocalization when the owner is out of sight or out of contact. When the motivation for the vocalization is anxiety or fear, desensitization and counter-conditioning will form the basis of the treatment programme. Social interactions are a vital part of parrot behaviour and, depending on the background of the parrot and the individual circumstances of the owner, it may be appropriate to introduce an avian companion in some cases. However, this decision can only be made on a case-by-case basis and it should be remembered that introducing another parrot may be detrimental in some situations. Ideally the situation of isolation should be avoided whilst teaching birds new coping strategies to deal with separation, but this is rarely practical for owners who need to go out to work and go about their daily business. It is possible to adapt techniques that are more commonly associated with separation-related behaviour in a canine context, such as teaching a safety signal or associating a favourite foraging toy with owner absence.

Excessive vocalization due to owner absence may also be caused by frustration. However, this would usually also be seen in contexts where the owner is present but the parrot is unable to access them, for example when the bird is caged or when the owner is in another room. The parrot may also manifest its frustration with redirected aggression, perhaps towards toys in the cage if it is confined. Frustration is caused by the parrot having expectation or desire of a reward that is not forthcoming, so behaviour modification in this case involves altering expectations and guiding the bird into more appropriate responses. It is also vital to address any shortages in enrichment, as a well-stimulated parrot is less likely to show frustration responses.

Inappropriate owner–bird bonding

Parrots have a strong instinct to pair-bond and this can have serious consequences in the domestic setting. When the bonding involves a parrot and a human a range of behavioural responses can occur, including feather plucking, aggression and separation-related vocalization. It is important to identify whether an inappropriate pair-bond is an influencing factor in any behavioural presentation. Once a pair-bond has formed it can be extremely difficult to break, particularly when the owner and bird need to continue to live together. For this reason, the emphasis of first aid advice is to prevent birds from forming pair-bonds with their human carers. Ideally all members of the household should be involved in looking after the bird, rather than allowing one individual to become the focus of caregiving. It is also important to avoid inappropriate interactions such as responding to regurgitation or stroking on the back, which have sexual connotations in the context of parrot behaviour. Often owners thoroughly enjoy the special bond they have with their bird and owner education is vital to ensure that they understand the implications of pair-bonding from a parrot's perspective. Altering the owner's perception and their behaviour toward their avian companion can be extremely challenging.

Feather damage

Feather damage is a frequent behaviour problem observed in captive parrots. The birds may chew at its feathers or it may pluck them. Some parrots, particularly cockatoos, may also cause severe damage to soft tissue, which can become life-threatening. Feather-damaging behaviour can respond well to treatment if it is dealt with swiftly and effectively in its early stages, but if it is not addressed it can develop into a severe compulsive disorder that is far more complex, if not impossible, to resolve fully (Fig. 20.3).

The possible underlying motivations for this form of self-mutilation are numerous and there are often several factors playing a role. A full medical work-up by an experienced avian veterinarian is essential in order to rule out underlying diseases, such as endoparasites, dermatological conditions, nutritional imbalance, prolonged reproductive behaviour or systemic disease. Assessment of the husbandry is also

Fig. 20.3. Male Eclectus with long-term feather plucking disorder. (Photo Clare Wilson.)

crucial in order to identify contributory factors, such as environmental cigarette smoke, irritation following a poor technique of wing trimming, lack of access to clean bathing water and lack of sleep, which can seriously affect feather condition and result in self-damage. The infliction of damage may occur during normal preening behaviour which becomes extended in time or escalated in intensity due to the presence of a stressor, but it can also occur as a specific activity where the bird interrupts another activity, such as eating or playing, to damage their feathers. Identification of the pattern of the behaviour can be useful in identifying the underlying motivation. Over-preening is more likely to be associated with triggers such as moulting, poor wing trimming or lack of stimulation (particularly limited foraging opportunities) while interruption of another behaviour would be more likely to occur if there is underlying pruritus (for example associated with dry skin due to lack of water bathing, or with internal parasites such as giardia) or the bird is suffering from a compulsive disorder. Although further research is warranted, it seems likely that there is a hereditary component to feather plucking (Garner *et al.*, 2006) as well as an underlying neuropathology and psychopathology, which may be related to developmental inadequacies such as maternal deprivation effects or environmental effects. The highly complex nature of this disease makes it very challenging to treat (Fig. 20.4a).

Research has shown a clear link between foraging opportunities and a decrease in feather-damaging behaviour (Meehan *et al.*, 2003) and therefore owners should be advised to focus on increasing environmental enrichment, especially in the form of foraging activities. Size of

Fig. 20.4. (a) Relinquished pet Greenwing macaws still feather plucking despite pair bonding and living as close to natural conditions as possible. **(b)** Providing nuts in shells increases the proportion of time spent feeding. (Photos Clare Wilson.)

food should also be considered, as Rozek and Millam (2011) found that Orange-winged Amazons preferred oversized pellets and were prepared to work harder to gain access to them. In addition, they spent longer periods manipulating these larger pellets with their beaks and feet and had a more similar foraging time budget to their wild counterparts. Similarly, simple changes such as feeding whole nuts in shells rather than shelled nuts greatly improved the parrot's experience in manipulating, destroying and foraging (Fig. 20.4b).

As mentioned in Chapter 11, recent research regarding the gut–brain axis and the connection between gut health and the general nervous system (Suchodolski, 2018) may also be highly relevant to resolving feather-damaging behaviour and this is an area for future research. Providing adequate and appropriate bathing opportunities can also be beneficial to help improve feather condition and encourage normal preening behaviour. Research by Murphy *et al.* (2011) on captive Orange-winged Amazons found that birds preferred to bathe approximately every 4 days and, in contrast with wild Amazons, preferred bathing in the mornings. One significant finding of this research was a reduction in preening for the hour following bathing as compared with the hour prior to bathing, as the birds dried off. This may be of benefit in discouraging feather-damaging behaviour. Ensuring that the bird has freedom to express all its natural behaviours is essential and, to this end, the introduction of a conspecific may be considered appropriate. However, it is important to ensure that owners are aware of the behavioural needs of parrots before considering taking on another avian companion and the approach may not be appropriate in some cases.

Owners should actively reinforce all desirable behaviours, such as relaxing on a perch, playing with toys and using foraging enrichments, by giving the bird quiet praise and attention. If the bird does start to damage feathers inappropriately it is best for the owner to ignore this behaviour in order to avoid the risk of accidental reinforcement but it is important to remember that birds are social and that, in cases where plucking is triggered by a lack of interaction from the owner and a lack of social enrichment, isolation is going to cause further stress rather than help to resolve the problem. The advice given in some texts that the owner should walk out of the room and only return when the bird has stopped damaging the feathers can be detrimental and should be treated with great caution. In cases of extreme self-trauma, it may be necessary to use barrier methods, such as jumpers or collars, to limit further damage but these should only be fitted by experienced veterinarians or behaviourists, as many birds will not tolerate such devices. Any intervention that increases negative emotional arousal should be avoided or used with extreme caution.

Research into the efficacy of medication for feather-plucking disorders is limited but therapy can be trialled with care on a case-by-case basis. When studying epilepsy, Amin *et al.* (2017) identified significant differences between dose rates required in different avian species, suggesting that care must be taken with dose rates for behaviour-modifying drugs. Administration of medication must also be considered and owners should be advised and assisted with training low-stress handling techniques (Cook, 2012; Speer *et al.*, 2018). In relation to administration techniques, methods of delivery must also be considered (Coutant *et al.*, 2018) and compounding of appropriate formulations (Powers and Davidson, 2018). As with all behaviour problems, medication must be considered an adjunct to behaviour modification and should ideally only be used in combination with referral to an avian behaviourist. One placebo-controlled trial showed that beneficial effects of clomipramine (Seibert *et al.*, 2004) and other potential medications would include serotinergic reuptake inhibitors such as fluoxetine. Haloperidol has been recommended by some but, in the author's experience, can cause severe sedation and its use is not advised. Van Zeeland (2018) provided an in-depth review of behaviour-modifying drug use in avian patients and readers are strongly advised to read further prior to prescribing, unless under the guidance of an avian veterinary behaviourist, as there is insufficient information available regarding efficacy, pharmacokinetics and dose rates to be able to make recommendations in a general text at this point in time.

Finally, owners must be advised that efficacy is variable, the use of such medications is off-licence and the bird is highly likely to relapse if medication is stopped.

Stereotypical behaviours (excluding compulsive feather plucking)

Stereotypies are defined as fixed patterns of behaviour performed repetitively and having no obvious goal or function which are rarely, if ever, observed in the wild and occur infrequently in large enriched captive environments (Engebretson, 2006). They are highly significant as an indicator of compromised welfare. Locomotor stereotypies are the most common type seen in pet parrots and might involve repetitive pacing up and down a perch or route tracing in the cage. There is a wealth of research in laboratory, farm and zoo settings demonstrating that lack of environmental enrichment relevant to the natural requirements of the species is highly correlated with the development of stereotypical behaviour. Enrichment for a social species such as parrots does not just refer to the provision of foraging opportunities and other physical and mental stimulation, but also must address opportunities for social interaction. Stereotypies usually respond well to improvements in enrichment but may never be fully resolved. An appropriately qualified behaviour practitioner may be able to give additional advice in the resolution of stereotypies.

Chewing and destructive behaviour

Parrots have an innate need to chew objects, not only during food manipulation but also during investigatory activity and play (Fig. 20.5).

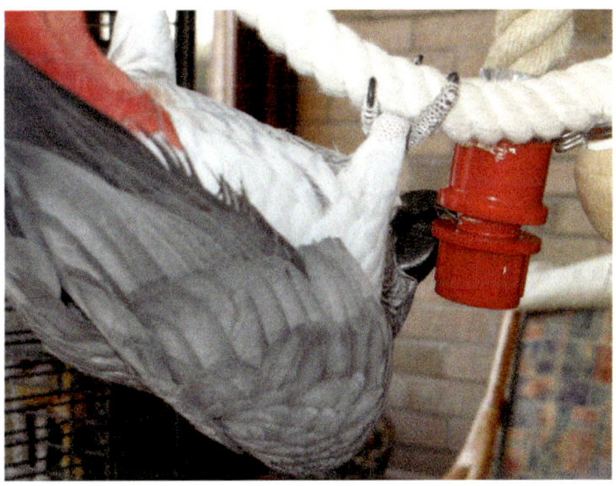

Fig. 20.5. African Grey enjoying a manipulation task of unscrewing plastic. (Photo A.G. Turner.)

If the parrot is causing inappropriate destruction around the home, such as chewing wooden door frames or skirting boards, the owner needs to provide opportunities for the bird to express this behaviour with more appropriate objects. Education plays a key role in solving this problem, as parrots cannot be expected to simply stop chewing but rather need unrestricted access to an appropriate outlet for this instinctive behaviour. There are numerous natural materials that are safe for parrots to chew and many parrots will thoroughly enjoy destroying a cardboard box, fruit-tree branches, wooden clothes-pegs or natural corks. Owners should carefully examine the toys that their parrot has access to and bear in mind the birds' destructive nature, as many commercial toys are of little value to parrots. There are an increasing number of appropriately made parrot toys on the market using wood, cardboard, leather and ropes and these can provide hours of safe chewing (Fig. 20.6).

However, these commercially available items are expensive and homemade toys can provide just as much fun for the bird without the cost (Fig. 20.7). Novelty is important for environmental enrichment to be effective and providing the bird with items such as new cardboard

Fig. 20.6. African Grey enjoying the destruction opportunities of a well-designed parrot toy. (Photo A.G. Turner.)

Fig. 20.7. Foraging in cardboard boxes provides outlets for destruction without the expense of toy shopping. (Photo Clare Wilson.)

boxes, empty cardboard tubes from toilet rolls, natural corks from empty wine bottles and paper bags from grocery shopping on a daily basis is cheap and straightforward. This will encourage a good compliance rate and encourage owners to provide novel toys at a far more frequent rate.

References

Amin, F., Dar, A.H., Osama, K. *et al.* (2017) A species dependent response to the pro-epileptic drug pentylentetrazole in birds. *Brain Research Bulletin* 134, 189–194.

Cook, E.K. (2012) Teaching avian patients and caregivers in the examination room. *Veterinary Clinics of North America: Exotic Animal Practice* 15(3), 513–522.

Coutant, T., Vergneau-Grosset, C. and Langloise, I. (2018) Overview of drug delivery methods in exotics, including their anatomic and physiologic considerations. *Veterinary Clinics of North America: Exotic Animal Practice* 21(2), 215–259.

Engebretson, M. (2006) The welfare and suitability of parrots as companion animals: a review. *Animal Welfare* 15, 263–276.

Garner, J.P., Meehan, C.L., Famula, T.R and Mench, J.A. (2006) Genetic, environmental, and neighbor effects on the severity of stereotypies and feather picking in Orange-winged Amazon parrots (*Amazona amazonica*): an epidemiological study. *Applied Animal Behaviour Science* 96, 153–168.

Jayson, S.L., Williams, D.L. and Wood, J.L.N. (2014) Prevalence and risk factors of feather plucking in African Grey Parrots (*Psittacus erithacus erithacus* and *Psittacus erithacus timneh*) and Cockatoos (*Cacathus* spp.). *Journal of Exotic Pet Medicine* 23, 250–257.

Meehan, C.L., Millam, J.R. and Mench, J.A. (2003) Foraging opportunity and increased physical complexity both prevent and reduce psychogenic feather picking by young Amazon parrots. *Applied Animal Behaviour Science* 80, 71–85.

Murphy, S.M., Bruan, J.V. and Millam, J.R. (2011) Bathing behaviour of captive Orange-winged Amazon parrots (*Amazona amazonica*). *Applied Animal Behaviour Science* 132, 200–210.

Powers, L.V. and Davidson, G. (2018) Compounding and extra-label use of drugs in exotic animal medicine. *Veterinary Clinics of North America: Exotic Animal Practice* 21, 195–214.

Rozek, J.C. and Millam, J.R. (2011). Preference and motivation for different diet forms and their effect on motivation for a foraging enrichment in captive Orange-winged Amazon parrots (*Amazona amazonica*). *Applied Animal Behaviour Science* 129, 153–161.

Seibert, L.M., Crowell-Davis, S.L., Wilson, G.H. and Ritchie, B.W. (2004) Placebo-controlled clomipramine trial for the treatment of feather picking disorder in cockatoos. *Journal of the American Animal Hospital Association* 40(4).

Shoemaker, N.J. (2018) Gonadotrophin-releasing hormone agonists and other contraceptive medications in exotic companion animals. *Veterinary Clinics of North America: Exotic Animal Practice* 21, 443–464.

Speer, B.L., Hennigh, M. and van Zeeland, Y.R.A. (2018) Low-stress medication techniques in birds and small mammals. *Veterinary Clinics od North America: Exotic Animal Practice* 2, 261–285.

Suchodolski, J.S. (2018) Gut brain axis and its microbiota regulation in mammals and birds. *Veterinary Clinics of North America: Exotic Animal Practice* 21, 159–167.

Van Zeeland, Y. (2018) Medication for behaviour modification in birds. *Veterinary Clinics of North America: Exotic Animal Practice* 21, 115–149.

21 Arranging Referral: When and Who to

Trudi Atkinson

There has been increasing recognition over recent years that an animal's behavioural and emotional well-being is no less essential to good welfare than its physical health (Boissy *et al.*, 2007a,b; CAWC, 2008). Admittedly it can be argued that some behaviour problems may be more of an issue for the owner than for the pet. However, even when this is the case, the animal's welfare may still be at risk because of the disruption to the pet–owner relationship and the potential for intentional or unintentional abuse by the owner, either due to frustration or through misguided attempts at punishment or training. In addition, there is the increased likelihood of relinquishment of the pet by re-homing or euthanasia. The prevention and treatment of behavioural disorders should therefore be considered to be of equal importance to the diagnosis and treatment of physical ill-health within the veterinary context. Sadly, this is not always the case and behaviour problems can sometimes be dismissed as being of minor importance or having little to do with veterinary care.

When behavioural issues are brought to the attention of the veterinary practice it is essential not to disregard them, even if the owner does not appear to rate the problem high on their list of concerns. Discussion may well reveal that the pet's behaviour, as well as being a welfare concern for the animal, is having a seriously damaging effect on the owners' lives. Many owners do not realize that help is available via the veterinary practice and may be embarrassed to mention these issues during a veterinary consultation. Lack of awareness of the link between emotional and physical health has led many owners to consider behaviour to be something that does not fall within the remit of their veterinary practice, which can explain why it is not uncommon for clients to fail to mention their pet's behaviour to the vet or nurse but then ask the practice receptionist for details of a local dog trainer or behaviourist.

Some owners may be reluctant to discuss their pet's behaviour or might 'play down' serious or potentially serious issues, especially aggression. This might be due to fear of prosecution or the belief that euthanasia will be the only solution that a veterinary practice has to offer. In order to increase the potential for clients to open up about their behavioural concerns, it is important for the practice to have an empathic and non-judgemental approach.

Questioning clients about their pet's behaviour during routine consultations and health checks is a good way of increasing awareness that the practice is interested and concerned about behavioural issues (Fig. 21.1). If these questions are part of every routine consultation, it will be easier for clients to talk about their pet's behaviour when issues arise. In addition, regular questioning may reveal the existence of actual or potential behaviour problems that may have been a major cause of distress for the owner or the animal for months or even years.

DOI: 10.1079/9781800621312.0021

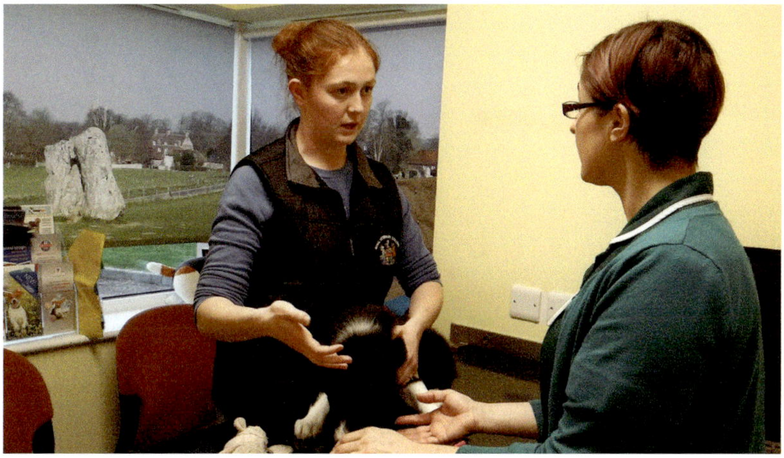

Fig. 21.1. Questioning owners about their pet's behaviour during routine consultations and health checks may reveal the existence of problems that might otherwise not be mentioned. (Photo: Trudi Atkinson.)

When to Refer

As with cases within any veterinary discipline, behaviour cases can vary greatly in severity and complexity. Offering 'in-house' behaviour consultations can have undoubted benefits for both the practice and the client but only if members of the practice staff have appropriate experience and qualifications. The quality of service that the practice is able to offer should always be considered along with the best option for the client and their pet. The Royal College of Veterinary Surgeons code of conduct states that a veterinary surgeon should recognize when a case or treatment is out of their area of competence and be prepared to refer to an individual or organization that they are satisfied is competent to carry out the investigations or treatment involved (RCVS, 2018). Although this primarily relates to medical or surgical cases, the same should also apply to behaviour cases.

Taking a short history of the most relevant factors and offering appropriate and often essential first aid advice is an important part of the initial response to a behaviour case. If the practice has sufficient resources, then treatment in-house can be an effective way forward. However, all cases must be closely monitored and if the behaviour problem increases or is not improved within the expected timescale, or the client wishes to investigate the issues at a level that the practice is unable to offer, then the option of referral should be offered.

The decision by any veterinary practice as to whether to refer or treat a behaviour case 'in-house' should be based on a number of important considerations, including: (i) education and experience of practice staff; (ii) time and space; (iii) the nature of the problem and associated risks; and (iv) species.

Education and experience of practice staff

Treatment for any behavioural problem should be based on correct identification of the underlying cause and this is no different from the appropriate veterinary approach to any clinical case. Failure to do so may result in a worsening of the current problem and even the development of further, possibly more serious behavioural issues. Treatment or advice offered should therefore be based on sound scientific evidence and any veterinary surgeon or nurse who is working with a patient with a behavioural presentation must possess sufficient knowledge based on both experience and study.

Ideally professional veterinary practice staff should have sufficient behavioural knowledge to enable them to correctly answer basic questions posed by clients and to provide preventive or first aid advice for behaviour problems (Fig. 21.2). There has been an increase in recent years in the behavioural medicine content of the undergraduate veterinary degree

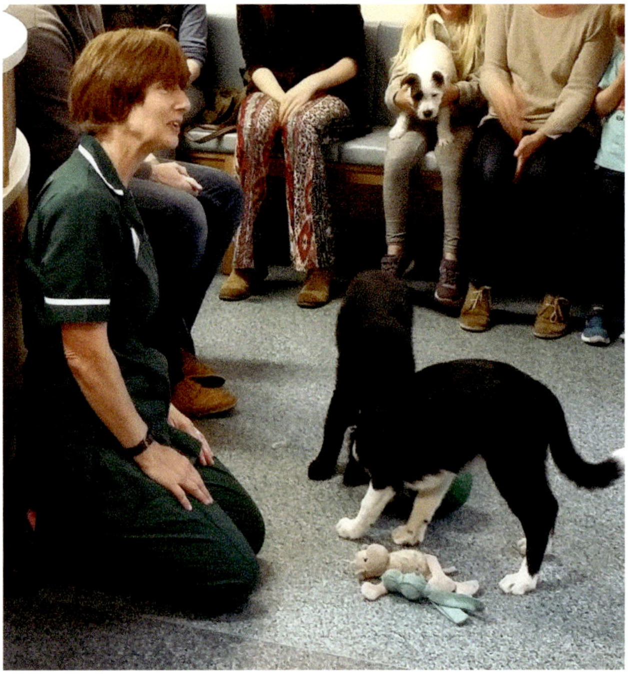

Fig. 21.2. Professional practice staff should have sufficient behavioural knowledge to answer basic questions and provide preventive or first aid behavioural advice. (Photo: Trudi Atkinson.)

course at some of the UK veterinary schools but this is not consistent. Likewise, there are some veterinary nursing courses that offer good education in this field but there is no standardization of that provision. For veterinary surgeons and nurses who are already qualified and feel that they have not received sufficient education during their training, a basic general practice level of knowledge can be obtained by reading relevant texts and by attending recognized continued professional development (CPD) events, including streams within veterinary congresses. As with any education, it is very important to check the qualifications of any CPD provider, as the standard of information can be highly variable.

The treatment of established behaviour problems can be more complex or serious than may at first appear and these cases often require a far greater level of experience and relevant education, usually to degree or post-graduate level. The science and study of companion animal behaviour is continually evolving and ongoing relevant CPD is essential. It should also never be assumed that gaining knowledge and experience of one species is sufficient to be able to deal with the behavioural issues of all other species.

Time and space

The treatment of companion animal behaviour problems can be extremely time consuming. Correctly identifying the underlying cause(s) requires a holistic approach to history taking, exploring all aspects of the animal's health, management and behavioural history. Once sufficient information regarding the animal's motivation and past influences on behaviour has been acquired, differential diagnoses explored and eventually the underlying cause(s) of the behaviour identified, this needs to be explained to the client, who will also need to be given advice as to how their pet's behaviour may be modified. This often includes the demonstration and teaching of relevant training techniques.

On average a standard behaviour consultation will last approximately 2–3 hours, but a practitioner does need to be prepared that some consultations can last longer.

A large amount of verbal information will have been passed to the client during the consultation, all of it essential to the client's understanding of the problem and subsequent treatment. It would be unreasonable to expect the client to remember it all; therefore, following each consultation, a full and individualized written report for the client needs to be prepared.

Follow-up communication to assess progress, provide support and sometimes to explore additional or related issues is also necessary. Follow-up consultations may also be required depending on the nature or complexity of the original problem and the number of presenting behaviour issues.

Another requirement for a behavioural consultation is adequate space. The demonstration and practice of training exercises can require a fair amount of both indoor and outdoor space (Fig. 21.3). Additionally, an animal's behaviour can be influenced by the people that it regularly has contact with, so it is sometimes

Fig. 21.3. The demonstration and practice of necessary training exercises may require more space than is available in many veterinary practices. (Photo: Natalie Light.)

necessary for all people who are involved in the animal's care and training, or who are directly affected by the pet's behaviour, to be present at the consultation. This may involve the whole family plus, in some cases, extended family members, friends, neighbours and even dog walkers and pet sitters. It is advisable to see other family pets as well, as these may also have an influence on the behaviour of the 'problem' animal. Taking all of this into account it is clear that the average small veterinary consulting room might not be the most practical environment for a full behaviour consultation.

The nature of the problem and associated risks

When considering whether to treat in-house or to refer on, it is important to be aware of the risks, immediate or potential, that could arise if the problem behaviour continues or worsens as a result of inappropriate or insufficient investigation or treatment. These include not only the risks to the animal, the owner, other people and other animals, but also potential risks to the person conducting the behavioural consultation. If client expectations are not met in relation to a behavioural problem that they have discussed with the veterinary surgeon or nurse, there can also be the potential risk of losing client confidence in the veterinary practice as a whole.

Specific considerations may be required in relation to certain behavioural presentations, such as: (i) aggression; (ii) destructive behaviour and house soiling; and (iii) excessive vocalization.

Aggression

Cases of reported aggression should never be dismissed or viewed lightly, even if the level of aggression appears mild, or if the practice suspects that the owner might be exaggerating or misinterpreting their pet's behaviour. It is always wise to remember that the behaviour you see in the practice might not be the same behaviour that the animal demonstrates in other contexts.

A risk assessment approach is needed in all cases of reported aggression. While some risks may be obvious, others may be less so. The risk of injury to the owner or other persons is a major factor to consider. This can be obvious in cases where aggression towards people has already been demonstrated but human risk can also be a factor in cases of aggression between animals, with owners being at risk of injury when attempting to separate fighting cats or dogs, and in situations where the owner or handler is attempting to restrain the animal and the interaction induces a frustration response and the risk of redirected aggression.

When advising owners of potentially aggressive animals, the safety of all concerned, including the professional dealing with the case, must be of primary concern. An appropriate level of education and experience is needed in order to be fully aware of the potential risks involved with various behavioural interventions and what is most suitable to the individual animal and owner. Care must be taken that any behaviour modification advice given is clearly understood by the client and is not likely to unintentionally exacerbate the animal's aggression by, for example, inducing or increasing frustration.

Particular care must be exercised if children are involved, either as a target for the animal's aggression, or likely to be involved in handling of the animal, or just likely to be close by when the animal becomes aggressive. When this is the case, the practitioner must have a good knowledge and understanding of how the relationship and signalling between animals and children (young children especially) can differ from that between animals and adults.

Aggression towards other animals should not be regarded as any less important. The welfare of other animals is, without doubt, a significant concern and, as previously mentioned, aggression towards other animals may also result in human injury. Another important factor is the welfare of the aggressive animal itself. Many animals that display aggression towards others do so as a means of defence because they feel fearful and threatened. This in itself should be a welfare concern. In addition, these animals can be at risk of physical injury as a result of defensive or retaliatory behaviour by people or other animals. The owner or handler, in an attempt to punish or establish control, might also be more inclined to use harsh corrective methods or devices that could cause the animal increased fear, pain or even injury.

Welfare and injury are not the only concerns when treating aggression cases. For dog owners especially, there is also the risk of prosecution (Fig. 21.4).

In England and Wales under Section 3 of the *Dangerous Dogs Act 1991* (DDA) (amended in 1997 and 2014), a criminal prosecution can be brought against the owner or the person in charge of any dog, regardless of breed or type, that is considered to be 'dangerously out of control'. This is defined as 'any occasion on which there are grounds for reasonable apprehension that it will injure any person or assistance dog, whether or not it actually does so' (section 10(3) Dangerous Dogs Act 1991). An offence may be committed no matter where the incident takes place, even if it is in the owner's home or garden. However, under section 3(1A) of the Act a person is not guilty of an offence if the dog attack is directed towards a trespasser entering their home. But this exception does not apply to dog attacks involving trespassers in gardens, driveways, or outbuildings.

Under the DDA the police have the power to seize the dog until the court's final determination. If found to be 'dangerously out of control' and especially if injury has been sustained, the dog is likely to be destroyed unless the owner can prove that the dog does not constitute a danger to public safety and that they are a fit and proper person to be in charge of a dog. If so, then a Contingent Destruction Order may be imposed, which requires that the dog is kept under proper control and, if not, it will be destroyed without any further investigation. Contingent destruction orders often have conditions imposed by the court, such as the dog must be on a lead and muzzled at all times when in a public place. In some cases, multiple and stringent conditions are imposed at the discretion of the magistrate.

Fig. 21.4. Under The Dangerous Dogs Act 1991 (amended in 1997 and 2014) a criminal prosecution can be brought against the owner or keeper of a dog that injures or gives rise to reasonable apprehension that it will cause injury. (Photo: Pixabay.com.)

If the owner, or person in charge of the dog at the time of the incident (the keeper), is found guilty they could also face a prison sentence, varying from a maximum of three years in cases involving the injury or death of an assistance dog, five years in cases involving injury to a person, and up to 14 years if a person dies as a result of their injuries. If it is considered to be a non-aggravated offence (where no injury has been sustained) the owner or keeper could still face imprisonment for a maximum of six months, a community order, or a heavy fine (Sentencing Council, 2016). Where destruction is ordered, the court may also order that the owner/keeper be disqualified from having custody of a dog for a time period as the court sees fit (pet consultant Candy D'sa, 2018, personal communication). In addition, anyone found guilty under the DDA will have a criminal record, which can have serious implications for many other aspects of their life.

The DDA is not the only legislation relevant to aggressive or potentially aggressive dogs. There is also Section 2 of the *Dogs Act 1871*. Under this act, which is civil legislation and not criminal, a dog can also be considered dangerous no matter where an incident takes place. But unlike the DDA, where a guilty verdict necessitates a destruction order or a contingent destruction order, the Dogs Act allows the Magistrate's Court full discretion over what happens to the dog if the owner is found guilty. While the options include possible destruction, in many cases the owner is issued with a Control Order and made to pay costs. Another difference is that under this Act proceedings are made against the dog's owner and not the person in charge of the dog at the time of the incident (Dogs Act, 1871).

In Scotland the *Control of Dogs (Scotland) Act* was introduced in April 2010 and the Dogs Act 1871 was repealed in February 2011. Under the Control of Dogs Act, power is given to an authorized officer to issue a Dog Control Notice if a dog is not considered to be under effective control or if its behaviour gives rise to alarm or apprehension. A breach of a Control Order can lead to a large fine, disqualification from having custody of a dog and an order for the dog to be destroyed if the Court considers the dog to be dangerous (Control of Dogs (Scotland) Act, 2010).

Dog control legislation may come under further review and while the information given in this chapter is correct at the time of writing, changes may be made at a later date. It is also important to be aware that the law will be different in countries other than the UK and can vary between local authorities within the UK as a result of local by-laws. It is therefore advisable for readers to check the current law in their area regularly and how it relates to dogs.

Destructive behaviour and house soiling

These behaviours can cause immense distress to the owner and be seriously damaging to the pet–owner relationship. The financial risk for the owner should also be considered, since they may need to replace household furniture and fittings damaged by digging, scratching, chewing, or soiling, or have these items repaired, sometimes at great cost. If the behaviour occurs when the animal is separated from the owners, the clients may be in a situation where they are unable to leave the animal alone in the house, which can have a seriously damaging effect on their social and work life.

Destructive or house-soiling problems can also have a damaging effect on an owner's social and family life if they no longer feel able to invite friends and family to their home because of odours, mess and damage caused by their pet. In addition, the welfare of the animal must be considered, as these behaviours can often be indicative that the animal is distressed (Fig. 21.5).

A full assessment of the underlying cause must be carried out and treatment based on the findings of that assessment. Measures such as crating may remove the animal from the problem and 'resolve' issues for the owner. However, restrictive confinement can cause extreme distress for some cats and dogs, increasing the risk of stress-related behavioural and health issues and even injury to the animal if it tries to escape.

Excessive vocalization

A dog may bark excessively for a variety of reasons, the most common being separation-related distress, learned attention-seeking or territorial

Fig. 21.5. Problems such as excessive barking and destructive behaviour are not only upsetting for the owner or caregiver but can also be financially damaging and could even result in prosecution. Such behaviours can also be a sign that the animal is distressed. (Photo: Trudi Atkinson.)

defence. It is essential to correctly determine and address the underlying cause rather than simply attempt to suppress the behaviour, as this may result in the development of alternative ways of coping that can include aggression, destructive behaviour and self-trauma.

A continually or excessively barking dog can be a source of distress not only for the owners but also for all those within earshot and is a common cause of neighbourhood disputes. If a local authority receives a complaint about a barking dog, it must investigate. An informal warning letter might be sent to the owner initially but if this is unsuccessful a noise abatement notice may be served. If the local authority decides that the owner or keeper of the dog is failing to comply, in other words if the problem continues, the defendant may then face prosecution under the Environmental Protection Act 1990 (EPA, 1990), potentially resulting in a heavy fine and a Criminal Behaviour Order requiring the offender to reduce the number of dogs owned (Cooper and Co., 2020).

Species

While much of the available behavioural literature relates to dogs, the behavioural problems of other species are of no less importance. The distress caused to the owner and the welfare implications for the animal are just as relevant for all species. Feline behaviour, especially the causes and treatment of stress in cats, is highly relevant to the veterinary practice since research has shown that chronic stress can be an underlying factor in a number of physical conditions (Bourdin, 1997; Cameron and Casey, 2004; Landsberg, 2008; Stella *et al.*, 2011; Westropp *et al.*, 2006) (Fig. 21.6).

Who to Refer to

Once it has been decided that a behaviour case needs to be referred, the next step is to decide who to refer to. Obviously, the person that the case is referred to should be able to offer more in terms of relevant education, experience and practical help than can be offered by the veterinary practice.

Referral to a veterinary surgeon specializing in behavioural medicine is the most familiar form of referral, as it will involve exactly the same process as referral to a veterinary colleague in any other discipline. The veterinary behaviourist will take on the duty of care of the patient in the usual way. If referral is made to a non-veterinary behaviourist the duty of care will remain with the referring veterinary surgeon, but this does not mean that the referral should be any less of a professional process.

The referring veterinary surgeon and the pet owner need to feel assured that the behaviourist will take all reasonable steps to ensure that the behaviour modification or training techniques that they use or advise will not cause the animal any unnecessary or avoidable pain or distress.

Identifying a suitable person to refer behavioural cases to can be a minefield and the fact that the referring veterinary surgeon will retain duty of care if the person they refer to is not a veterinary surgeon necessitates careful selection. It is not only inappropriate from an animal welfare perspective to refer cases to someone who is not suitably qualified, but it is also potentially detrimental to the veterinary surgeon's professional standing. Referring to an appropriately qualified behaviourist is important in order to ensure a competent, professional service.

Fig 21.6. Stress has been shown to be a contributory factor in a number of feline disease conditions. Behavioural assessment and treatment aimed at reducing stress in cats is therefore highly relevant to the veterinary practice. (Photo: Pixabay.)

If the client has pet insurance cover for their animal it is possible that the cost of the behavioural consultation may be covered, at least in part, by some insurance companies, but cover may depend on the accreditation of the practitioner involved.

Designation

There are several titles that individuals may use to indicate that 'their job' is to treat companion animal behaviour problems. These include terms such as whisperer, listener, psychologist, behaviourist, behaviour counsellor, behaviour therapist and many more.

None of these titles are protected and currently there is no statutory requirement as to who can treat companion animal behaviour problems or what a practitioner can call themselves. Just because an individual may claim to be able to treat behaviour problems, or has an impressive title, it is not a guarantee that they have the education, experience, or practical resources to do so safely and effectively.

Qualifications

Before referring a behaviour case it is essential that a veterinary practice investigates the qualifications of any potential behaviourist thoroughly. It is important to find out not only which qualifications the person claims to hold but also where the qualifications were obtained from and whether they are current, relevant and sufficient. This might not be an easy task, as there are numerous qualifications that might be cited and a range of course providers offering qualifications at varying standards in the field of companion animal behaviour. Post-nominal letters may give some indication as to an individual's qualifications but are not always a reliable endorsement that the qualifications achieved are relevant or sufficient. In addition, post-nominals may relate not to qualifications achieved but to memberships of groups or organizations that might be irrelevant or have minimal or no entry requirements.

A reputable behaviourist should be willing to share details of their academic and clinical achievements and provide anonymized examples of reports to clients which, whilst using scientifically based and ethical techniques, should be written in clearly explained, simple terms that can be easily understood and applied.

Continued education

As with any other science-based profession, a clinical animal behaviourist needs to keep up to date with the latest research and to revise past education in order to maintain and continue to develop their professional skills and knowledge. Some assurance that the individual maintains a level of continued professional development (CPD) is therefore desirable. Evidence of current and up-to-date accreditation should provide this assurance, since it is dependent on demonstration that a certain level of annual CPD has been undertaken.

Experience

The experience of an individual is clearly important and needs to be taken into account when considering referral. However, experience without initial education and CPD can be of little worth, as this might just mean that the same mistakes and inappropriate methods are being repeated.

The species to be referred

The behavioural repertoire and motivations to perform behaviours can vary greatly between different species and it is crucial that each species is dealt with according to its species-specific needs. The veterinary practice therefore should feel confident that the person an animal is to be referred to has sufficient knowledge and experience of that species. Just because a person has adequate behavioural knowledge and experience of one species, it should not be assumed that they are as knowledgeable and experienced with the behavioural issues of other species.

Under the Certificated Clinical Animal Behaviourist (CCAB) scheme run by the

Association for the Study of Animal Behaviour (ASAB), clinicians can only be accredited for species in which they have been fully assessed.

Insurance

It is essential that any behaviourist that a veterinary practice refers to holds full public liability and professional indemnity insurance so that they are able to meet any liabilities that they may incur.

Ongoing support for clients

The behavioural consultation will involve identification of the underlying cause of the problem and education of the owner as to how they can work to alter their pet's emotional reactions and resulting behaviours. Ongoing support following the consultation is essential, both in terms of phone or electronic support and in terms of practical application of the advice. It is therefore important to select a behaviourist who is either able to offer the ongoing practical support themselves or is able to recommend the services of a competent trainer or animal behaviour technician who, under the supervision of the behaviourist, can assist the client in carrying out the recommended behaviour modification programme.

Membership of groups or organizations

There are a number of companion animal behaviour groups and organizations. Competent and reputable behaviour practitioners will belong to an organization that sets strict academic and clinical standards for their members and ensures that they are insured, adhere to a code of conduct, and regularly update their knowledge by undertaking continuing professional development.

Not all organizations have strict selection procedures. Membership of some bodies may be granted after completion or attendance at a specific educational course. The content and teaching of such courses can vary tremendously and although some are of a high standard, others may provide only basic or in some cases inappropriate and non-scientifically-based content. Some organizations require no evidence of education or experience at all and only ask for the payment of a membership fee. Included in this category are organizations that, although reputable and highly acclaimed, do not represent qualified professionals but are interest groups open to anyone, regardless of academic qualification or experience in companion animal behaviour. These include the British Veterinary Behaviour Association (BVBA) and the Association for the Study of Animal Behaviour (ASAB). Even so, some individuals may cite membership of such bodies that could be incorrectly read as evidence of their education and abilities. (Membership of ASAB should not be confused with ASAB accreditation (CCAB), which does require evidence of high-level academic qualification in relevant subjects and practical experience, plus theoretical and clinical assessment.)

It is no surprise that pet owners and members of the veterinary profession can find it confusing when trying to find a reputable and suitably qualified companion animal behaviourist.

The Animal Behaviour and Training Council

In 2008 the Companion Animal Welfare Council (CAWC) presented a report outlining the need for regulation of behaviour therapy and animal training (CAWC, 2008). In response to this the Animal Behaviour and Training Council (ABTC) was developed in 2010 as a regulatory body and umbrella organization that recognizes practitioner organizations that set the necessary standards required for their members' practitioner roles.

The ABTC recognizes and holds registers for five separate practitioner roles:

- **Clinical Animal Behaviourist** (ABTC-CAB). This standard relates to the clinical application of animal behaviour and the modification of problematic, inappropriate, or dangerous behaviour. This can also include behaviours that have a link to physical changes and pathologies that require diagnosis in collaboration with a veterinary surgeon. Individuals must have an extensive understanding of clinical animal behaviour and related scientific and clinical literature.

- **Veterinary Behaviourist** (ABTC-VB). Register for Clinical Animal Behaviourists who are also qualified and registered Veterinary surgeons.
- **Animal Behaviour Technician** (ABTC-ABT). This standard relates to the provision of prophylactic and first aid behavioural advice and the implementation of behaviour or environmental modification following assessment by a CAB.
- **Animal Training Instructor** (ABTC-ATI). This standard relates to general animal training or the training of animals to undertake specific tasks. It covers direct interactions with the animal as well as teaching the animal's owner or handler how to introduce and reinforce desirable behaviours by way of training.
- **Animal Trainer** (ABTC-AT). This standard also relates to general animal training or the training of animals to undertake specific tasks. It covers direct interactions with the animal but does not cover the teaching of training skills to the animal's owner or handler.

All activities by all standards should be carried out within the constraints of the *Veterinary Surgeons Act 1966*. Activities covered by the Act are therefore limited unless practitioners are also qualified veterinary surgeons.

Conclusion

An animal's behavioural well-being should be considered to be equally as important as its physical health and is therefore a necessary part of veterinary care. As with any physical condition, when behaviour problems arise the best available treatment should be provided. This might not always be accessible within the veterinary practice and referral might need to be arranged. The task of finding a reputable and suitably qualified behaviour practitioner can be daunting and difficult. At the time of writing there is no regulation as to who can claim to treat animal behaviour problems. However, in addition to the RCVS registers of Advanced Practitioners in Companion Animal Behaviour and Specialists in Behavioural Medicine, a regulatory body, the Animal Behaviour and Training Council, is now available and this allows veterinary professionals to search for a suitably qualified clinical animal behaviourist (CAB) with the necessary skills, education and experience to offer a service to clients with behavioural concerns about their pets.

Links to Relevant Websites

Association for the Study of Animal Behaviour – CCAB Accreditation
https://asab.org/ccab
Animal Behaviour and Training Council
https://abtc.org.uk

Website links for legislation (accessed April 2020):

Control of Dogs (Scotland) Act 2010
www.legislation.gov.uk/asp/2010/9/contents
Crown Prosecution Service (CPS)
https://www.cps.gov.uk/legal-guidance/dangerous-dog-offences
Dangerous Dogs Act 1991
http://www.gov.uk/ukpga/1991/65/section/3
Dogs Act 1871
www.legislation.gov.uk/ukpga/Vict/34-35/56
Environmental Protection Act 1990
http://www.legislation.gov.uk/ukpga/1990/43/part/III

References

Boissy, A., Manteuffel, G., Jensen, M.B., Moe, R.O., Spruijt, B. *et al*. (2007a) Assessment of positive emotions in animals to improve their welfare. *Physiology and Behavior* 92, 375–397.

Boissy, A., Arnould, C., Chaillou, E., Duvaux-Ponter, C., Greiveldinger, L. *et al*. (2007b) Emotions and cognition: a new approach to animal welfare. *Animal Welfare* 16(S), 37–43.

Bourdin, M.C. (1997) Feline psychogenic alopecia and behavioural disorders. In: *Proceedings of the First International Conference on Veterinary Behavioural Medicine*, pp. 127–132.

Cameron, M.E. and Casey, R.A. (2004) A study of environmental and behavioural factors that may be associated with feline idiopathic cystitis. *Journal Small Animal Practice* 45, 144–147.

CAWC (2008) *The Regulation of Companion Animal Services in Relation to Training and Behaviour Modification of Dogs*. Companion Animal Welfare Council. Available at: https://www.reg-council-dtbp.uk/CAWC%20Report%202008%20onT&B%20Services.pdf (accessed 30 April 2020).

Cooper and Co. (2020) *Dog Law*. Available at: https://www.doglaw.co.uk/barking/ (accessed 30 April 2020).

Landsberg, G.M. (2008) Is it medical or is it behavioural? In: *Proceedings of CABTSG Study Day*, pp. 6–19.

RCVS (2018) *Code of Professional Conduct for Veterinary Surgeons. Referrals and Second Opinions*. Royal College of Veterinary Surgeons. Available at: https://www.rcvs.org.uk/setting-standards/advice-and-guidance/code-of-professional-conduct-for-veterinary-surgeons/supporting-guidance/referrals-and-second-opinions (accessed 30 April 2020).

Sentencing Council (2016) *Dangerous Dog Offences Definitive Guidelines*. Available at: https://www.sentencingcouncil.org.uk/wp-content/uploads/Dangerous-Dog-offences-definitive-guideline-Web.pdf (accessed 27 April 2020).

Stella, J.L., Lord, L.K. and Buffington, T. (2011) Sickness behaviors in response to unusual external events in healthy cats and cats with interstitial cystitis. *Journal of the American Veterinary Medical Association* 238(1), 67–73.

Westropp, J.L., Kass, P.H. and Buffington, T. (2006) Evaluation of the effects of stress in cats with idiopathic cystitis. *American Journal of Veterinary Research* 67(4), 731–736.

22 The Role of the Referring Vet in Case Management

Emma Brown

Introduction

More than perhaps any other field of veterinary medicine, the management of behavioural cases is a team effort involving the client, their family, trainers, behaviour counsellors, veterinary staff and sometimes others such as groomers, kennel/cattery staff and pet sitters.

Behaviour clinicians will generally be recommending a range of measures, including environmental management, behaviour modification and, in some cases, in discussion with the referring veterinary surgeon, medication. More information on the referral process and what to expect can be found in Chapter 21.

There may be specific protocols to be carried out in key locations such as the veterinary surgery or groomers and the behaviour clinician may work alongside rehabilitation trainers who will help the clients to implement the protocols.

To achieve the desired outcome, it is essential that all members of the team work effectively together. By establishing a good working relationship with the behaviour clinician, the referring veterinary surgeon should be able to receive regular updates, allowing them to monitor progress and discuss any concerns with both the owner and the behaviour clinician.

Follow-up After Behavioural Referral

Following a behaviour consultation, the referring vet should receive a full report from the behaviour clinician outlining their findings and recommendations. On some occasions the behaviourist may wish to discuss medication or nutraceuticals to support the behaviour modification protocol. The responsibility for prescribing medication remains with the veterinary surgeon. The use of psychoactive medications will be discussed later in this chapter.

Medical conditions can cause, mask, or exacerbate behavioural signs. Behaviourists often spend several hours observing an animal in its own environment and have long conversations with clients about the animal's behaviour and how it has changed. This means that behaviourists may become aware of indicators of medical concerns that may not have been apparent in the initial brief veterinary consultation. In these cases, the behaviour clinician should recommend that the animal returns to the veterinary practice for further investigation and contact the veterinary team to explain their concerns.

The veterinary team must listen to these concerns and, if appropriate, carry out further investigations. By working together, the veterinary and behaviour teams can provide the best service for clients and their animals.

DOI: 10.1079/9781800621312.0022

Identification of a medical concern should not mean that behavioural assessment or treatment is terminated but the vet and behaviour clinician will need to ensure that they communicate regularly to manage the case effectively.

An example would be a dog with a sudden onset of aggression during handling which, after assessment, is found to have orthopaedic pain. Clearly investigation and management of the pain is essential for the welfare of the animal and this may result in a reduction in aggressive behaviour. However, it is unlikely that the animal will immediately be pain free at all times (despite our best efforts) and there will be a learnt component to the animal's response (handling = pain; and showing an aggressive response has reduced handling in the past), so the animal may continue to exhibit the behaviour. The behaviourist will be able to give advice to owners on management to reduce risk and demonstrate how to start to form new positive associations with sensitive handling, but this needs to be done with support from the veterinary team to guide them as to what level of handling is appropriate.

In some cases, the link between the behaviour and the medical condition may not be so immediately obvious. The veterinary surgeon should make every effort to investigate and manage any medical conditions and keep the behaviour clinician informed of any new diagnoses.

It would be usual for a clinical animal behaviourist to remain in regular contact with a client for several months and they may carry out further visits or refer to a trainer for ongoing support. The veterinary surgeon should ensure that they know what follow-up is offered by the behaviourist that they refer to. This will allow them to help clients select a service that fits their needs and help them plan ongoing management.

If the animal has been prescribed medication as part of a behaviour modification protocol, then it is important that any response to this is monitored, so that doses can be adjusted to maximize response and any side effects can be identified and responded to appropriately. Regular prescription checks will be needed as for any medication and may include blood tests, depending on the medication prescribed.

Behaviour clinicians use a variety of techniques to monitor response to treatment, for example some clients will keep weekly or daily diaries. There may be scheduled follow-up visits, phone calls or emails. The veterinary surgeon should check with the behaviour clinician what protocol they use and agree a schedule for follow-up contact. At a minimum the referring veterinarian and behaviourist should be in contact if there are any significant changes in the animal's behaviour, if there are any changes to medication and when making significant changes to the behaviour modification protocol.

Medication Use

Pharmacological intervention can be helpful and, in some cases, essential. It is important that it is used appropriately. Both veterinarians and owners can be reluctant to use medication for behaviour cases, meaning that it is often reserved for 'severe or longstanding' cases. This can have a negative impact both on the animal's welfare and on the outcome of the case. Veterinary surgeons should not be afraid to use psychopharmacology, but it should always be used alongside appropriate behaviour modification and environmental management.

The decision to use medication in behavioural cases will involve input from the owner, veterinary surgeon and clinical animal behaviourist. Owners will need to be able to administer any medication prescribed safely. The veterinary surgeon is best placed to decide whether a medication is appropriate, given the animal's medical history, and should be aware of any potential interactions with medications the animal may already be receiving. In complex cases advice can be sought from a veterinary behaviourist. A list of these can be found on the Royal College of Veterinary Surgeons (RCVS) website (Specialist and Advanced Practitioner lists) (https://findavet.rcvs.org.uk/home/) and the Animal Behaviour & Training Council (ABTC) website (http://www.abtcouncil.org.uk/veterinary-behaviourists.html).

Medication should be considered in all cases where the animal is experiencing distress that cannot be prevented by avoiding the trigger. It should also be considered in cases where the animal's emotional state is such that it is likely to impact on the outcome of the behaviour modification protocol.

In some cases, it will be appropriate to implement management changes, start the behaviour modification protocol and then review the need for medication. It is therefore important that the veterinary surgeon and behaviour clinician continue to discuss progress in order to make this decision.

It may be that an animal makes initial good progress in response to the behaviour management protocol but then progress plateaus without reaching the stated goal. In these cases, adding in medication may allow further progress.

Resistance to medication use

It is fairly common for owners to express concerns about using psychoactive medications for their pets. Owners may worry that the medication will change their pet's personality or cause them to be sleepy.

It is important that veterinary surgeons discuss the advantages and risks associated with these medications. They should discuss potential side effects (including the potential for increased aggression) and complications but also why they feel that medication is appropriate for the animal.

Although as with all medications there are potential risks, used appropriately behavioural medications are generally safe and can be life changing for some animals and their owners. It is important, however, that clients have realistic expectations. For example, it may take 4–6 weeks to see full effects of selective serotonin re-uptake inhibitor/tricyclic antidepressant (SSRI/TCA) medications. Additionally, situational use of medications such as benzodiazepines may cause sedation but, as these wear off fairly rapidly and the medications are used infrequently, this is unlikely to have a significant impact on the dog's daily life.

Prior to prescribing a psychoactive medication, ensure that:

- Clients are aware of what to expect in terms of effect – how long to onset, likely duration of effect, what changes they should expect.
- Clients have been given realistic expectations of how long the animal is likely to need to remain on medication.
- Clients are aware of what side effects can occur and what action to take if they see these. Some of the commonly used medications can result in transient side effects in the early days of administration and it is good practice to warn owners that these may occur, how long they might be expected to last and when they should contact the prescribing vet for advice.
- Clients are aware of the signs of serious complications such as serotonin syndrome and what action to take if they suspect these.
- Informed consent has been obtained for off-licence use of medications – see section on licensing and the cascade.

Long-term or context-specific medication

It is essential that the clinician prescribing psychoactive medication has a clear idea of what they aim to achieve. For example, some animals will require daily medication to support treatment of, for example, a generalized anxiety or abnormal repetitive behaviour. In other cases, the animal may only need support at very specific times, such as during a thunderstorm or fireworks display for an animal with a sound phobia. Different medications will be better suited to these different uses and the veterinary surgeon must ensure that the medication selected is appropriate.

Medications are generally used in one of three ways.

1. Long-term. In some cases, an animal will need ongoing psychoactive medication administered daily for months or years.
2. Situational or context-specific use. Usually, a small number of doses given prior to or in response to an environmental trigger, such as vet visits or storm. These can be used:
   - prior to a predicted adverse event; and
   - during or following an adverse effect.

3. A combination of long-term and situational use.

Combinations of medications

In some cases, a combination of medications may be required. For example, a dog on long-term treatment with a tricyclic antidepressant for separation anxiety might require context-specific medication for a sound phobia with a licensed medicine or benzodiazepine for fireworks night.

There is limited research into the effects of combining medications in behavioural medicine. Before considering using a combination of medications the veterinary surgeon must have a good understanding of how each medication works, in order to minimize the risk of toxic effects.

Sometimes a case will be presented in which the animal's behaviour is so serious that the welfare and safety of the animal or owner are severely compromised, and waiting for the relatively slow onset of a TCA or SSRI medication would not be appropriate. In these cases, combination therapy until the long-acting medication starts to take effect is indicated.

Considerations prior to prescribing

Before prescribing psychoactive medication, it is important to carry out a full physical examination to identify any ongoing medical conditions (Table 22.1). It is important to note that identifying a medical condition that may be impacting on the animal's behaviour does not negate the need for behavioural support. For example, finding orthopaedic pain in a dog that has been showing aggression when handled should of course lead to further investigation and treatment of the pain, but a behavioural assessment and treatment protocol may still be appropriate as there will often also be a learnt component to the response.

The animal's medical history should be reviewed and blood and urine samples analysed to give a baseline prior to starting medication. Most psychoactive medications are metabolized by the liver or kidneys and therefore identification of any underlying concerns as well as obtaining a pre-treatment baseline is important. In some cases, it may be impossible to obtain these baseline samples without significant risk to staff or distress to the animal. In these cases, a risk/benefit analysis must be carried out considering

Table 22.1. Considerations prior to prescribing psychoactive medication.

Factors to consider	Specific considerations
Have a full medical examination and baseline blood/urine tests been carried out?	Identify any related/concurrent health concerns Carry out a risk/benefit analysis prior to these interventions
Have you reviewed the medical history?	Age, general health and any ongoing conditions that may impact behaviour
Is there an accurate behavioural diagnosis?	Allows selection of appropriate medication
What is the time to response for the medication considered?	Medications used situationally need to have a rapid onset of action
Formulation	Consideration needs to be given to the practicality, safety and patient impact of administration
Duration of action	How frequently will the medication need to be administered? Is this practical?
Interactions with other medications	Be aware of possible interactions with existing medications that the animal is taking, including supplements or herbal medications
Is there a need for ongoing monitoring?	How practical will this be to achieve? Have you carried out a risk/benefit analysis?
Licensing/cascade[a]	Is there a licensed product? Can you justify use of a medication within the cascade system? Do you have informed consent?
Realistic expectations	Is the client aware of the expected timeframe before changes are seen and what changes to expect?
Are you aware of potential side effects and has the client been informed of these?	Both medical and behavioural
Consider the potential for human abuse	Especially with benzodiazepines

[a] More information on prescribing and the cascade system can be found at: https://www.gov.uk/guidance/the-cascade-prescribing-unauthorised-medicines

the age and health status of the animal, previous medical history, safety profile of the medication considered, risks to owner and veterinary staff and the likely level of distress experienced by the animal. It may be appropriate, after a risk benefit discussion with the client, to start medication prior to carrying out these tests.

Administering medication

Consideration must be given to how easy the medication is to administer. Cats in particular can be difficult to administer medication to. If an animal is very difficult to tablet, then consideration should be given to a liquid or capsule formulation. Transdermal administration has been considered for veterinary medicine and, although there are limited areas to which these medications could be applied on animals and research suggests that absorption may be limited, this route may become useful in the future (Crowell-Davis *et al.*, 2019). As yet there is no licensed formulation for transdermal administration in the UK.

Many of the medications used in behavioural medicine will be off-licence for the species being treated and therefore achieving the correct dose may be difficult. Pill cutters may help with accurate splitting of tablets (if not coated or very bitter when split).

Medication may be taken if disguised in high-value food, for example meat/fish paste, soft cheese and Lick-e-Lix cat treats often work well. Distraction by giving several tasty treats, then the medication and then following with several tasty treats can work well for some animals.

Consideration can also be given to training animals to accept a tablet (see Appendix I, 'Training Animals to Accept Medication').

Some medications (such as SSRIs) are associated with a relatively high incidence of side effects in the early days of administration. If these are significant, then consider restarting at a lower dose and then increasing after 2–3 weeks. Warning clients of possible side effects allows them to prepare and increases compliance. In some cases where the risk of side effects is considered significant (for example in an animal prone to gastrointestinal disease) the veterinary surgeon may choose to start at a lower dose and then increase after 2–3 weeks if there are no adverse effects.

How psychoactive medications work

The majority of psychoactive medications used in veterinary behaviour cases act by modulation of neurotransmitters and their metabolites. The neurotransmitters of particular interest are the monoamine neurotransmitters dopamine, noradrenaline and serotonin and the amino acid neurotransmitters glutamate and gamma-aminobutyric acid (GABA) (Table 22.2).

Psychoactive medications may increase neurotransmitter production or release, inhibit re-uptake of neurotransmitters, inhibit neurotransmitter breakdown or act on neurotransmitter receptors (as agonists or antagonists).

Psychoactive medications used in veterinary behavioural medicine

The use of psychoactive medication in veterinary behaviour medicine is a very exciting area with lots of new research and new products being introduced. Clinicians with an interest in animal behaviour must ensure that they keep up to date with the latest research.

There are now several excellent books available, which discuss the use of psychoactive medication and doses can be found in these (Horwitz and Mills, 2012; Landsberg *et al.*, 2012; Overall, 2013; Crowell-Davis *et al.*, 2019). The BSAVA small animal formulary (Allerson, 2020) provides an excellent and continually updated resource for medication doses. The intention in this chapter, therefore, is not to discuss the use of individual compounds in detail but instead to give an overview of the use of behaviourally active medication, when to consider this, what options are currently available and how to make decisions regarding medication use.

Veterinary surgeons working in first-opinion practice should ensure that they are familiar with the indications and therapeutic use of the most frequently used medications in behavioural medicine. There are a large number of medications that may potentially be used, but in many cases it is appropriate to use one of the small number of medications for which we have good databases.

At the time of writing the following medications are most readily available or most commonly used.

Table 22.2. Overview of the major neurotransmitters modulated by veterinary psychoactive medications.

Neurotransmitter	Modulates	Notes
Serotonin	Mood Cognitive function Impulsivity Social behaviours	In humans, serotonin depletion increases irritability, hostility and impulsiveness
Noradrenaline	Arousal Mood Learning and memory Stress response	In humans, depletion is associated with depression while elevated levels are associated with mania
Dopamine	Cognitive function Coordination of movement Mood	In humans, reduced dopamine has been associated with depression Increased levels have been associated with stereotypical behaviours
GABA	Arousal Seizure activity Anxiety Memory formation Muscle tone	GABA dysregulation is associated with fears and phobias
Glutamate	Arousal Seizure activity Anxiety Memory formation Muscle tone	In humans, high levels can be associated with aggression, impulsive and schizophrenic disorders

- **Selective serotonin reuptake inhibitors (SSRIs)**, highly selective blockade of the reuptake of serotonin leading to prolonged action of serotonin molecules. Response to treatment may not be seen for several weeks and treatment should be continued for 6–8 weeks before making decisions about efficacy. These medications are not suitable for situational use and should be given regularly. SSRIs must never be given with monoamine oxidase inhibitors.
- **Tricyclic antidepressants (TCAs)** primarily inhibit reuptake of serotonin and noradrenaline. Many also have antihistaminic and anticholinergic effects and act as alpha 1 adrenergic antagonists. Like SSRIs, onset of clinical effect can take several weeks and these medications should be used regularly. They are not suitable for situational use. Must not be given with monoamine oxidase inhibitors.
- **Benzodiazepines** potentiate the effects of GABA by binding to GABA-A receptors. Relatively rapid onset and limited duration of activity make these compounds suitable for situational use. There have been reports of hepatic necrosis in cats treated with benzodiazepines, and diazepam in particular should be used with caution and careful monitoring in this species.

There is significant individual variation in response to these medications and in some animals a paradoxical excitation can be seen following administration. It is therefore advisable to administer a test dose whilst the owners can observe the response.

Long-term use can result in tolerance and dependence. If longer-term (ongoing medication for weeks rather than just isolated situational dosing) use is necessary, this should be withdrawn gradually to prevent rebound of symptoms as the medication is withdrawn.

- **Monoamine oxidase inhibitors (MAOs)** block oxidative deamination of brain amines. The compound most commonly used in veterinary medicine is selegiline. Selegiline is fairly specific for dopamine, having limited effect on serotonin. In the UK selegiline has a licence for treatment of behavioural disorders of emotional origin but in the USA it is licensed for cognitive dysfunction.

Selegiline may be effective for fear-related behaviours but has limited anxiolytic effect. Use of MAOs for behavioural conditions has to a large extent been superseded by newer products.

One of the metabolites of selegiline is amphetamine and therefore it may be contraindicated in animals that show over-arousal or increased activity.

MAOs are not suitable for situational use and should be used regularly for at least 4 weeks before assessing whether they are effective.

Very significant and potentially fatal side effects can occur when MAOs are combined with a number of other medications, including SSRIs, TCAs, Tramadol and Amitraz. Veterinary surgeons should take great care to check for possible contraindications when prescribing medication alongside an MAO. A 2-week washout period should be observed when changing from an MAO to a TCA or SSRI. When changing from an SSRI to an MAOI a longer washout period may be required (up to 5–6 weeks).

- **Asapirones**, partial serotonin agonists. Busiprone has been used in veterinary behavioural medicine. May be of benefit in anxiety disorders or phobia. Can facilitate pro-social behaviour in cats. Changes may be seen within the first week, but treatment will need to be continued for several weeks for maximum efficacy to be seen. These medications often require dosing multiple times per day.
- **Serotonin 2A antagonist/re-uptake inhibitors (SARIs)** antagonize serotonin receptors inhibiting serotonin reuptake. Trazodone has been used in veterinary behavioural medicine both as a long-term and situational medication. It has also been used in combination with SSRI and TCA medications but care must be taken when using combinations of medications that increase serotonin levels (see 'Serotonin syndrome', below).
- **Imepitoin,** low-affinity partial agonist for the benzodiazepine binding site of GABA receptors, now licensed in the UK as Pexion for the reduction of anxiety and fear associated with noise phobia in dogs, as well as control of seizures. Rapid onset of effect but it is advised that administration should be started a few days prior to the anticipated phobic event.
- **Alpha-2 adrenoceptor agonists**, inhibit release of noradrenaline. Dexmedetomidine is licensed in the UK as Sileo for the alleviation of acute anxiety and fear associated with noise in dogs. The licensed formulation is a gel that must be applied to the oral mucosa, which can be difficult to administer to some dogs. Clonidine has been used in the treatment of panic, anxiety and impulsivity in dogs, often as an adjunctive medication alongside long-term medication with an SSRI or TCA, although care must be taken with TCAs that also impact noradrenaline levels.
- **Gabapentin**, GABA analogue. The precise mechanism of action for Gabapentin remains unclear. Despite its name, Gabapentin does not bind to GABA receptors but instead is believed to act by binding to voltage-gated calcium channels, resulting in inhibition of the release of excitatory neurotransmitters. Used in veterinary medicine to treat seizures and neuropathic pain. Has been used in the treatment of anxiety and compulsive disorders. Rapid onset of action means that Gabapentin can be suitable for situational as well as long-term use and it has been found useful prior to veterinary visits, especially in cats. If Gabapentin is administered in the long term, it should be withdrawn gradually, as sudden termination of treatment can result in seizures.
- **Beta-adrenergic receptor antagonists** (beta blockers). Propranolol has been used in veterinary medicine usually as a situational medication alongside long-term medication in cases of anxiety or phobia.
- **Pain relief**. We are increasingly becoming aware of the impact that pain can have on animal behaviour. In a recent review of behaviour cases seen at referral clinics, pain was felt to be a factor in between 28% and 80% of cases (Mills *et al.*, 2020). Animals presenting with behavioural conditions can be difficult to examine and if there is any suspicion of pain as component of the clinical presentation then a full work-up should be carried out and a pain relief trial considered. In order to fully assess the impact of pain on behaviour, the trial should be continued for a minimum of 5–6 weeks. We must remember that central sensitization

can occur and there will be a learned component to a response; therefore the behaviour may not disappear as soon as the pain is reduced.

- **Cannabidiol** (CBD oil), extracted from leaves and flowers of the hemp plant. Binds to CB2 endocannabinoid receptors to give anti-inflammatory but potentially also antidepressant and anxiolytic effects. Recent studies have looked at the efficacy of CBD for chronic pain and epilepsy and reported weak positive results (Gamble *et al.*, 2018; McGrath *et al.*, 2019). In the UK the VMD state: 'We consider that veterinary products containing Cannabidiol (CBD) are veterinary medicines and should be regulated as such.' There are no authorized veterinary products currently available in the UK.
- **Cabergoline**, dopamine agonist, licensed in the UK for the treatment of false pregnancy in bitches, including the behavioural changes associated with this. The datasheet (NOAH, 2020) recommends a 4–6-day course but a review of treatment protocols suggests that treatment for 14 days may be needed for behavioural signs including aggression (Root *et al.*, 2018).
- **Deslorelin acetate**, synthetic gonadotrophin-releasing hormone (GnRH) analogue, available in the UK as Suprelorin for the induction of temporary infertility in healthy, entire, sexually mature male dogs for up to at least 6 months. Although not specifically indicated for behaviour management, deslorelin acetate has been used to assess the likely effect of castration on the behaviour of male dogs. Clinicians using deslorelin acetate in this way must be aware that there is an initial increase in testosterone levels (as the pituitary initially responds to the increase in circulating GnRH, prompting an LH and FSH surge). This lasts for an average of 7–10 days, but in some dogs could last for up to 3 weeks (Lucas, 2014).

Licensing and the cascade system

Very few prescription medications are licensed in the UK for use in behavioural conditions (Table 22.3). At the time of writing there are five licensed products available for behavioural conditions in dogs and none for cats. This means that use of psychoactive medications is often off-licence use of a medication. In these cases,

Table 22.3. Medications authorized for behavioural conditions available in the UK (at the time of writing).

Trade name	Active ingredient	Manufacturer	Licence indication
Clomicalm	Clomipramine	Elanco UK AH Ltd	As an aid in the treatment of separation-related disorders in dogs manifested by destruction and inappropriate elimination (defecation and urination) and only in combination with behavioural modification techniques
Sileo	Dexmedetomidine	Zoetis UK Ltd	Indicated for the alleviation of acute anxiety and fear associated with noise in dogs
Pexion	Imepitoin	Boehringer Ingelheim Animal Health UK Ltd	For the reduction of anxiety and fear associated with noise phobia in dogs
Reconcile	Fluoxetine	Forte Healthcare Ltd	As an aid in the treatment of separation-related disorders in dogs manifested by destruction and inappropriate behaviours (vocalization and inappropriate defecation and/or urination), and only in combination with behavioural modification techniques
Selgian	Selegiline hydrochloride	CEVA Animal Health Ltd	1. Treatment of behavioural disorders of purely emotional origin: depression, anxiety 2. In association with behaviour therapy, treatment of signs of emotional origin observed in behavioural conditions such as over-activity, separation problems, generalized phobia and unsociable behaviour

veterinary surgeons in the UK must follow the prescribing cascade (Gamble *et al.*, 2018).

Informed consent should always be obtained for off-licence use of medication. In the UK, template forms for off-licence consent are available from the British Small Animal Veterinary Association (BSAVA).

Supplements and nutraceuticals

There is an ever-increasing array of supplements, diets and complementary therapies marketed for the alleviation of behavioural disorders in animals in the UK. Although some of these products may be beneficial and many have been researched and quality controlled, we must remember that the regulations governing these products are very different to those covering pharmaceuticals. At the current time good-quality, placebo-controlled research is lacking for many of these products. This does not mean that they will not prove in time to have value.

Clients often like the idea of a 'natural' product and may not realize that some of these products can cause toxicity, interact adversely with other medication or may not be suitable for all animals.

Behavioural conditions can cause significant distress to animals, their owners and those they come into contact with. Veterinary surgeons have a responsibility to advise based on the best available scientific evidence. It is not appropriate simply to suggest a supplement without fully investigating the behavioural condition.

Table 22.4 outlines the most common ingredients, their indications and contraindications.

Table 22.4. Supplements and nutraceuticals commonly available in the UK.

Name	Action	Additional information
Alpha-casozepine (Zylkene)	Protein found in milk purported to have affinity for GABA receptors. Anxiolytic effects	No known contraindications There have been studies suggesting positive effects (Beata *et al.*, 2007a,b)
Tryptophan	Precursor of serotonin. Uptake is complex and may require reduction in other dietary large amino acids to minimize competition for carrier proteins Suggested for anxiety	Care when combining with other products that enhance serotonin effect Minimal evidence to support use
L-theanine (Anxitane)	Structural analogue of glutamate. Has been suggested to treat fear, anxiety and stress	Generally palatable There have been studies suggesting positive effects (Michelazzi *et al.*, 2010; Crowell-Davis *et al.*, 2019)
Valerian	Thought to bind to a beta subunit on GABA-A receptors Sedative and anxiolytic effects	Minimal studies into effect in animals
Passiflora extract	Extracted from plants of the *Passiflora* genus. The extract contains flavonoids and glycosides Claimed to have sedative and anxiolytic properties	Minimal studies into effect in animals
St John's Wort	Mode of action not fully understood. Proposed to act as a reuptake inhibitor for serotonin, noradrenaline, dopamine and glutamate Has been suggested for anxiety, sedation and to treat compulsive disorders	Must not be given with SSRIs, TCAs or MAOIs. Care when using alongside any product that might enhance serotonin Minimal studies into effect in animals
Skullcap	Mode of action poorly understood. Possible action on GABA receptors	Minimal studies into effect in animals Risk of side effects at high doses Possible hepatotoxicity
Melatonin	Melatonin is synthesized from serotonin and plays a role in the regulation of circadian rhythms. Has been used for sleep cycle disorders and anxiety	Minimal studies into effect in animals

Many of the products on the market contain combinations of compounds. Veterinary surgeons considering recommending these products should check the level of active ingredient actually contained in the preparation and carefully consider the evidence for both efficacy and safety.

Pheromones

The following definition for pheromones has been proposed (Mills *et al.*, 2013): 'Chemical signals used in intraspecific communication, which are typically detected by the vomeronasal organ and which appear to have an intrinsic effect on the emotional processing of the receiver.'

At the time of writing, synthetic analogues of several pheromones are commercially available in the UK in a variety of formulations (sprays, diffusers, impregnated collars) (Table 22.5).

There is research to suggest that some of these products may be helpful in some situations. It is important that they are used alongside other measures such as environmental management, behaviour modification and, if appropriate, medication.

Serotonin syndrome

Serotonin syndrome is the name given to the potentially fatal toxicity that can result when excess levels of serotonin-increasing medications are administered. This can potentially occur with any medication that increases serotonin levels, but the risk is increased in combination therapy.

Clinical signs are neurological, autonomic and gastrointestinal and can include restlessness, hyperthermia, shaking, tachycardia, tachypnoea, diarrhoea, vomiting, hypersalivation, abdominal pain, tremors, seizures, coma and death.

It should be remembered that some commonly used medications also increase serotonin and clinicians must ensure that medications are reviewed when animals are placed on psychoactive medication and that they check for any incompatibilities before starting any additional medication for an animal receiving psychoactive medication.

It should also be remembered that some supplements, such as St John's Wort and tryptophan, may increase serotonin levels. The prescribing veterinary surgeon must check whether owners are giving any additional supplements.

How to select a psychoactive medication

As in all branches of veterinary medicine, selection of an appropriate medication will require an accurate diagnosis. Reaching a behavioural diagnosis requires a full physical examination and a thorough behavioural history and evaluation. Veterinary surgeons (unless they are also qualified behaviourists) must work alongside the behaviour clinician to get to this point.

It is important to understand that the unwanted behaviours that are observed are non-specific. For example, cats showing aggression towards humans may be motivated by fear or frustration or showing inappropriate play behaviour. Even if we can narrow the diagnosis to fear, we must still consider that many factors will have been involved in the development, expression and ongoing maintenance of the behaviour.

We must also consider that behaviours that owners consider undesirable may be a normal part of the behavioural repertoire of the species involved.

The behaviours expressed are the result of genetic predisposition, neurochemical balance, learning through prior experience, disease states and emotional state. It is not appropriate to base

Table 22.5. Commercially available pheromone products.

Pheromone	Brand names	Indications
Feline Facial Fraction F3	Feliway	Urine marking Management of stressful situations
Feline Facial Fraction F4	Felifriends	Encourage acceptance of new cats or unfamiliar people Multi-cat households
Feline Pheromone Complex (FPhC)	Feliway Optimum	Urine Marking, Scratching, Multicat Households
Feline interdigital semiochemical (FIS)	Feliscratch	Encourages cats to scratch in desired locations
Dog-appeasing pheromone	Adaptil	Anxiety, fear and management of stressful situations

the selection of medication on the phenotypic behaviour displayed. Instead, we must choose a medication based on our understanding of the neurochemistry and the effect we are hoping to achieve, such as a reduction in anxiety or arousal.

What if the response to medication is suboptimal?

Animals are individuals and there can be significant variation in response to psychoactive medications. If, following an appropriate period to assess response, this is deemed to be suboptimal, then the clinician has three options:

1. Increase the dose. If there has been a favourable response and the dose is not at the maximum recommended level, then a dose increase should be considered.
2. Add an additional medication. There is very little data on combining psychoactive medications in veterinary medicine. The clinician must consider how each medication acts and whether the effects are likely to be synergistic or risk toxicity. There are textbooks that discuss possible combinations in more detail (Overall, 2013; Crowell-Davis *et al.*, 2019). If unsure, clinicians should seek guidance from a qualified veterinary behaviourist (lists available at http://www.abtcouncil.org.uk/veterinary-behaviourists.html and the Royal College of Veterinary Surgeons registers at https://www.rcvs.org.uk).
3. Change the medication. In some cases it may be necessary to change to an alternative medication.

Changing medication

There will be times when it is necessary to change from one psychoactive medication to another. This may be because the side effects are significant or because the medication has not resulted in the desired effect.

Before changing medication, ensure that the course has been long enough to truly assess efficacy. The effects of TCAs and SSRIs may not be seen for 4–6 weeks and full effects may take several months. Clients should be encouraged to continue giving these medications for 6–8 weeks before deciding that they have not been effective. If some mild improvement has been seen, then consideration could be given to an increased dose.

If you do need to change medication, then this must be done carefully. A washout period of 2 weeks will be needed when changing from a TCA/SSRI to an MAOI, or vice versa, and up to 6 weeks should be left between an SSRI and an MAOI. Consideration will need to be given to whether the animal will need additional support in this period with management changes or the addition of short-term medication.

Stopping medication

Clients should be given reasonable expectations when commencing treatment as to how long this is likely to be required for. With appropriate environmental management and behaviour modification, many animals will be able to be weaned off medication, but others will continue to require medication for the long term.

When starting one of the longer-term medications, clients should be advised that unless there is a reason to stop earlier, they should expect their animal to require medication for 6–12 months before considering weaning off. In most cases medication should be continued beyond the point at which no further improvements are being seen. At this point a gradual weaning-off process can be started. Generally, a reduction of 25% every 2–4 weeks is a good starting point for weaning off most long-term medications. In some cases, it may not be possible to withdraw the medication completely, but the gradual reduction may allow identification of the lowest effective dose.

If there is a recurrence of the unwanted behaviour during the dose reduction protocol, clients should be advised to return to the last effective dose and continue this for a further period alongside behaviour modification before trying once more to reduce the dose.

The Role of Veterinary Facilities and Staff in the Behaviour Modification Protocol

As well as providing ongoing monitoring, veterinary practices can become involved in behaviour programmes at a practical level.

Behaviourally aware practice

Increasingly we are aware of the responsibilities that veterinary practices have both to advise clients on behavioural conditions, including where to seek help if needed, and to ensure that every experience is as positive as possible for patients and clients.

There are now excellent schemes to encourage and acknowledge behaviourally aware practice. These include:

- **Cat Friendly Clinics** – worldwide programme from International Society of Feline Medicine (ISFM) veterinary division of the charity International Cat Care (ICC), based in UK. Available at: https://catfriendlyclinic.org/ (accessed July 2020).
- **Dog Friendly Practices** – scheme created by Dogs Trust and British Veterinary Behaviour Association (BVBA). Available at: https://www.dogfriendlypractice.com/resources (accessed July 2020).
- **Fear Free Certification** – issued by Fear Free Pets, based in Denver, Colorado, and providing online education for veterinary and pet professionals and pet owners. Available at: https://fearfreepets.com (accessed July 2020).

It is important that all members of staff are aware of the practice's behaviour policies and receive training to help them meet these standards. This should ensure a consistent message to clients and a consistent level of care for patients.

Forming positive associations with the veterinary facility

Many animals with behavioural conditions will be anxious in a veterinary setting and working to change their response to veterinary visits can be hugely beneficial. Animals may find veterinary visits difficult for a variety of reasons. They may have had previous adverse experience, be fearful of interaction with strangers, uncomfortable with handling or worried by other animals that they may encounter within the practice.

The impact of an animal's behaviour within the clinic on clients should not be overlooked. Many owners are distressed by seeing their animals frightened or behaving aggressively and if not addressed this may lead to reluctance to bring the animal in for routine appointments.

When animals are relaxed and cooperative, staff safety is improved, patient welfare improves, clinical examinations are improved and clients feel more comfortable.

Working with clients to improve an animal's experience of the veterinary clinic can be hugely rewarding for staff, improves patient welfare and staff safety and is great for practice–client relationships. It does, however, take time and requires dedicated knowledgeable staff.

The principles of desensitization and counterconditioning can be used to gradually change the animal's emotional response to their veterinary visits whilst also helping clients to relax. Details of the protocol used by the author for dogs can be found in Appendix II ('Veterinary Staff Protocol for Building Positive Associations with Veterinary Visits: Dogs'). Figures 22.1–22.3 illustrate some of the stages in this process.

For cats, protocols should initially focus on forming strong positive associations with the cat carrier so that the cat arrives relaxed as shown in Fig. 22.4 (see Appendix III, 'Feline Behaviour Handout: Cat Carriers'). Then on attending a clinic at a very quiet time where the cat is simply offered high-value food within the carrier and then taken home. This can then be built upon, gradually increasing confidence in the practice environment (see Appendix IV, 'Veterinary Staff Protocol for Reducing Feline Anxiety at the Vet').

These processes can be carried out by veterinary staff alongside behaviour modification protocols under the guidance of a behaviour clinician.

Managing behaviour modification protocols within the practice

In an ideal scenario every veterinary practice would have a behaviour advocate – a member of staff with training in animal behaviour to coordinate the practice's management of behaviour cases and implement training protocols, as well as being available to give behavioural advice as needed.

In the ideal scenario, clients would be approaching their veterinary surgeon as the first port of call for all issues relating to an animal's

Fig. 22.1. Forming positive associations with the car park and practice surroundings. (Image: author's own.)

Fig. 22.2. Search games in the waiting area build confidence. (Image: author's own.)

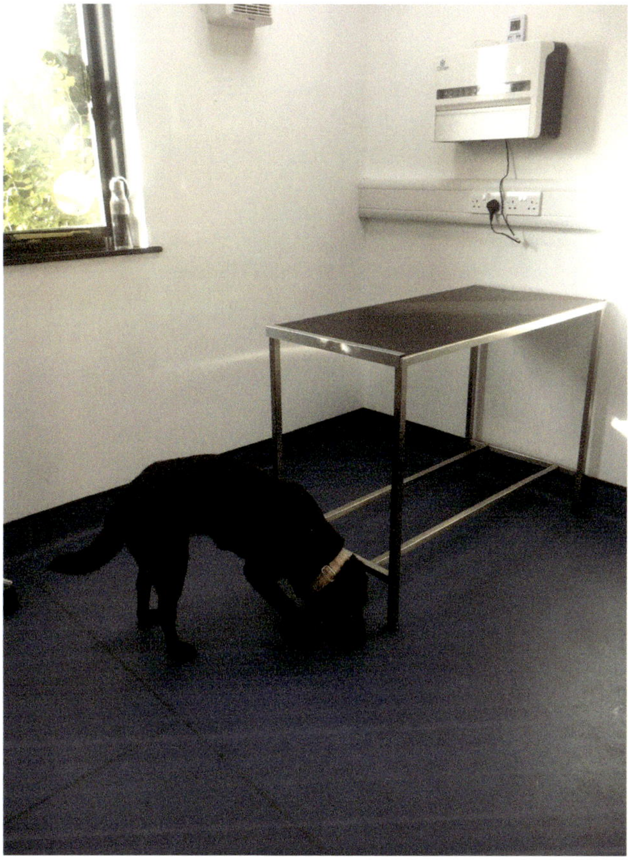

Fig. 22.3. The dog is allowed to explore the consulting room whilst food rewards are intermittently scattered for her to find. (Image: author's own.)

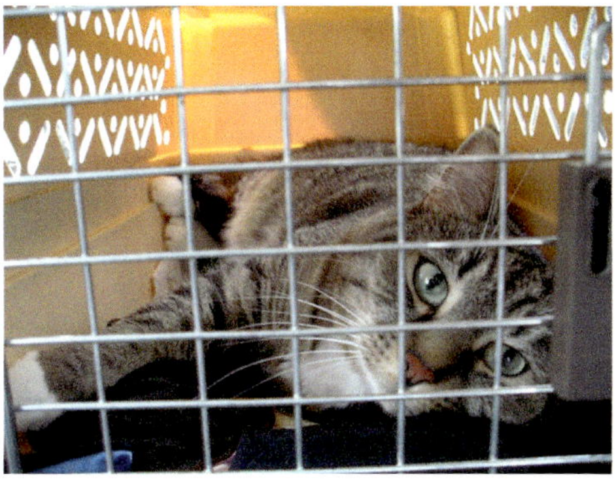

Fig. 22.4. A cat relaxed in the carrier. ('Are we there yet?' by Pjmorse, licensed under CC BY-SA 2.0.)

physical or emotional well-being. A behaviour advocate can advise on selection of an appropriate animal, early management, local registered trainers, timing of neutering, preparing for veterinary visits and helping manage animals following surgery as well as giving initial first aid advice on many other issues and knowing who to refer on to in more complex cases (see Chapter 21).

This staff member can also advise on practice design/layout (see Chapter 6) and offer training and support on low-stress handling and managing difficult patients within the practice.

Many animals benefit enormously from forming a relationship with a member of staff and clients appreciate a point of contact. We routinely assign a behaviourally aware veterinary nurse to each case requiring behaviour modification within the clinic. This allows both client and animal to build trust with a member of the team.

We try to ensure regular meetings between this nurse (behaviour advocate) and the client

Fig. 22.5. A cat exploring the consulting room. ('Ozu at the vet' by HeyRocker, licensed under CC BY 2.0.)

and their animal. These meetings may initially be carried out at the client's home or outside in the practice grounds. At first these are simply focused on forming a relationship (both with the client and with the animal) and will include discussion of favoured food rewards, games and what handling (if any) the animal enjoys.

As time progresses the behaviour advocate will work together with the client to address specific concerns. This can work very well if the practice's behaviour advocate communicates effectively with the behaviour clinician. It can be difficult sometimes for clients to appreciate the progress that their animal has made. Having an experienced staff member who can objectively assess progress and identify any potential concerns can help ensure that progress continues and the behaviour clinician can adjust the protocol taking this feedback into account.

Ensuring wherever possible that routine appointments are booked at times when the animal's assigned behaviour advocate can be present can be very beneficial. The behaviour advocate will know the animal's behavioural history, what level of handling they are likely to be able to cope with and what rewards they will find most salient. They will be a familiar point of contact with whom owners should feel confident to discuss any concerns or suggestions and a familiar person with whom the animal will have built a history of positive interactions.

Assisting with specific aspects of the behaviour protocol

Sometimes a behaviour clinician may recommend work on a specific task such as building confidence with claw clipping, muzzle training or learning to accept handling. The experience and training of the veterinary team means that they are well placed to assist owners with these protocols if required.

For example, animals can be booked for an initial session with their nurse to start work on muzzle training. This allows the nurse to demonstrate the technique, observe owners carrying out the initial stages of training and give feedback on timing and technique. In this way we hope to ensure that owners fully understand the stages in the protocol, how the training should look and the importance of working at a pace guided by the animal. Owners should then be encouraged to work on the task at home but contact their nurse if they have any questions. Repeat appointments should be booked to assess progress and give further advice if needed.

It is possible that the animal being treated for a behaviour problem will need a veterinary appointment during the behaviour modification protocol. For some animals this will not be a concern but this will need very careful management for those who struggle with handling, close contact with strangers or have specific vet fears.

Planning is essential. Non-urgent procedures and appointments should be delayed while the practice works with the owner to improve the experience. Non-urgent but essential appointments should be carefully planned to ensure the best and safest outcome.

If an animal undergoing behaviour therapy for problems within the practice environment needs to be seen, having an emergency protocol in place is essential. Every member of the team, including the out-of-hours service, must be able to access the protocol.

Consider the following points.

- Medication pre-appointment if possible and appropriate (see Table 22.6, below).
- Muzzle training for dogs (see Appendix V, 'Canine Handout: Muzzle Training').
- Timing: can they attend at a quiet time when there are fewer people and animals in the building?
- If the animal is nervous within the practice, ask the owner to book in at reception and then wait in the car with their animal.
- Are members of staff who are familiar with the animal available to carry out the examination or to assist?
- Could the consultation be carried out safely at home or would the situation be appropriate for a telemedicine consultation?
- Use high-value food rewards generously if not contraindicated.
- Take your time – long appointments should always be booked.
- Can cats remain within the lower section of their carrier, or under a blanket or towel from home, for part or most of the examination?
- Consider sedation early rather than risk distress to the animal and owner or injury to personnel or owners.
- Use appropriate low-stress handling techniques (Yin, 2009; Hedges, 2014).

The protocol should be specific to the patient. Include details of the behavioural concerns and how this will impact the animal's response to veterinary treatment. It should include information on any medications that the animal is currently receiving as well as details of medications to facilitate handling that have worked well in the past and handling techniques that have been useful. Include information about the stage of any behaviour modification protocol the animal has reached, who is overseeing the case and which staff have been working with the animal as well as any ongoing medication (see Appendix VI, 'Protocol for Veterinary Visits for an Animal Undergoing Behavioural Therapy').

After each visit this protocol should be updated and any relevant information should be passed on to the behaviour clinician overseeing the case.

Medication prior to appointments

If an animal needs a veterinary examination and this will potentially be stressful for the animal, consideration should be given to pre-consultation medication for the client to administer at home. The level of medication required will vary according to both the animal's level of anxiety or fear and the procedure intended. For example, a quick weight check may be accomplished at a quiet time with minimal concern, but blood sampling might need more intervention.

No medication is specifically licensed for use prior to veterinary examination but there are several medications that have been found to

Table 22.6. Medication prior to veterinary visits.

Medication	Dose	Other information
Gabapentin	Dogs – 10–30 mg/kg (lower doses for geriatric animals) Cats – 100 mg per cat (50 mg for very petite or geriatric cats and up to 150 mg for very big cats	Give 2–3 hours prior to arrival at the clinic For dogs, give a dose the evening before the visit as well Generally effective and well tolerated in cats particularly
Trazodone	Dogs – 5–10 mg/kg. Lower doses (up to 3 mg/kg) should be considered for animals also receiving other medications that increase Serotonin	Care if administering other medications that enhance serotonin Can be given the evening before and again on the morning of the visit
Benzodiazepines	Alprazolam: Dogs – 0.1 mg/kg Cats – 0.25 mg per cat Diazepam: Dogs – 2 mg/kg	Risk of paradoxical excitation, therefore test dosing is essential Risk of disinhibition of aggressive behaviour Risk of hepatic necrosis in cats (especially diazepam) Administer 30–60 minutes prior to arrival
Dexmedetomidine	Dogs – Sileo oral gel 125 micrograms per m^2	Licensed in the UK for alleviation of acute anxiety and fear associated with noise in dogs Administer 20 minutes prior to arrival Minimal sedation at this dose
Clonidine	Dogs – 0.01–0.05 mg/kg	Administer 60 min prior to arrival
Acepromazine (ACP)	Dogs – tablets 1–2 mg/kg or injectable acepromazine 0.01–0.05 mg/kg oral transmucosal	Minimal anxiolytic effect. For this reason, ACP should not be used as a sole agent. Generally, the use of ACP should be avoided in behaviour cases. It may have a role in protocols when combined with medications that more reliably provide anxiolysis. Consideration should be given to the duration of action of each of the medications used to ensure that anxiolytic effect continues throughout the period of sedation
Buprenorphine	Cats – 0.03 mg/kg oral transmucosal	To be given 30–60 min prior to arrival

be useful (Table 22.6). In all cases prior to prescription, veterinary surgeons should review the medical history if possible.

For some animals, a combination of medications may be required and it can take a few trials to find the combination and dose that best suits an individual. Pre-consultation sedation can reduce anxiety and allow safe and low-stress examination of some animals. Their use must be combined with low-stress handling and good planning (plenty of time, quiet room, experienced staff). You should be ready to administer injectable sedatives rather than stress the animal further if they are resisting.

Some animals will present a significant risk to veterinary staff and in these cases you should plan for administration of an injectable sedative.

Conclusion

Management of behaviour cases is a team effort and the veterinary surgeon plays a vital role in coordinating that team. Forming good working relationships with behaviour clinicians brings benefits to all involved. Working with behaviour cases is hugely rewarding but can be emotionally demanding and take time.

Although never a replacement for behaviour modification and environmental management, judicious use of psychoactive medication can help cases progress and this must always be prescribed by a veterinary surgeon based on the best possible behavioural diagnosis.

Every member of staff within the veterinary practice has a role to play in ensuring the optimum experience for all patients, including those undergoing behaviour modification. From the receptionist who listens to the client's concerns and ensures they are booked in at a quiet time and allowed to wait in a quiet place, to the nurse who handles the animal gently and appropriately, and gives high quality advice, to the veterinary surgeon who oversees the referral and prescribes medication: every team member can help to ensure that both clients and patients have the best possible experience.

References and Further Reading

Allerson, F. (ed.) (2020) *BSAVA Formulary*, 10th edn, *Part A: Canine and Feline*. BSAVA, Gloucester, UK.

Beata, C., Beaumont-Graff, E., Coll, V., Cordel, J., Marion, M. *et al.* (2007a) Effect of alpha-casozepine (Zylkene) on anxiety in cats. *Journal of Veterinary Behavior* 2, 40–46.

Beata, C., Beaumont-Graff, E., Diaz,C., Marion,M., Massal,N. *et al.* (2007b) Effects of alpha-casozepine (Zylkene) versus selegiline hydrochloride (Selgian, Anipryl) on anxiety disorders in dogs. *Journal of Veterinary Behavior* 2, 175–183.

Bowen, J. and Heath, S. (2005) *Behaviour Problems in Small Animals – Practical Advice for the Veterinary Team*. Elsevier Saunders, Philadelphia.

Crowell-Davis, S.L, Murray, T.F. and Mattos de Souza Dantas, L. (2019) *Veterinary Psychopharmacology*, 2nd edn. Wiley Blackwell, Hoboken, New Jersey.

Dramard, V., Kern, L., Reme, C., Nicolas, C., Chala, V. and Navarro, C. (2018) Effect of L-theanine tablets in reducing stress-related emotional signs in cats: an open-label field study. *Irish Veterinary Journal* 71, Article 21.

Engel, O., Masic, A., Landsberg, G., Brooks, M., Mills, D.S. and Rundfeldt, C. (2018) Imepitoin shows benzodiazepine-like effects in models of anxiety. *Frontiers in Pharmacology* 9, 1225.

Gamble, L.-J., Boesch, J., Frye, C., Schwark, W., Mann, S. *et al.* (2018) Pharmacokinetics, safety, and clinical efficacy of cannabidiol treatment in osteoarthritic dogs. *Frontiers in Veterinary Science* 5, Article 165.

Hedges, S. (2014) *Practical Canine Behaviour for Veterinary Nurses and Technicians*. CAB International, Wallingford, UK.

Horwitz, D.F. and Mills, D.S. (eds) (2012) *BSAVA Manual of Canine and Feline Behavioural Medicine*, 2nd edn. BSAVA, Gloucester, UK.

Landsberg, G., Hunthausen, W. and Ackerman, L. (2012) *Behaviour Problems of the Dog and Cat*, 3rd edn. Elsevier Saunders, Edinburgh.

Love, L. (2019) Oral sedatives and anxiolytics for veterinary visits. North American Veterinary Anesthesia Society. Available at: https://www.mynavas.org/post/oral-sedatives-and-anxiolytics-for-veterinary-visits#:~:text=A%20common%20combo%20includes%20trazodone,3%20hours%20prior%20to%20travel.&text=Timing%20is%20very%20important%20with,when%20to%20give%20the%20medications. (accessed July 2020).

Lucas, X. (2014) Clinical use of Deslorelin (GnRH agonist) in companion animals: a review. *Reproduction in Domestic Animals* 49 (Suppl. 4), 64–71.

Martin, K. (2016) Psychopharmacology – which drug and why? In: *Proceedings of the Canadian Veterinary Medical Association (CVMA) and the Society of British Columbia Veterinarians (SBCV)*, Niagra Falls, New York.

McGrath, S., Bartner, L., Rao, S., Packer, R. and Gustafson, D. (2019) Randomized blinded controlled clinical trial to assess the effect of oral cannabidiol administration in addition to conventional antiepileptic treatment on seizure frequency in dogs with intractable idiopathic epilepsy. *Journal of the American Veterinary Medical Association* 254(11), 1301–1308.

Michelazzi, M., Berteseli, G., Minero, M. and Cavallone, E. (2010) Effectiveness of L-theanine and behavioural therapy in the treatment of noise phobias in dogs. *Journal of Veterinary Behavior* 5(1), 34–35.

Mills, D., Braem Dube, M. and Zulch, H. (2013) *Stress and Pheromonatherapy in Small Animal Clinical Behaviour*. Wiley-Blackwell, Oxford.

Mills, D.S., Demontigny-Bedard, I., Gruen, M., Klinck, M., McPeake, K. *et al.* (2020) Pain and problem behavior in cats and dogs. *Animals* 10(2), 318.

NOAH (2020) Compendium of data sheets for animal medicines. Available at: https://www.noahcompendium.co.uk/ (accessed July 2020).

Ogata, N. (2015) Behaviour: Psychopharmacology. In: *40th Congress of the World Small Animal Veterinary Association Proceedings* May 15–18, Bangkok, Thailand.
Overall, K.L. (2013) *Manual of Clinical Behavioural Medicine for Dogs and Cats*. Elsevier, St Louis, Missouri.
Pinel, J.P.J. (2009) *Biopsychology*, 7th edn. Pearson, London.
Reisner, I. (2014) The use of medications in canine behaviour therapy. *Today's Veterinary Practice* July/August, pp. 63–67.
Rodin, I. and Heath, S. (2016) *Feline Behavioural Health and Welfare*. Elsevier, St Louis, Missouri.
Root, A.L., Parkin, T.D., Hutchinson, P., Warnes, C. and Yam, P.S. (2018) Canine pseudopregnancy: an evaluation of prevalence and current treatment protocols in the UK. *BMC Veterinary Research* 14, Article 170.
Self, I. (ed.) (2019) *BSAVA Guide to Pain Management in Small Animal Practice*. BSAVA, Gloucester, UK.
Seskel, K. (2012) Antidepressants and small animals: What should you know? In: *Proceedings WSAVA / FECAVA / BSAVA World Congress*, Birmingham, 11–15 April. BSAVA, Gloucester, UK.
Van Haaften, K.A., Eichstadt, L.R., Pharm, D.F., Stelow, E.A. and Bain, M.J. (2017) Effects of a single preappointment dose of gabapentin on signs of stress in cats during transportation and veterinary examination. *Journal of the American Veterinary Medical Association* 251(10), 1175–1181.
VMD (Veterinary Medicines Directorate) (2015) The cascade: prescribing unauthorised medicines. Available at: https://www.gov.uk/guidance/the-cascade-prescribing-unauthorised-medicines (accessed July 2020).
VMD (Veterinary Medicines Directorate) (2018) VMD statement on veterinary medicinal products containing Cannabidiol. Available at: https://www.gov.uk/government/news/vmd-statement-on-veterinary-medicinal-products-containing-cannabidiol (accessed July 2020).
Yin, S. (2009) *Low Stress Handling: Restraint and Behaviour Modification of Dogs and Cats* – techniques for developing patients who love their visits. Cattle Dog Publishing, Davis, California.

Appendix I

Training Animals to Accept Medication

Some animals will simply swallow a tablet if offered, but sometimes a course of medication can cause significant stress both to you and your pet. There are a variety of techniques to consider to help make this easier for you both:

- Discuss the issue with your veterinary team – is there an alternative formulation that might be easier? A spot-on or an oral liquid, or is there a more palatable tablet formulation?
- Try hiding the medication in a small amount of strong-smelling food that you know your pet enjoys. Canned fish, cheese, liver or fish paste and Lick-e-Lix can all work well.
- For some animals offering several small treats, then one containing the medication and then several more without can work well.

It is possible to train some animals to take a tablet. This is a brief outline of the techniques we use.

Dogs

The idea will be to teach the dog to swallow a less palatable item in order to earn a high-value reward. Consider what is your dog's favourite reward. Then consider what safe food items they do not enjoy so much (pieces of vegetable can work well for many dogs).

- We like to have a marker that shows the dog that we are training. You can simply use a mat or blanket that you encourage your dog to stand on for each training session.
- Start by offering high-value rewards. Feed two in quick succession, praising your dog for taking each. Repeat a couple of times until they are fully engaged with the training.
- Now offer a less exciting but still enjoyed food such as a piece of dry food, praise and immediately follow with the high-value reward. Again, repeat this several times. Gradually extend the time between the less exciting and high-value food.
- Now move on to offering the less preferred food (maybe a piece of a vegetable they do not usually enjoy). Offer this and if they eat it praise and immediately offer the high-value food. Be generous with the high-value rewards at this point (several tiny pieces is fine). If they do not take it, try

offering both the less preferred and high-value food together a couple of times, then try again.
- Once they are reliably eating the less preferred food you can try with a tablet (you may be able to get empty gelatine capsules to practise with).

Cats

We want to teach your cat to readily accept the plastic pill giver so that you can easily give a tablet.

- Start by smearing high-value soft food on the outside of the pill giver, then place it in a bowl and move away. Hopefully, your cat will then eat the food. If your cat has had a previous bad experience with a pill giver, this may take time so keep trying over several days.
- Once your cat is happily eating the food from the pill giver, try gradually staying closer while they eat. Build up until you can hold on to the pill giver while they eat the food. Try reloading the pill giver a few times.
- Once your cat will eat while you hold the pill giver, try using it to feed them soft food. It is important that you keep the pill giver still and let the cat choose to interact. What we are looking for is them starting to grab the pill giver in their mouth.
- Now try placing a small piece of dry food in the pill giver and again smearing the outside with wet food. Hopefully, your cat will grab the pill giver and take the dry food as well. Follow up with more soft food.
- Repeat this frequently. The aim is that when you need to give a tablet you will be able to use the same technique.

If you have any questions contact your veterinary team or a qualified trainer/behaviourist (http://www.abtcouncil.org.uk/).

Appendix II

Veterinary Staff Protocol for Building Positive Associations with Vet Visits: Dogs

Prior to First Session

- Look at notes from previous vet visits – these may give a clue as to how severe the issue is.
- Check whether the dog has had a behaviour consultation – if so check whether there are any specific concerns and discuss these with the owner/behaviour clinician.
- Contact the dog's owner and request that they bring with them high-value treats (ideally soft, smelly and lots of them!). Avoiding feeding the dog prior to the session will help ensure the rewards have value.
- Ask about the dog's reaction to people, other dogs, etc. If the dog is uncomfortable with people in other situations or with other dogs, then you will need to factor this into your protocol – for example you may need to meet outside standard consult times. If you have any concerns, advise a conversation with a behaviourist to assess whether a full consultation is required prior to starting work.

First Session

- Meet in the car park, walk around, encourage the dog's owner to give treats (it is usually best to allow the owner to give the rewards as if you encourage the dog to approach you with food it may be tempted into coming closer than it is really comfortable with).
- Assess the dog's behaviour – look for signs of fear or anxiety: arousal, panting, lip licking, ear/tail position.
- If the dog appears relaxed, approach the door and assess the reaction to this. If still relaxed, enter and walk around reception – still giving treats. If at any point the dog looks worried, exit, allow to settle and then repeat the process a few times. Once the

dog is comfortably entering the building, high-value food should then only come while inside, but it should remain the dog's choice whether to stay or leave. You can have a second pot of lower-value food and continue to give this outside – we want these dogs to love every part of coming to the practice.

Homework:

- If the dog is very anxious, suggest frequent visits to initially feed the dog in the car, and then walk around car park and reward with treats before arranging to meet again, once the dog is more relaxed around the building.
- If the dog is okay with the owner handling them, suggest that they practise handling at home – building up positive associations (give the appropriate handouts and links).
- If appropriate, advise muzzle training. Give the owners a copy of the Muzzle Training handout.
- Book an appointment for a second session if appropriate, or refer to a behaviourist if needed.
- Explain to the owner the importance of the dog not having a bad experience and how this can be avoided.
- Make sure that you record on the dog's clinical record how far in the protocol the dog has got and what it is likely to be able to cope with. Discuss each case with a specific behaviourally aware vet and complete a 'Protocol for Veterinary Visits for an Animal Undergoing Behaviour Therapy' form (see Appendix IV).

Second and Subsequent Sessions

- Start as before and build on what was achieved in the last session. Always start from basics again, but you may be able to progress more rapidly. Always be guided by the dog's response.
- Once the dog is happily walking into the practice you can try arranging a visit at a quiet time between consult blocks so that you can play find-it/search games in the waiting room.
- Try propping open a consulting room door and scattering treats on the floor in the room. Allow the dog to choose whether or not to go in.
- Build up to having the dog comfortably in a consulting room and then to starting an examination. Never continue if the dog looks worried – back off to the previous level.
- Gradually up the level of behaviour that you require for high-value reward.
    - Carpark
    - Reception
    - Consulting room
    - Handling
- Low-value treats can continue to be offered in other locations. We want these dogs to love coming to the vets!
- Try to reward calm appropriate behaviour, not excitement.
- Once the dog is happily approaching you, start offering treats yourself.
- Once you start the examination process it is important that each stage is followed with a treat, for example touch ear, treat, touch ear, treat, touch ear for a bit longer, treat, etc. This ensures that the handling becomes the predictor of good things!
- When the dog is comfortable with you, introduce other members of staff.
- Set appropriate homework for the owner after each session, if this is possible.
- Make sure that there is a plan in place for an emergency visit that may need to take place before the handling protocol is completed.

For a list of qualified behaviourists visit the Animal Behaviour and Training Council Website (http://www.abtcouncil.org.uk/).

Appendix III

Feline Behaviour Handout: Cat Carriers

Sadly, many visits to the vet or cattery start with a fight to get the cat into their carrier, and it is not unusual for vet visits to be cancelled because the cat has disappeared at the first sight of the cat carrier. If we want to reduce stress for our cats and us, we need a little advance planning. We want your cat to learn to love their carrier.

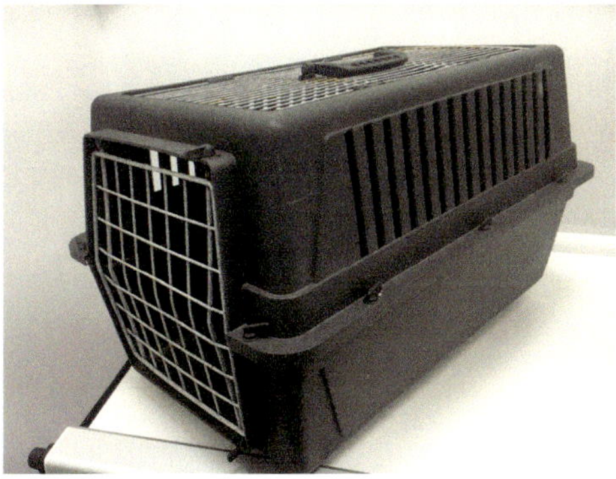

Cat carrier. (Image: author's own.)

1. Cat carrier choice

Try to think practically rather than based on appearance. For many cats this sort of cat box is a good choice.

The carriers that come apart in the middle are excellent for nervous cats as they can be opened up at the vets, allowing the cat to sit in the security of their own box for much of the examination.

The top opening allows cats to be gently lifted in and out, which can be easier and less upsetting than trying to push them through a small door into the box.

If you already have a cat carrier, think about whether it can be made more cat friendly. For example, the top-opening wire boxes mean that the cat cannot hide at all from other people and animals. The addition of a towel over the box can make a huge difference.

2. Advance preparation

We all know that cats love a box so why does the carrier cause such a panic? Could it be because it only appears just before a visit to the vet/cattery? Rather than hide the carrier away, make it part of your cat's safe space. Put the carrier in a preferred position (you know your cat – do they like a cosy hideaway under furniture or do they like to be up high keeping watch? Would they love a cosy bed by a radiator?) and add comfy bedding. If your cat likes an enclosed bed, then you can start with a towel over the box. If they prefer an open bed, start with the top section off, using the bottom as a basket.

Scent is very important to cats. They mark their home territory, familiar objects and even us with scent. All that rubbing their heads against us is about marking us with pheromones from glands on their cheeks. We can help them to feel more secure by ensuring that the carrier smells right. We can put in a piece of bedding from their own bed and also use a cloth to gently rub around your cat's face and then rub this over the box. Another option is to use one of the commercial feline pheromone sprays to spray the inside of the carrier. If you do this, allow at least 15 minutes and ideally half an hour before allowing the cat near the box.

Once the cat carrier is in position, add a few of your cat's favourite treats and let them explore. Keep adding a few treats every day and encouraging your cat to use the box.

3. Travelling

A trip in the car is unlikely to be top of any cat's wish list. We can make it less difficult by getting them secure in their box, which should by now be a safe familiar place, and then perhaps covering this with a towel to provide additional sound insulation, as long as you ensure that there is adequate ventilation on hot days. You can spray this towel with pheromones and, if needed, it can be used at the vet for the cat to hide under while they are examined.

Make sure the box is well secured in the car and that it is ventilated but not in a draught.

Time vet visits for quiet times or in cat-only sessions if your vet has these.

On arrival at the vet use a cat waiting area if there is one, find a quiet corner or ask if you can wait in the car. Most cats find being down at floor level close to unknown dogs, but unable to move away, very stressful. If there is a safe raised surface on which to place the carrier, do so and consider covering it with a towel, leaving just a small gap so that your cat can peep out if they want to.

4. Removing your cat from their carrier

Ideally you will simply open the carrier and allow your cat to come out in their own time. Placing their towel over the vet's table so they have a familiar surface with a familiar scent can help. It is surprising how many cats will venture out if given the chance and this gives you the opportunity to chat with the vet about any concerns you have prior to the examination. A few of their favourite treats might help to tempt them out.

If your cat is not keen and you have one of the carriers that comes apart in the middle, then this is a good option. Take the top section off and allow your cat to sit in the bottom section. You can now place the towel gently over them so that they still feel like they can hide. Try to avoid dragging your cat out of the carrier. It is very important to cats to feel like they have some control over the situation, so allowing them to make the choice to enter and leave the carrier can really help them to feel more secure.

If your cat really struggles with travelling or visits to the vet, seek advice from your veterinary surgeon or a qualified behaviourist (http://www.abtcouncil.org.uk/).

Appendix IV

Veterinary Staff Protocol for Reducing Feline Anxiety at the Vet

Veterinary visits can be very difficult for cats. They have to deal with confinement, loss of control, change in routine, handling by unfamiliar people, new smells and experiences. They may encounter other species that they are not comfortable with and may feel unwell or be in pain. It is not surprising then that many cats become fearful of vet visits. We can help to reduce this anxiety.

Prior to a Visit

- Make sure the owner has a copy of the cat carrier handout and is working on this.
- Discuss what food rewards the cat really enjoys and ensure these will be available on the day of the visit.
- Check the cat's medical and behavioural records. If you have any concerns, discuss these with a behaviourist.
- Ask the owner not to feed the cat on the day of the visit and to bring a familiar towel from home that has been in the cat's bed and smells of them.

On the Day

- Ask the client to call in to let reception know that they have arrived. If appropriate, they should then be asked to bring the cat into the cat room in their carrier and wait there. If this is not possible, they should be asked to wait in the car until you can collect them.
- Open the carrier and simply allow the cat to decide whether to exit while you chat to the owner. Ask the owner to place a small amount of the cat's favourite food in the carrier. If they eat this, add more. If the cat chooses to exit the carrier, allow them to explore in their own time and place high-value food around the room for them to find.
- If the cat seeks interaction with you, stroke very briefly then wait to see whether they choose to interact again. At this stage we want the cat to have as much choice as possible.
- Discuss with the owner what the cat finds difficult. Is it travel, handling or specific procedures? Discuss techniques for minimizing stress on future visits – managing travel, finding handling techniques that suit the cat, desensitization and counterconditioning to common veterinary procedures, discussing medication with a vet, etc.
- Some good resources can be found on the Fellowship of Animal Behaviour Clinicians (FACBCs) website and YouTube channel. (https://www.youtube.com/c/FellowshipofAnimalBehaviourClinicians/playlists?view=50&sort=dd&shelf_id=3).

Follow-up

- Complete or update a protocol for veterinary visits for an animal undergoing behavioural therapy form.
- Discuss the case with a veterinary surgeon – would the cat benefit from medication prior to veterinary visits?
- Book a further appointment if appropriate.

Appendix V

Protocol for Veterinary Visits for an Animal Undergoing Behaviour Therapy

Animal's name ....
Species ....
Age ....
Sex / neuter status ....

Client's name ....
Address ....
....
....
Telephone number ....
Email address ....

Behavioural assessment ....
Treatments implemented ....
Stage of muzzle training reached (if relevant) ....

Psychoactive medications including current dose ....
Supplements currently received ....
Other ongoing medication ....
Members of staff currently working with the patient ....

Plan for veterinary visits ....
Medication to be given pre-visit ....
Will sedation be required for examination? ....
What doses have been effective in the past (if known) ....
Any other comments ....

Index

Note: The page numbers in italics and bold represents figures and tables respectively.